Textbook of
Finite Element Analysis

P. Seshu

Director
Indian Institute of Technology Dharwad, Karnataka
and
Professor
Department of Mechanical Engineering
Indian Institute of Technology Bombay, Mumbai

PHI Learning Private Limited

Delhi-110092
2025

*In fond memory of **Shri Asoke K. Ghosh** (October 1942 – February 2024), Founder Chairman and Managing Director of PHI Learning, whose vision endlessly inspires.*

The Legacy Continues....

Published by Pushpita Ghosh, PHI Learning Private Limited, Rimjhim House, 111, Patparganj Industrial Estate, Delhi-110092 and Printed by Syndicate Binders, A-20, Hosiery Complex, Noida, Phase-II Extension, Noida-201305 (N.C.R. Delhi).

₹725.00

TEXTBOOK OF FINITE ELEMENT ANALYSIS
P. Seshu

ISBN-978-81-203-2315-5 (Print Book)
ISBN-978-93-90544-35-6 (e-Book)

The export rights of the book are vested solely with the publisher.

Respectfully Dedicated to

My Parents and Teachers

Contents

Preface   *ix*

1. Introduction   **1–15**

1.1 Typical Application Examples 4

  1.1.1 Automotive Applications 4

  1.1.2 Manufacturing Process Simulation 7

  1.1.3 Electrical and Electronics Engineering Applications 8

  1.1.4 Aerospace Applications 14

Summary 14

2. Finite Element Formulation Starting from Governing Differential Equations   **16–65**

2.1 Weighted Residual Method—Use of a Single Continuous Trial Function 16

2.2 The General Weighted Residual (WR) Statement 28

2.3 Weak (Variational) Form of the Weighted Residual Statement 33

2.4 Comparison of Differential Equation, Weighted Residual and Weak Forms 36

2.5 Piece-wise Continuous Trial Function Solution of the Weak Form 41

2.6 One-dimensional Bar Finite Element 48

2.7 One-dimensional Heat Transfer Element 57

Summary 61

Problems 61

3. Finite Element Formulation Based on Stationarity of a Functional   **66–88**

3.1 Introduction 66

3.2 Functional and Differential Equation Forms 67

3.3 Principle of Stationary Total Potential (PSTP) 73

  3.3.1 Rayleigh–Ritz Method 75

3.4   Piece-wise Continuous Trial Functions—Finite Element Method   81
    3.4.1   Bar Element Formulated from the Stationarity of a Functional   81
    3.4.2   One-dimensional Heat Transfer Element Based on the Stationarity of a Functional   83
3.5   Meaning of Finite Element Equations   84
Summary   87
Problems   88

4.   One-dimensional Finite Element Analysis   89–144

4.1   General Form of the Total Potential for 1-d   89
4.2   Generic Form of Finite Element Equations   90
4.3   The Linear Bar Finite Element   93
4.4   The Quadratic Bar Element   101
    4.4.1   Determination of Shape Functions   101
    4.4.2   Element Matrices   102
4.5   Beam Element   117
    4.5.1   Selection of Nodal d.o.f.   117
    4.5.2   Determination of Shape Functions   118
    4.5.3   Element Matrices   119
4.6   Frame Element   125
4.7   One-dimensional Heat Transfer   132
Summary   138
Problems   138

5.   Two-dimensional Finite Element Analysis   145–231

5.1   Introduction—Dimensionality of a Problem   145
5.2   Approximation of Geometry and Field Variable   148
    5.2.1   Simple Three-noded Triangular Element   149
    5.2.2   Four-noded Rectangular Element   152
    5.2.3   Six-noded Triangular Element   153
5.3   Natural Coordinates and Coordinate Transformation   156
    5.3.1   Alternate Methods of Deriving Shape Functions   157
    5.3.2   Natural Coordinates—Quadrilateral Elements   159
    5.3.3   Natural Coordinates—Triangular Elements   164
5.4   2-d Elements for Structural Mechanics   167
    5.4.1   Generic Relations   167
    5.4.2   Three-noded Triangular Element   171
    5.4.3   Four-noded Rectangular Element   179
    5.4.4   Compatibility of Displacements   181
    5.4.5   Four-node Quadrilateral Element   183
    5.4.6   Eight-node Quadrilateral Element   188
    5.4.7   Nine-node Quadrilateral Element   190
    5.4.8   Six-node Triangular Element   192

5.5   Numerical Integration   194
    5.5.1   Trapezoidal Rule   195
    5.5.2   Simpson's 1/3 Rule   196
    5.5.3   Newton–Cotes Formula   197
    5.5.4   Gauss Quadrature Formula   198
    5.5.4   Gauss Quadrature in Two Dimensions   201
5.6   Incorporation of Boundary Conditions   205
5.7   Solution of Static Equilibrium Equations   206
5.8   2-d Fluid Flow   220
Summary   225
Problems   226

6. Dynamic Analysis Using Finite Elements   **232–294**

6.1   Introduction   232
6.2   Vibration Problems   232
6.3   Equations of Motion Based on Weak Form   235
    6.3.1   Axial Vibration of a Rod   235
    6.3.2   Transverse Vibration of a Beam   237
6.4   Equations of Motion Using Lagrange's Approach   240
    6.4.1   Formulation of Finite Element Equations   242
    6.4.2   Consistent Mass Matrices for Various Elements   245
6.5   Consistent and Lumped Mass Matrices   246
    6.5.1   HRZ Lumping Scheme   247
6.6   Form of Finite Element Equations for Vibration Problems   253
6.7   Some Properties of Eigenpairs   255
6.8   Solution of Eigenvalue Problems   257
    6.8.1   Transformation Based Methods   258
    6.8.2   Vector Iteration Methods   264
6.9   Transient Vibration Analysis   272
    6.9.1   Modelling of Damping   272
    6.9.2   The Mode Superposition Scheme   275
    6.9.3   Direct Integration Methods   279
6.10 Thermal Transients—Unsteady Heat Transfer in a Pin-Fin   289
Summary   293
Problems   293

7. Application Examples   **295–307**

7.1   Finite Element Analysis of Crankshaft Torsional Vibrations   295
    7.1.1   Beam Element Model of Crankshaft Assembly   296
    7.1.2   Results and Discussion   299
    7.1.3   Dynamic Response Analysis   301
7.2   Axisymmetric Finite Element Analysis of a Pressure Vessel   303
    7.2.1   Finite Element Formulation for Axisymmetric Loads   304
    7.2.2   Stress Analysis of a Pressure Vessel   305

Appendix A—Suggested Mini-Project Topics   *309–320*

Project 1: Thermal Analysis of a Pressure Vessel   309
Project 2: Structural Dynamic Analysis of a Pressure Vessel   310
Project 3: Dynamics of a Scooter Frame   312
Project 4: Automotive Chassis Dynamics   313
Project 5: Analysis of a Turbine Disk   316
Project 6: Dynamic Analysis of a Building   317
Project 7: Thermal Analysis of an IC Engine Cylinder   318
Project 8: Stress Concentration   319
Project 9: Dynamics of a Hard Disk Drive Read/Write Head Assembly   319

Appendix B—Review of Preliminaries   *321–323*

B1.1   Matrix Algebra   321
B1.2   Interpolation   322

Appendix C—Typical Finite Element Program   *324–328*

Index   *329–330*

Preface

Many excellent books have been written on Finite Element Analysis, for example, books authored by R.D. Cook, D.S. Malkus and M.E. Plesha; J.N. Reddy; K.J. Bathe; O.C. Zienkiewicz, K. Morgan, et al. I have immensely benefitted from these books, both as a student and as a teacher. Most books, however, present finite element method (FEM) primarily as an extension of matrix methods of structural analysis (except, for example, the ones by J.N. Reddy and Huebner). Present-day applications of FEM, however, range from structures to bio-mechanics to electromagnetics. Thus the primary aim of the present book, based on several years of teaching the course at Indian Institute of Technology (IIT) Bombay, is to present FEM as a general tool to find approximate solutions to differential equations. This approach should give the student a better perspective on the technique and its wide range of applications. Finite element formulation, based on stationarity of a functional, is also discussed, and these two forms are used throughout the book.

Over the past few years, several universities and engineering institutes in India have introduced a one-semester course on finite element method at the senior undergraduate level. The material presented in this book covers the syllabus of most such courses. Several worked-out examples, drawn from the fields of structural mechanics, heat transfer and fluid flow, illustrate the important concepts. Some of the issues are in fact brought out through example problems. At the same time, some "problems to investigate" given at the end of each chapter encourage the student to think beyond what has been presented in the book.

FEM is a technique (numerical tool), and the various nuances are best mastered by attempting challenging real-life problems. While teaching the course at IIT Bombay, I have successfully used two types of term projects (running through the semester)—one where a near-real-life problem is modelled using a commercial finite element analysis software and the other where the student attempts to develop his own code and verify the same on simple text-book type problems. Some topics for the former are suggested in Appendix A and a sample code is given in Appendix C for isoparametric, nine-node quadrilateral element.

I am indebted to IIT Bombay and my colleagues (particularly, Profs. C. Amarnath, Bharat Seth, and Kurien Issac) in the Mechanical Engineering Department for providing a congenial work environment. The financial support provided by the Quality Improvement

Programme (QIP) at IIT Bombay for the preparation of the manuscript is gratefully acknowledged.

My students Anurag Ganguli, Vignesh Raja, Gaurav Sharma, Kartik Srinivasan and V.K. Gupta did a splendid job by studiously going through the entire manuscript and offering their critical and incisive comments. I am also thankful to my student Pankaj Langote for assisting me in getting the diagrams drawn and to Mrs. Padma Amin for keying in the whole manuscript. I would like to thank my wife Uma and my little jewels Soumya and Saket for their continuous support and encouragement. Preparation of the manuscript took longer than I expected and I am grateful to my wife for helping me stay focussed till the very end. Finally, I wish to thank the Publishers, PHI Learning—their sales, editorial and production team—for patiently pursuing the scheduling of the manuscript and for its careful processing.

P. Seshu

CHAPTER 1

Introduction

Finite element analysis has now become an integral part of Computer Aided Engineering (CAE) and is being extensively used in the analysis and design of many complex real-life systems. While it started off as an extension of matrix methods of structural analysis and was initially perceived as a tool for structural analysis alone, its applications now range from structures to bio-mechanics to electromagnetic field problems. Simple linear static problems as well as highly complex nonlinear transient dynamic problems are effectively solved using the finite element method. The field of finite element analysis has matured and now rests on rigorous mathematical foundation. Many powerful commercial software packages are now available, enabling its widespread use in several industries.

Classical analytical methods consider a differential element and develop the governing equations, usually in the form of partial differential equations. When applied to real-life problem situations, it is often difficult to obtain an exact solution to these equations in view of complex geometry and boundary conditions. The finite element method (FEM) can be viewed simply as a method of finding approximate solutions for partial differential equations or as a tool to transform partial differential equations into algebraic equations, which are then easily solved. Some of the key ideas used in finite element formulation are now summarised:

➢ Since the solution for the field variable satisfying both the boundary conditions and the differential equation is unknown, we begin with an **assumed trial solution**. The trial solution is chosen such that the boundary conditions are satisfied.

➢ The trial solution assumed, in general, does not satisfy the differential equation exactly and leaves a **domain residual** defined as the error in satisfying the differential equation.

➢ In general, the domain residual varies from point to point within the domain and cannot be exactly reduced to zero everywhere. We can choose to make it vanish at select points within the domain, but we prefer to render the residual very small, in some measure, over the entire domain. Thus, **the weighted sum of the domain residual** computed over the entire domain is rendered zero.

➢ The accuracy of the assumed trial solution can be improved by taking additional, higher order terms, but the computations become tedious and do not readily render themselves for automation. Also, for complex real-life problems, choosing a single continuous trial function valid over the entire domain satisfying the boundary conditions is not a trivial task. We therefore prefer to **discretise the domain into several segments** (called **finite elements**) and use several **piece-wise continuous trial functions**, each valid within a segment (finite element).

➢ Trial functions used in each segment (finite element) are known as **element level shape functions**. These are defined in the form of interpolation functions used to interpolate the value of the field variable at an interior point within the element from its value at certain key points (called the **nodes**) in the element. Typical elements commonly used in finite element analysis are shown in Figure 1.1.

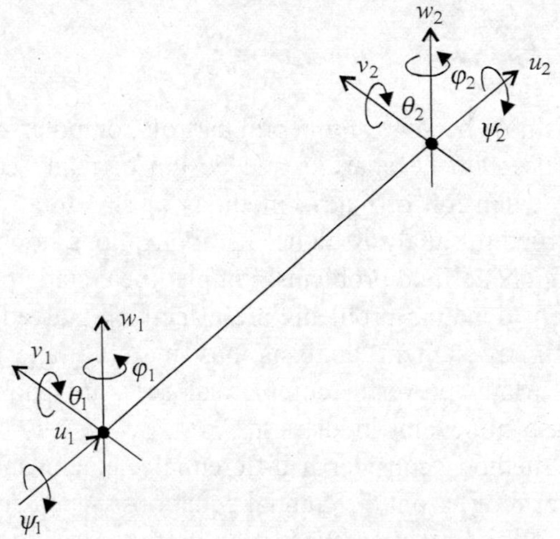

(a) General frame element (Six d.o.f. Per node)

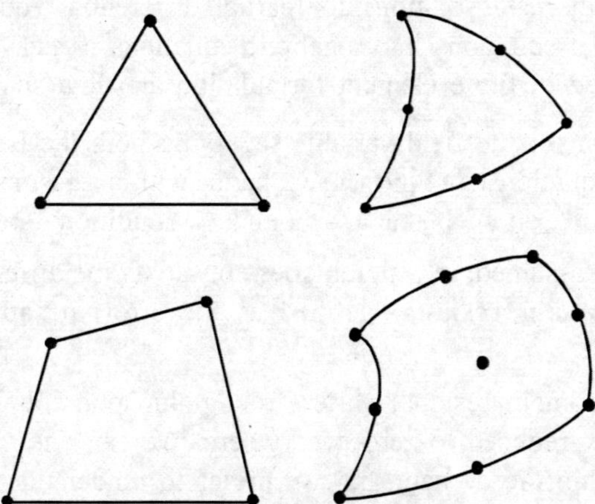

(b) Common 2-d elements

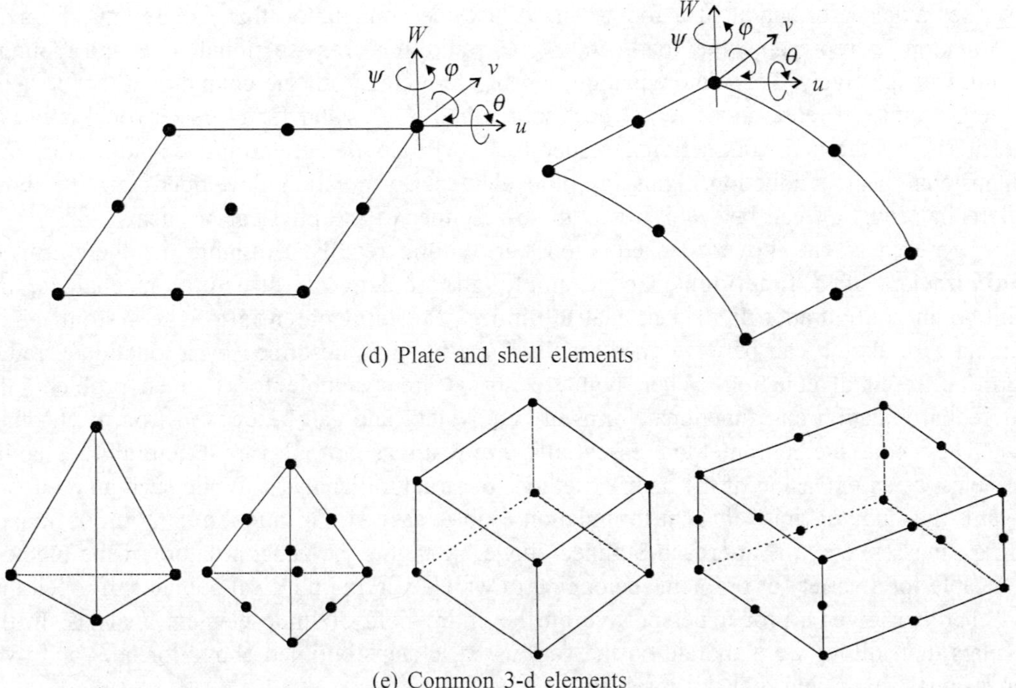

(d) Plate and shell elements

(e) Common 3-d elements

Fig. 1.1   Typical finite elements

➢ With these shape functions, the weighted sum of the domain residual is **computed for each element** and then **summed up over all the elements** to obtain the weighted sum for the entire domain.

➢ For all elements using the same shape functions, the computations will be identical and, thus, for each type of element we have **element level characteristic matrices**. These characteristic matrices for several types of elements are derived a *priori* and programmed into a finite element software such as ANSYS, NASTRAN, IDEAS, etc. The user can choose to discretise (model) his domain with a variety of different finite elements. The computer program sets up the characteristic matrices for each element and then sums them all up for the entire finite element mesh to set up and solve the system level equations.

The basic steps of finite element analysis, as outlined above, are quite generic and can be applied to any problem—be it from the field of structural mechanics or heat transfer, or fluid flow or electromagnetic fields, given the appropriate differential equation and boundary conditions. In view of the similarity in the form of governing differential equations, the finite element formulation for a particular *type* of differential equation can be used to solve a *class* of problems. For example, a differential equation of the type

$$AC \frac{d^2 f}{dx^2} + q = 0$$

describes axial deformation of a rod when we use the connotation that f represents the axial deformation, q represents the load, and A, C stand for cross-sectional area and Young's modulus, respectively. The same equation, when interpreted with the connotation that f stands for temperature, q represents internal heat source and A, C stand for cross-sectional area and coefficient of thermal conductivity, respectively will be the governing equation for one-dimensional heat conduction. Thus, a finite element formulation developed for the above differential equation can be readily used to solve either of the physical problems.

Sometimes, the governing equations are more readily available in the form of **minimization of a functional**. For example, in problems of structural mechanics, the equilibrium configuration is the one that **minimizes the total potential of the system**. Finite element formulation can be developed readily for a problem described by a functional, rather than a differential equation. When both the forms are available for a given problem, the differential equation and functional forms are equivalent and can be derived from each other.

The finite element method essentially grew up as a tool for structural mechanics problems, as an extension of the matrix methods of structural analysis. While such an approach towards the study of finite element formulation enables easy visualisation in the form of lumped springs, masses, etc., the approach outlined above highlights the generic nature of the method, applicable for a variety of problems belonging to widely varying physical domains. It is felt that this approach gives a proper perspective on the entire field of finite element analysis. In the chapters that follow, we elaborate on the various basic steps outlined above for one- and two-dimensional, static and dynamic problems.

We now present several examples of application of finite element analysis to real-life problems, to give an overview of the capabilities of the method. Our application examples are drawn from the fields of structural mechanics, aerospace, manufacturing processes, electromagnetics, etc.

1.1   Typical Application Examples

1.1.1   Automotive Applications

In a vehicle having monocoque construction, the body itself is connected to the suspension. Therefore, the body panels are subjected to road loads. Hence, stresses and strains in these body panels are of interest. Figure 1.2 shows a FE mesh of a floor panel from the rear end of the vehicle. Provision for spare wheel as well as the various depressions used as stiffeners can be seen in the figure. A total of about 13,000 quadrilateral and triangular shell elements have been used to perform modal analysis, torsional stiffness analysis, and service load analysis. The same finite element mesh is also used for crash analysis using LS-DYNA software.

An automotive engine cylinder block experiences severe pressures and temperature gradients and other transient loads. It is essential to predict accurately the stresses and the vibration levels for further correlation with noise predictions. Figure 1.3 shows a typical finite element (shell element) model of a four cylinder in-line diesel engine cylinder block. Such a model is used to predict the system natural frequencies and mode shapes, response to combustion gas pressure, etc.

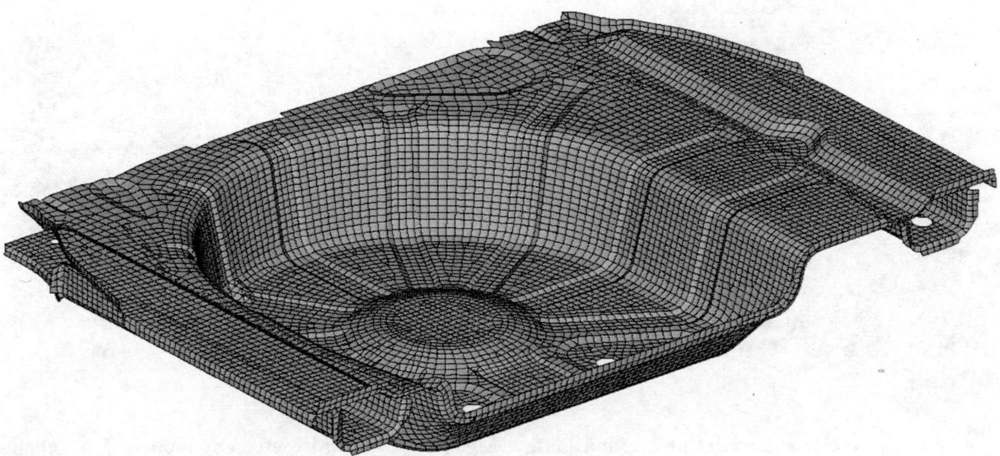

Fig. 1.2   Finite element model (MSC/NASTRAN) of the floor panel of an automobile.
(*Courtesy:* TELCO, *Pune.*)

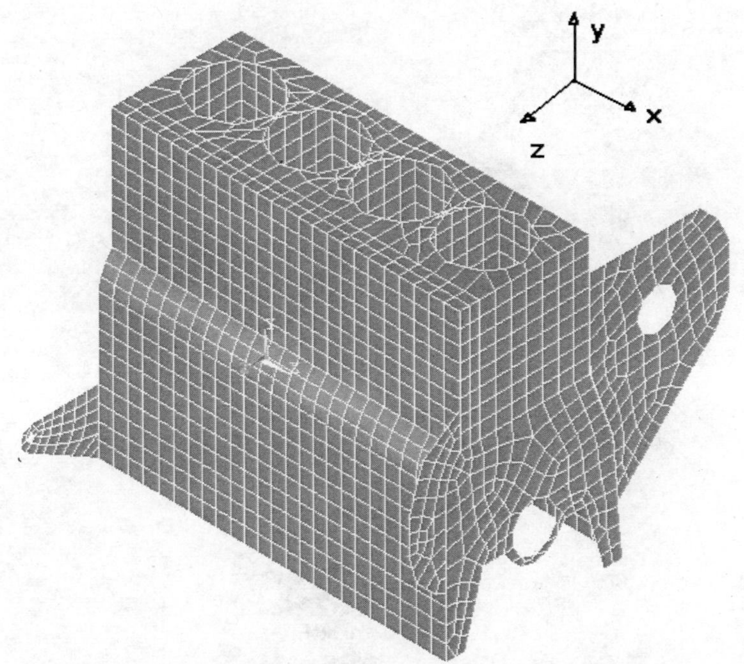

Fig. 1.3   Finite element model of an automotive engine cylinder block.
(*Courtesy: Mahindra & Mahindra Ltd., Nasik.*)

Figures 1.4–1.7 show representative finite element models of various components of a driveline where the gears have been modelled as friction wheels. Such a model can be used for studying the dynamic response of the entire driveline.

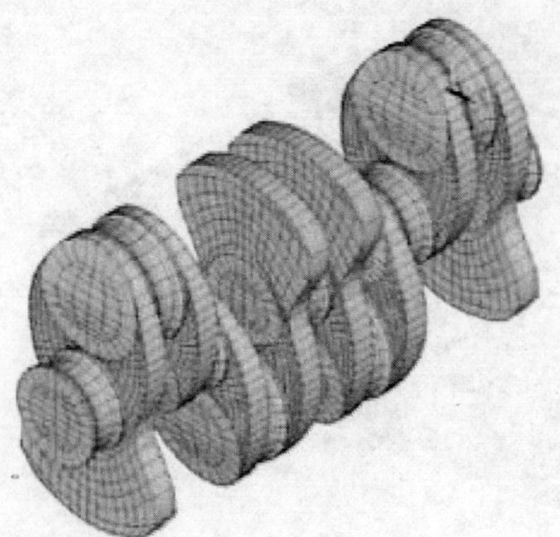

Fig. 1.4   3-d Finite element model of a crankshaft.
(*Courtesy: Mahindra & Mahindra Ltd., Mumbai.*)

Fig. 1.5   3-d Finite element model of a gearbox.
(*Courtesy: Mahindra & Mahindra Ltd., Mumbai.*)

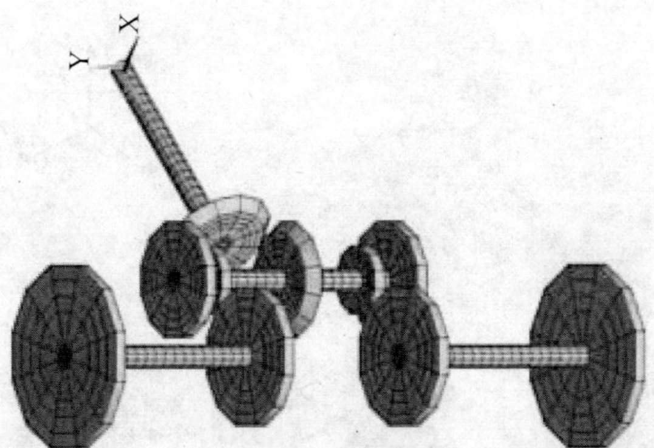

Fig. 1.6 3-d Finite element model of a differential.
(*Courtesy: Mahindra & Mahindra Ltd., Mumbai.*)

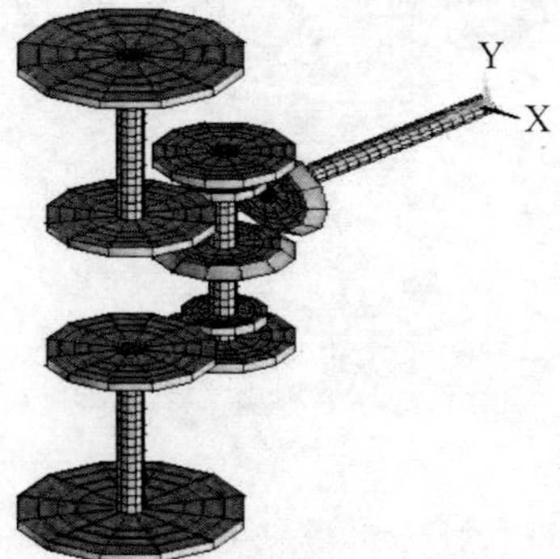

Fig. 1.7 3-d Finite element model of a rear axle.
(*Courtesy: Mahindra & Mahindra Ltd., Mumbai.*)

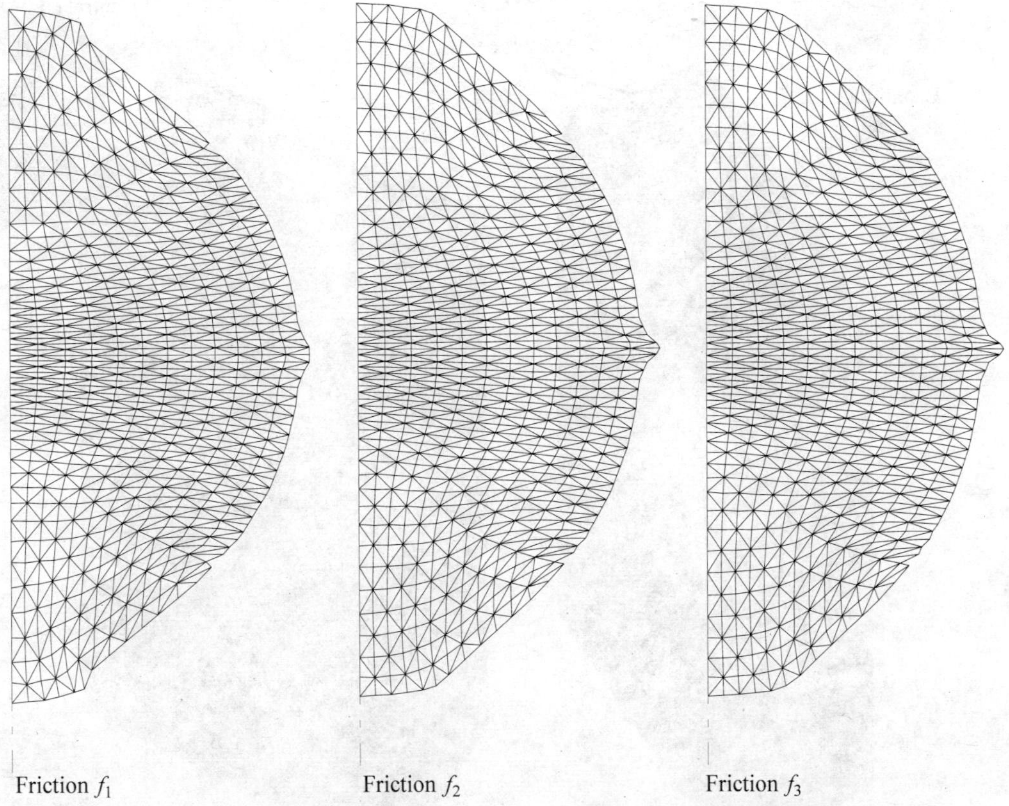

Fig. 1.8   Finite element modelling of cold heading process on a cylindrical slug.
(Courtesy: TRDDC, Pune.)

1.1.2   Manufacturing Process Simulation

Figure 1.8 is an illustrative example in which finite element tools have been used for inverse analysis. A cylindrical slug is cold headed into the form of a near spherical ball. The coefficient of friction plays an important role in the final ball shape. However, it is difficult to measure the value of coefficient of friction. To overcome this difficulty, the operation is simulated for various values of coefficient of friction and the pole diameter (diameter of near flat portion near the poles) is compared with the measured pole diameter in the actual heading process for one case. This value is used for further analysis and optimization of the process.

Figure 1.9 shows stress distribution in a fusion cast ceramic block. A transformation behaviour of one of the components of the material mixture causes anomalous expansion during cooling in a particular temperature range. This leads to tensile stress build-up and failure on the cast block. A 3-d thermomechanical FE analysis is used to study the solidification, thermal field and evolution of stress, and the cause for failure. This information is further used to change the processing conditions so as to eliminate these high tensile stresses.

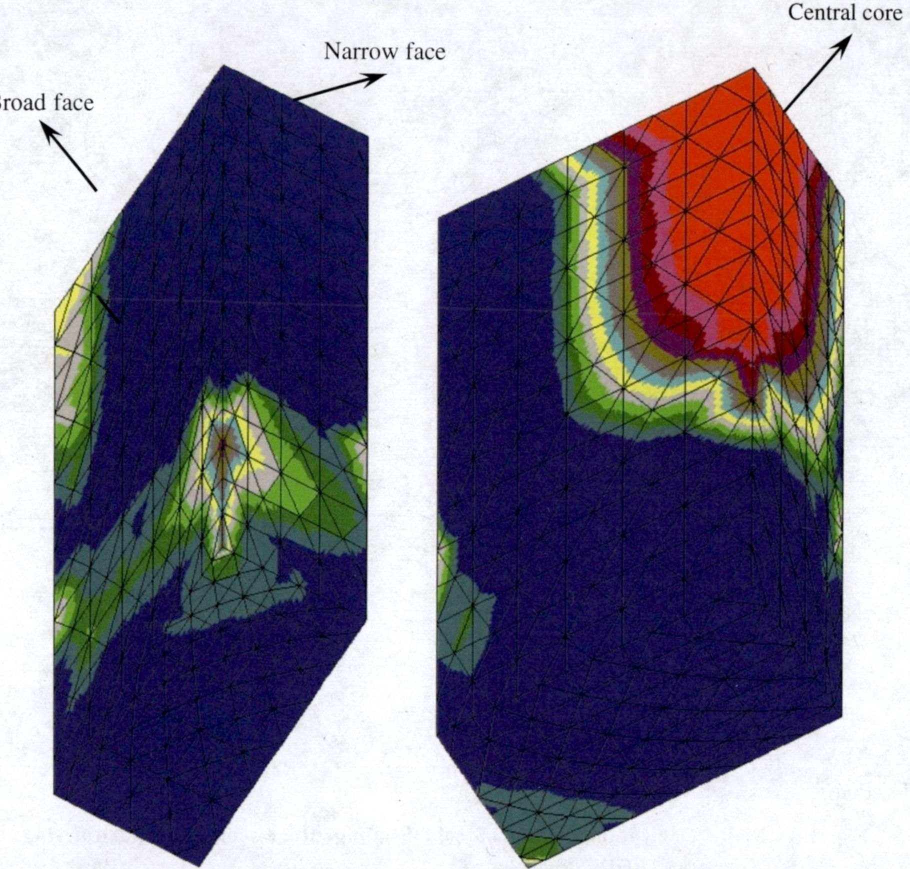

Fig. 1.9   Finite element model of fusion cast ceramic block.
(*Courtesy: TRDDC, Pune.*)

1.1.3   Electrical and Electronics Engineering Applications

FEA can be used for reliability enhancement and optimization of insulation design in high voltage equipment by finding accurately the voltage stresses and corresponding withstands. For complex configuration of electrodes and dielectric insulating materials, analytical formulations are inaccurate and extremely difficult, if not impossible. The FEA can be effectively used in such cases. A typical equipotential field plot for a high voltage transformer obtained from finite element simulation is shown in Figure 1.10. An analysis of eddy currents in structural conducting parts and minimization of stray losses in electrical machines is possible using FEM. A typical mesh for one such analysis is depicted in Figure 1.11, where it is aimed at estimating the eddy current losses in structural conducting plate supporting high current terminations.

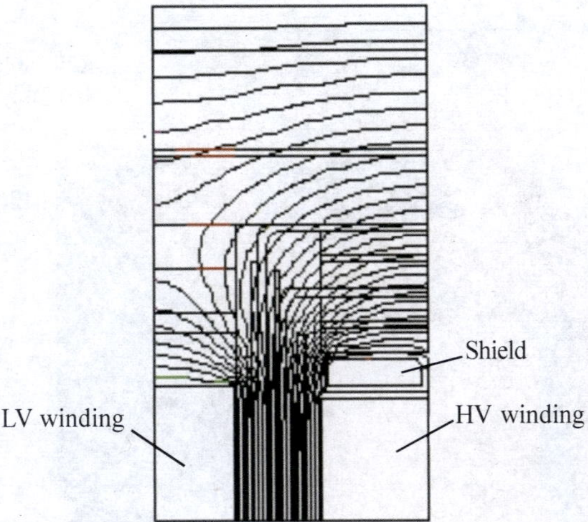

Fig. 1.10   High voltage insulation design of transformer using FEA.
(*Courtesy: Electrical Engg. Dept., Indian Institute of Technology Bombay.*)

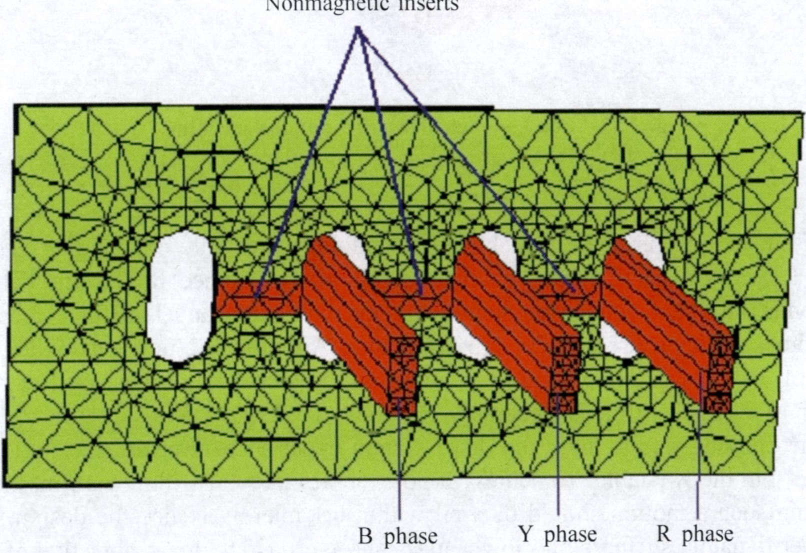

Fig. 1.11   Analysis of eddy currents using FEM.
(*Courtesy: Electrical Engg. Dept., Indian Institute of Technology Bombay.*)

Thermosonic wire bonding is one of the preferred processes for completing an electrical interconnection between a semiconductor chip and a leadframe by a thin metal wire. The wire bonder machine (solid model shown in Figure 1.12) consists of a linear motor driven precision XY-table on which a precision Z-axis assembly integrated with an ultrasonic transducer is mounted. The silicon chip is mounted on a hot plate called *heater block assembly* which is maintained at 200°C. The ultrasonic transducer helps in welding the gold wire to the silicon chip. The combined motion of the X, Y and Z-axis helps achieve a curvilinear profile to produce the required shape to the bonded wire.

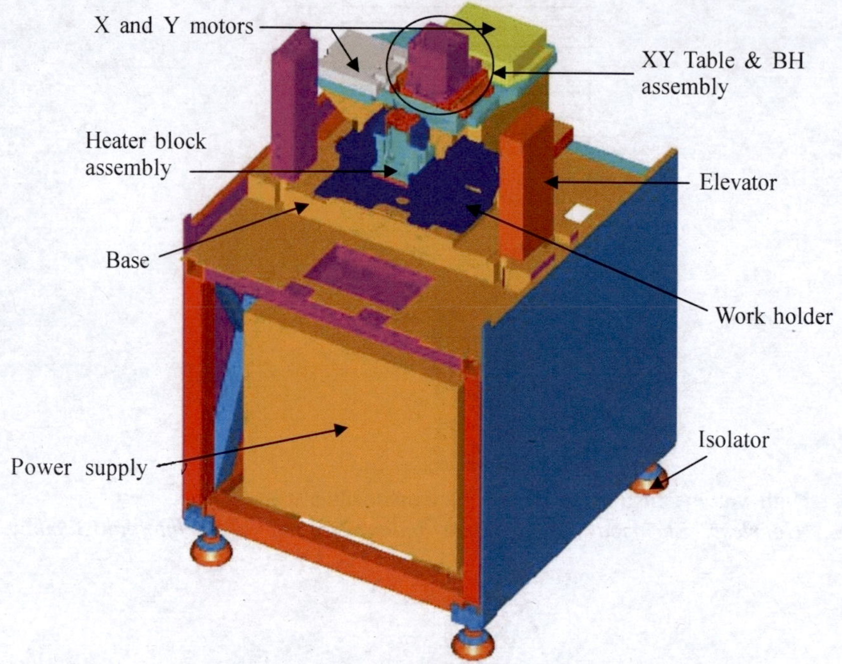

Fig. 1.12   Solid model of a wire bonder machine.
(*Courtesy: ASM International Pte Ltd., Singapore.*)

The present day needs of the wirebonder technology are quite challenging and have to meet stringent process specifications such as small pitch distance of 35 μm between the successive bonds, stringent bond placement accuracy of ±2 μm, and loop height consistency of 5 μm within a bonding area of 70 mm × 70 mm. Additionally, to meet higher output, a bonding rate of 6–12 wires per second, depending on the wire length range of 2 to 8 mm and wire diameter of 20 to 75 μm, is necessary.

Since the bonding rate is high, the drive system experiences a peak acceleration of 13 g (about 130 m/s^2) in the XY-table. To achieve a position accuracy of within ±2 μm, the residual vibration during such a motion should be well within 0.1 micron. Hence the design should be aimed at higher dynamic stiffness and lower mass and inertia. The main objective of the finite element simulation (a typical mesh is shown in Figure 1.13) is to analyse the dynamic rigidity of the system. A typical bending mode of the machine is depicted in Figure 1.14.

The next example relates to the dynamic analysis of a one-axis linear motor. The one-axis table (the FE model in Figure 1.15) consists of a three-phase linear motor that directly drives a 5 kg mass in rectilinear motion. The table is designed for a peak acceleration of about 130 m/s^2. Further, the design would be used in a typical semiconductor packaging machine such as the wire bonder (described above) which involves bonding gold wires at a high speed of 12 wire interconnects per second. Hence the finite element analysis aims at finding the resonant frequencies and the mode shapes very accurately. This will help in achieving a high dynamic rigidity with low mass. Also, knowledge of the mode shape helps in illustrating the

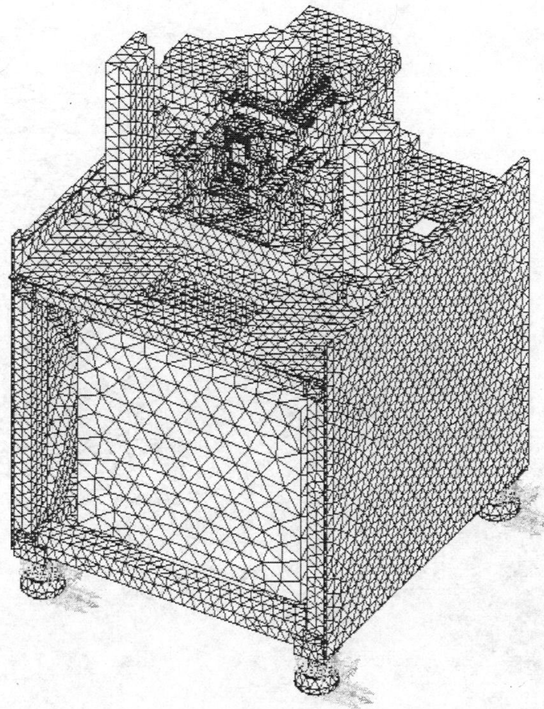

Fig. 1.13   Finite element model of a wire bonder machine.
(*Courtesy: ASM International Pte Ltd., Singapore.*)

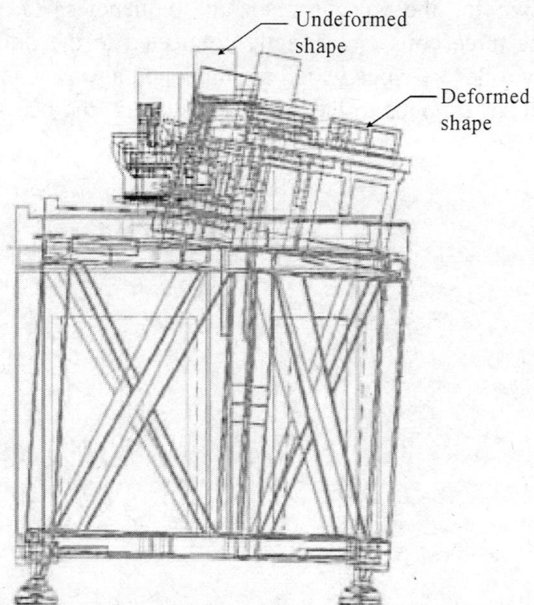

Fig. 1.14   Second mode shape (114 Hz) of wire bonder machine.
(*Courtesy: ASM International Pte Ltd., Singapore.*)

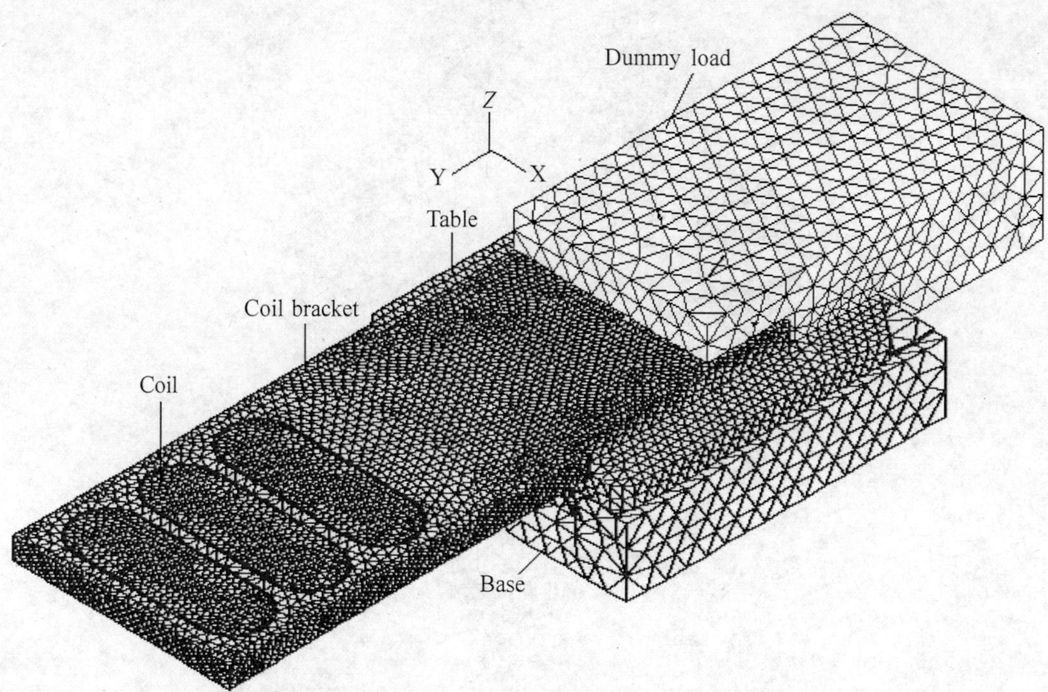

Fig. 1.15   Finite element model of a one-axis table.
(*Courtesy: ASM International Pte Ltd., Singapore.*)

weakness in the design for the various resonant frequencies. The model consists of a coil bracket containing the three coils and directly connected to the dummy mass by four screws. The table is guided by roller bearings which are modelled as equivalent springs with appropriate spring constant which corresponds to the applied preload in the bearings. A typical mode shape is shown in Figure 1.16.

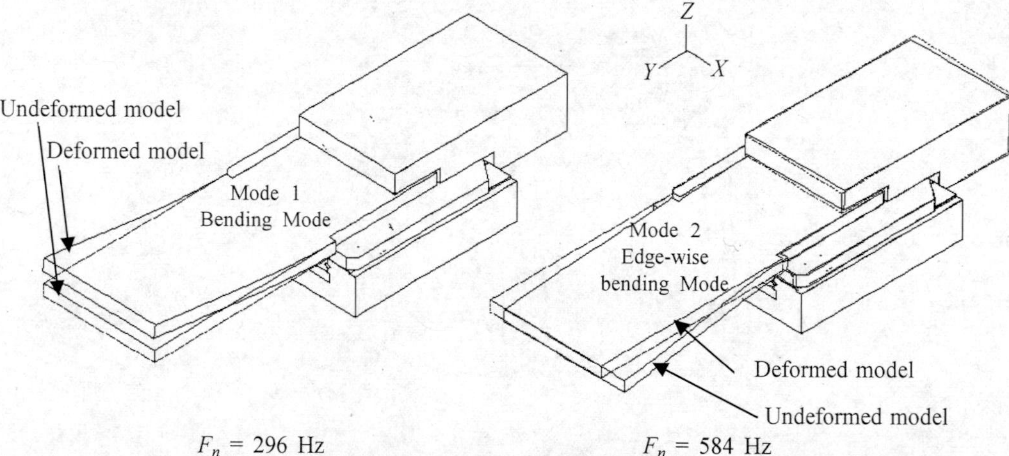

$F_n = 296$ Hz $F_n = 584$ Hz

Fig. 1.16   Natural frequencies and mode shapes of one-axis table.
(*Courtesy: ASM International Pte Ltd., Singapore.*)

Encapsulation of integrated circuits (IC) is commonly made using the transfer moulding process where a thermoset epoxy is heated and transferred under pressure so as to encapsulate the silicon IC chip. The moulding machine consists of a top and a bottom mould mounted on large platens. The top platen is supported by four steel columns, while the bottom platen guided by the four steel columns moves in a straight path using a motor-driven four bar mechanism. A very high pressure (equivalent load of about 100 tons) is maintained between the top and the bottom mould faces, so as to avoid liquid mould compound to leak at the interface. Heaters are placed in the moulds to produce the heat necessary for maintaining a constant uniform mould surface temperature of 175°C. Such a uniform temperature enables uniform melting of epoxy and its smooth flow into the mould cavities.

A typical finite element model of the complete system is illustrated in Figure 1.17. The top and bottom mould surfaces are coated with surface contact elements in order to readily determine the contact pressure. Further, thermal analysis of the model is conducted to predict the temperature distribution on the mould surfaces. Results also include the heat losses to the surroundings in the form of convection and radiation heat transfer as well as conduction heat transfer to the base of the machine. Knowledge of these heat losses helps in deciding the number of heaters and their locations as well as their corresponding wattage.

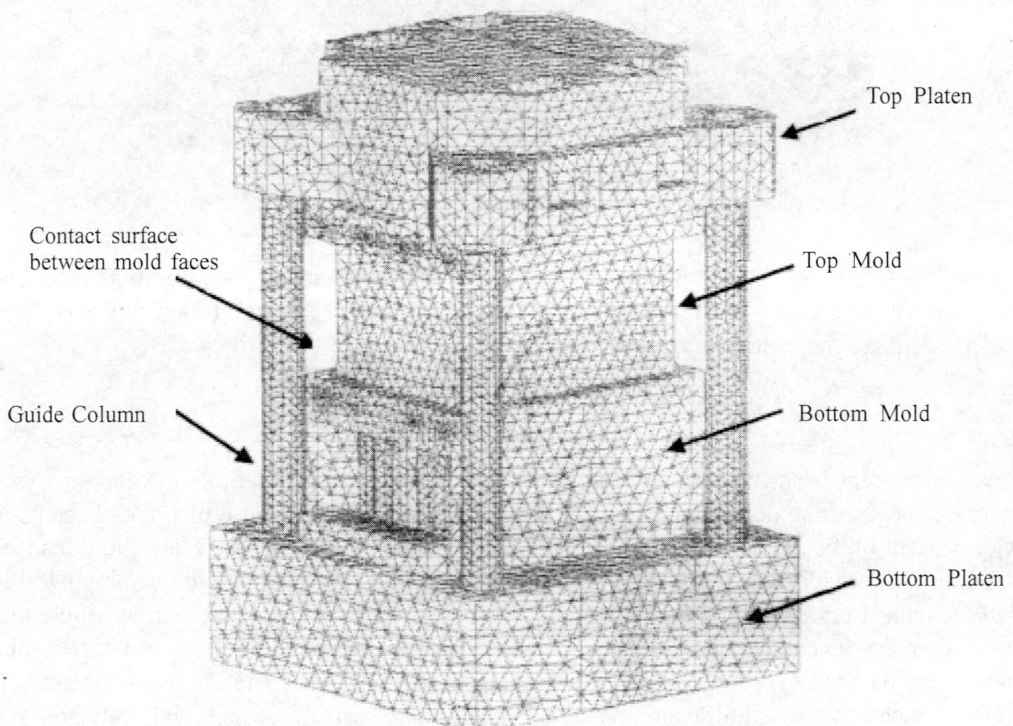

Fig. 1.17   Finite element model of a moulding machine.
(*Courtesy: ASM International Pte Ltd., Singapore.*)

1.1.4   Aerospace Applications

In typical aerospace applications, finite element analysis is used for several purposes, viz., structural analysis for natural frequencies, mode shapes, response analysis, aero-servo-elastic studies, and aerodynamics. For example, Figure 1.18 shows a typical combat aircraft designed

Fig. 1.18   Typical combat aircraft.
(*Courtesy: Aeronautical Development Agency, Bangalore.*)

and developed in India. Figure 1.19 depicts a typical deformed shape of the aircraft as predicted using finite element tools. To model the complete aircraft, various types of elements (viz., rod, shear panel, plate/shell, etc.) are used.

SUMMARY

As may be evident from the above examples of real-life application of finite element analysis, present day engineering design based on CAE tools involves extensive use of finite elements in a wide variety of fields. The question that naturally arises is how a single technique can be applied to such a wide variety of application domains. Once we recognize the commonality in the mathematical representation of these various problems (for example, the partial differential equations or the functional expression), we realize that a tool to solve a *type* of differential equation can be used to solve a *class* of problems. Knowledge of finite element principles is therefore crucial in two significant ways: to aid *intelligent* use of commercial software, and to lay a strong foundation for further research in this field. Thus this text focusses on emphasising the fundamental principles used in the formulation of finite element method in a lucid manner.

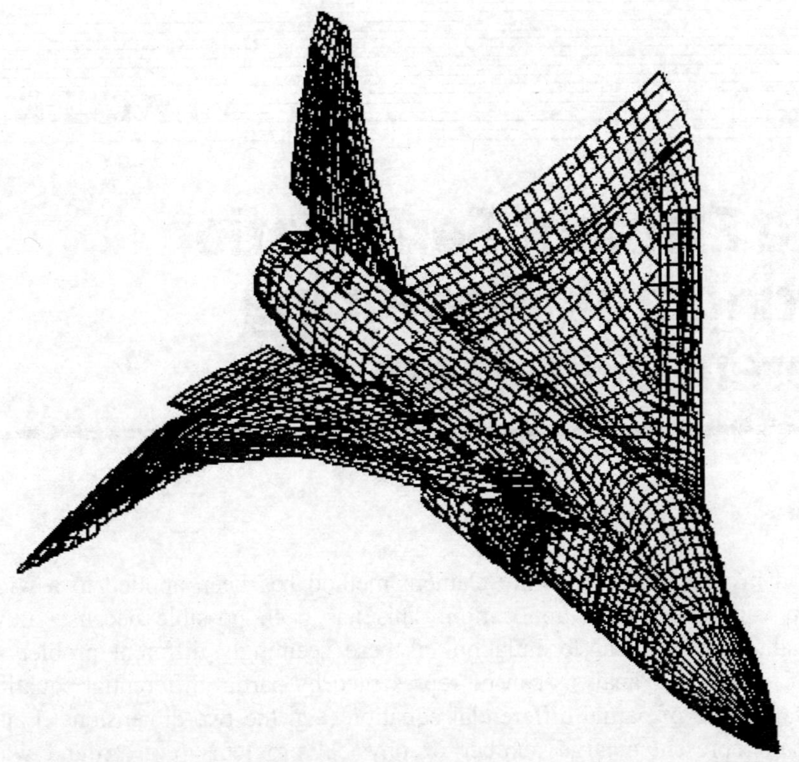

Fig. 1.19   **Finite element simulation of deformation of a typical combat aircraft.**
(*Courtesy: Aeronautical Development Agency, Bangalore.*)

CHAPTER 2

Finite Element Formulation Starting from Governing Differential Equations

As discussed in Chapter 1, the finite element method has been applied to a wide variety of problems in various fields of application. This has been possible because of the common features in the mathematical formulation of these seemingly different problems, e.g. many problems of engineering analysis can be represented by partial differential equations. In some cases, the same *type* of partial differential equation (e.g. the two-dimensional Laplace/Poisson equation) can represent a large number of physical problems (e.g. ground water seepage, torsion of bars, and heat flow). Thus, to gain a proper perspective of the method of finite elements, we would like to present it in this chapter as a method for finding an approximate solution to differential equations.

The Weighted Residual (WR) method is a powerful way of finding approximate solutions to differential equations. In particular, The Galerkin Weighted Residual formulation is the most popular from the finite element point of view. Piece-wise trial function approximation of the weak form of the Galerkin weighted residual technique forms the basis of the finite element method. In what follows, we will first introduce the general weighted residual technique and the Galerkin form of the weighted residual technique, using a set of trial functions, each of which is valid over the entire solution domain. We will then introduce the weak form of the same. Finally, we will present the piece-wise trial function approximation concept, wherein each of the trial functions used is valid only over a small part of the domain. This leads us to the formulation of the finite element method.

2.1 Weighted Residual Method—Use of a Single Continuous Trial Function

Let us consider a general problem of engineering analysis described in the form of a differential equation (to be valid within a particular domain Ω), while satisfying the prescribed boundary conditions on the boundary Γ. Our scheme of finding approximate solution to differential equations consists of the following steps:

16

➢ Assume a guess (or trial) solution to the problem. For example, for a one-dimensional problem, we may choose a trial solution as

$$f(x) = c_0 + c_1 x + c_2 x^2 + \cdots \tag{2.1}$$

➢ In general, the function so assumed will satisfy neither the differential equation within the domain (Ω) nor the boundary conditions (on Γ). By substituting the assumed function in the differential equation and the boundary conditions of the problem, find the error in satisfying these (we will call these "domain residual" and "boundary residual").

➢ Determine the unknown parameters (c_0, c_1, c_2, ...) in the assumed trial function in such a way as to make these residuals as low as possible.

In the process, if we can make the domain and boundary residuals identical to zero everywhere, we will get the exact solution to the problem itself. In general, we expect to get a reasonably accurate solution to the problem at hand. **In the context of the finite element method, we will limit our discussion to trial solutions that satisfy the applicable boundary conditions and hence, only domain residual remains.** The choice of trial solutions that implicitly satisfy the differential equation but not the boundary conditions (thus resulting in nonzero boundary residual) leads to the "boundary element method". As a detailed discussion on the boundary element method is beyond the scope of this text, the interested reader may refer to standard texts for details of this technique.

We will use the following simple example to illustrate the above method of finding approximate solutions to differential equations.

Example 2.1.   Consider a uniform rod subjected to a uniform axial load as illustrated in Figure 2.1. It can be readily shown that the deformation of the bar is governed by the differential equation

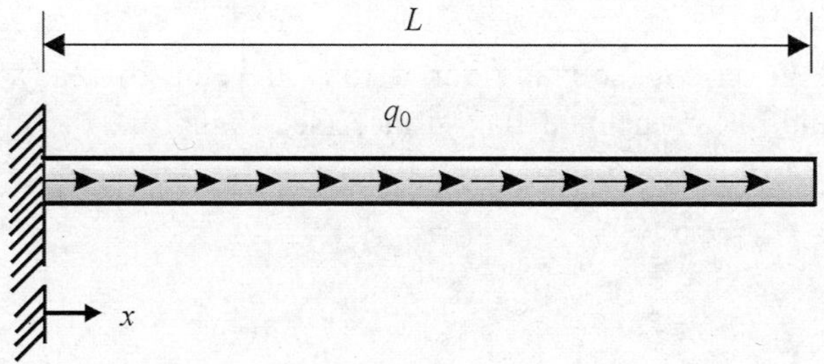

Fig. 2.1   Rod subjected to axial load.

$$AE \frac{d^2 u}{dx^2} + q_0 = 0 \tag{2.2}$$

with the boundary conditions $u(0) = 0$, $\left. \dfrac{du}{dx} \right|_{x=L} = 0$.

Let us now find an approximate solution to this problem using the method just discussed.

Step 1: *Assume a trial or guess solution.*   Let

$$u(x) \approx \hat{u}(x) = c_0 + c_1 x + c_2 x^2 \tag{2.3}$$

where the constants c_0, c_1, c_2 are yet to be determined. In order to satisfy the first boundary condition that $\hat{u}(0) = 0$, we have $c_0 = 0$. To satisfy the second boundary condition, we have $c_1 = -2c_2 L$. Thus we now have, for our trial solution,

$$\hat{u}(x) = c_2(x^2 - 2Lx) \tag{2.4}$$

Since the trial solution contains only one free parameter c_2, it is often referred to as a "one-parameter solution".

Step 2: *Find the domain residual.*   Substituting in the governing differential equation

$$R_d = AE\frac{d^2\hat{u}}{dx^2} + q_0 = AE(2c_2) + q_0 \tag{2.5}$$

Step 3: *Minimise the residual.*   Since there is one residual to be minimised and one parameter to be determined, we can readily solve for the undetermined coefficient by setting the residual to zero, i.e., $R_d = 0$, yielding

$$c_2 = \frac{-q_0}{2AE} \tag{2.6}$$

Thus our final solution is

$$\hat{u}(x) = \left(\frac{q_0}{2AE}\right)(2xL - x^2) \tag{2.7}$$

For this simple example, since we could make the residual identically zero everywhere, our final solution tallies with the exact solution.

Example 2.2. The governing equation for a fully developed steady laminar flow of a Newtonian viscous fluid on an inclined flat surface (see Figure 2.2) is given by

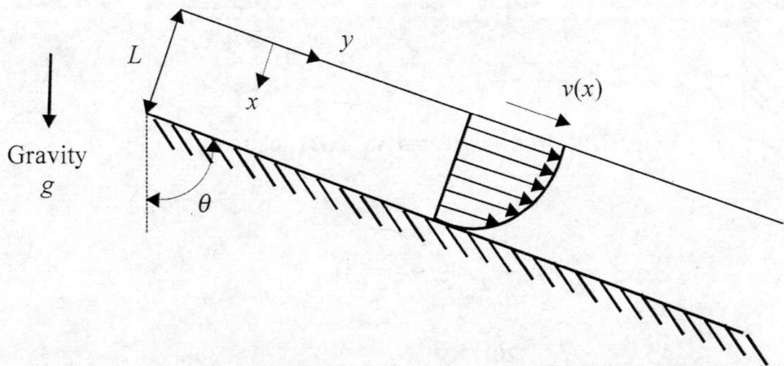

Fig. 2.2   Laminar flow on an inclined surface.

$$\mu \frac{d^2v}{dx^2} + \rho g \cos \theta = 0 \qquad (2.8)$$

where

μ = coefficient of viscosity,
v = fluid velocity,
ρ = density,
g = acceleration due to gravity,
θ = angle between the inclined surface and the vertical.

The boundary conditions are given by

$$\left. \frac{dv}{dx} \right|_{x=0} = 0 \qquad \text{(zero shear stress)} \qquad (2.9)$$

$$v(L) = 0 \qquad \text{(no slip)}$$

Let us find the velocity distribution $v(x)$ using the weighted residual method.

Step 1: *Assume a trial solution.* Let

$$v(x) \approx \hat{v}(x) = c_0 + c_1 x + c_2 x^2 \qquad (2.10)$$

Hence,

$$\frac{d\hat{v}}{dx} = c_1 + 2c_2 x \qquad (2.11)$$

From the boundary conditions, $c_1 = 0$, $c_0 = -c_2 L^2$. Therefore,

$$\hat{v}(x) = c_2(x^2 - L^2) \qquad (2.12)$$

Step 2: *Find the domain residual*

$$R_d = \mu(2c_2) + \rho g \cos \theta \qquad (2.13)$$

Step 3: *Minimise the residual.* R_d is a constant and can therefore be set to zero. Hence,

$$c_2 = \frac{-\rho g \cos \theta}{2\mu} \qquad (2.14)$$

Therefore,

$$\hat{v}_{(x)} = \frac{\rho g \cos \theta}{2\mu}(L^2 - x^2) \qquad (2.15)$$

It is readily verified that our solution matches the exact solution as we can make the domain residual identically zero.

Example 2.3. Consider the problem of a cantilever beam under uniformly distributed load q_0 as shown in Figure 2.3. The governing differential equation is given by

$$EI \frac{d^4v}{dx^4} - q_0 = 0 \qquad (2.16)$$

and the boundary conditions are given by

$$v(0) = 0, \qquad \frac{dv}{dx}(0) = 0$$

$$\frac{d^2v}{dx^2}(L) = 0, \qquad \frac{d^3v}{dx^3}(L) = 0 \tag{2.17}$$

where the first two boundary conditions enforce zero displacement and slope at the fixed end and the last two conditions prescribe zero bending moment and shear force at the free end.

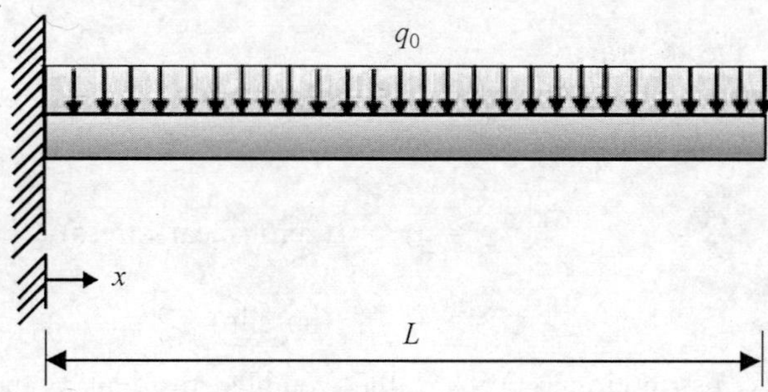

Fig. 2.3   Cantilever beam under load.

Step 1:   *Assume a trial or guess solution.*   We observe that it is not easy to select a trial function that satisfies all the boundary conditions. Let us choose $\hat{v}(x) = c_0 + c_1x + c_2x^2 + c_3x^3 + c_4x^4$.

From the boundary conditions that $\hat{v}(0) = 0$ and $d\hat{v}/dx\,(0) = 0$, we have $c_0 = 0 = c_1$. In order to satisfy the boundary conditions at $x = L$, we should have

$$c_2 = -3c_3L - 6c_4L^2, \qquad c_3 = -4c_4L \tag{2.18}$$

Substituting and rearranging the terms in the trial solution, we get

$$\hat{v}(x) = c_4(x^4 - 4Lx^3 + 6x^2L^2) \tag{2.19}$$

We thus observe that finding trial solution functions that satisfy all the boundary conditions could, in general, be cumbersome.

Step 2:   *Find the domain residual.* Substituting in the differential equation, we get the domain residual as

$$R_d(x) = 24EIc_4 - q_0 \tag{2.20}$$

Step 3:   *Minimise the residual.* Since there is one residual to be minimised and one parameter to be determined, we can readily solve for the undetermined coefficient c_4 by setting the residual to zero, i.e., $R_d = 0$, yielding thereby

$$c_4 = q_0/(24EI) \tag{2.21}$$

Thus our trial solution is

$$\hat{v}(x) = q_0/(24EI)\,[x^4 - 4Lx^3 + 6x^2L^2]$$

which can be readily verified to be the exact solution itself. This is to be expected since we were able to make the residual identically zero within the entire domain.

Example 2.4.   Let us consider the example of a simply supported beam under uniformly distributed load as shown in Figure 2.4. The governing differential equation and the boundary conditions are given by

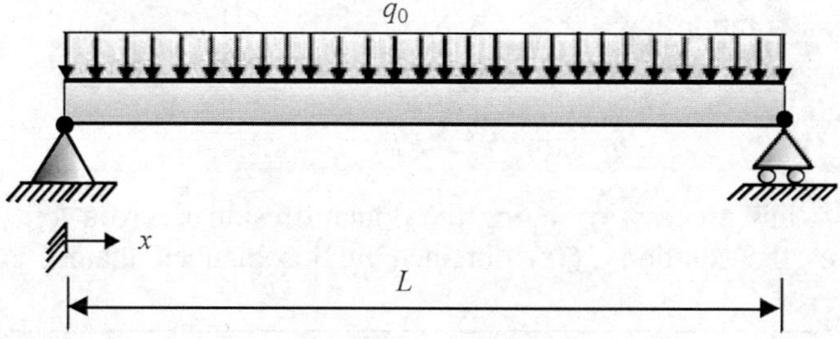

Fig. 2.4   Beam on simple supports.

$$EI\frac{d^4v}{dx^4} - q_0 = 0$$

$$v(0) = 0, \qquad \frac{d^2v}{dx^2}(0) = 0 \tag{2.22}$$

$$v(L) = 0, \qquad \frac{d^2v}{dx^2}(L) = 0$$

Step 1: *Assume a trial solution.* We can use the approach of Example 2.3 and find a polynomial trial solution satisfying all boundary conditions. However, in view of the special boundary conditions of simple supports, we can make the process simple by choosing trigonometric functions. Let

$$v(x) \approx \hat{v}(x) = c_1 \sin(\pi x/L) \tag{2.23}$$

This one-parameter trial solution satisfies all boundary conditions.

Step 2: *Find the domain residual.* Substituting the trial solution $\hat{v}(x)$ in the governing differential equation, the domain residual is obtained as

$$R_d = c_1(\pi/L)^4(EI) \sin(\pi x/L) - q_0 \tag{2.24}$$

Step 3: *Minimise the residual.*   We observe that, unlike in the previous examples, the domain residual is now varying from point to point within the domain ($0 < x < L$). Since we have only one coefficient to be determined, we can set the residual zero only at any one point of our choice within the domain if we follow the approach of previous examples. This technique is called the **point collocation technique**, wherein we set the residual (in general, a function of x) to zero at chosen points within the domain—the number of points being equal to the number of coefficients in the trial function that need to be determined. We will illustrate this procedure now. In this procedure, however, there is a danger that the residual might be unduly large at some other points within the domain. We may thus want to "minimise" the residual in an overall sense over the entire domain rather than setting it identically zero at only few selected points. This will be discussed in the next section.

Solution by point collocation.   Let us make $R_d = 0$ at $x = L/4$, i.e.

$$c_1(\pi/L)^4 (EI) \sin(\pi/4) - q_0 = 0 \tag{2.25}$$

yielding thereby

$$c_1 = \frac{\sqrt{2}}{\pi^4}\frac{q_0 L^4}{EI} \tag{2.26}$$

and the resulting trial solution is

$$\hat{v}(x) = \frac{\sqrt{2}}{\pi^4} \frac{q_0 L^4}{EI} \sin \frac{\pi x}{L} \tag{2.27}$$

We may repeat this process by setting the domain residual zero at other axial locations. Figure 2.5 compares the solutions $\hat{v}(x)$ obtained in this manner against the classical exact

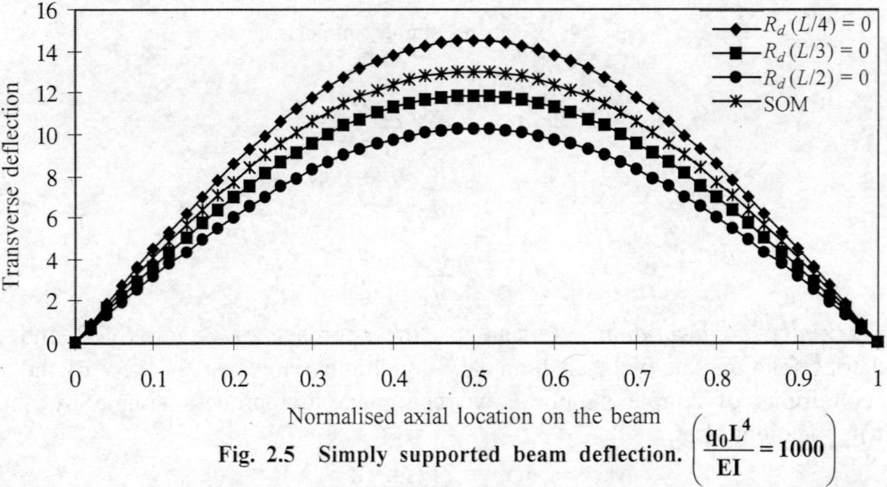

Fig. 2.5   Simply supported beam deflection. $\left(\dfrac{q_0 L^4}{EI} = 1000 \right)$

solution. Figure 2.6 shows a plot of the domain residual in each case. In this way, we are able to generate different approximate solutions to the problem but each of these solutions deviates appreciably from the exact solution. We can improve our trial solution by adding one more term to the Sine series. Since the problem at hand is symmetric about $x = L/2$, we modify our trial solution as follows:

$$v(x) \approx \hat{v}(x) = c_1 \sin (\pi x/L) + c_3 \sin (3\pi x/L) \tag{2.28}$$

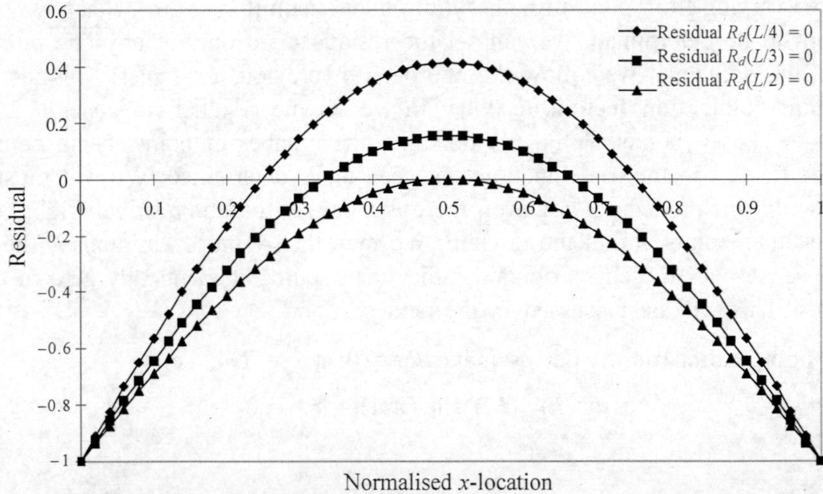

Fig. 2.6   Plot of residual-simply supported beam (Example 2.4).

With this two-term trial function, we obtain the domain residual as

$$R_d = c_1(\pi/L)^4(EI) \sin(\pi x/L) + c_3(3\pi/L)^4(EI) \sin(3\pi x/L) - q_0 \tag{2.29}$$

Since we have two constants to be determined, we can set the residual zero at two selected points. Let us set

$$R_d\left(\frac{L}{4}\right) = R_d\left(\frac{L}{3}\right) = 0 \tag{2.30}$$

i.e.,

$$c_1\left(\frac{\pi}{L}\right)^4 EI\left(\frac{1}{\sqrt{2}}\right) + c_3\left(\frac{3\pi}{L}\right)^4 EI\left(\frac{1}{\sqrt{2}}\right) - q_0 = 0 \tag{2.31}$$

$$c_1\left(\frac{\pi}{L}\right)^4 EI\left(\frac{\sqrt{3}}{2}\right) + c_3\left(\frac{3\pi}{L}\right)^4 EI(0) - q_0 = 0 \tag{2.32}$$

Solving these equations, we obtain

$$c_1 = \frac{2}{\sqrt{3}\,\pi^4}\frac{q_0 L^4}{EI} \tag{2.33}$$

$$c_3 = 0.00003289\,\frac{q_0 L^4}{EI} \tag{2.34}$$

Figure 2.7 compares the solution $\hat{v}(x)$ obtained in this manner against the classical exact solution. Figure 2.8 shows a plot of the domain residual. We observe that this approximates the solution perhaps better than the earlier one-parameter solution, but it still deviates appreciably from the exact solution.

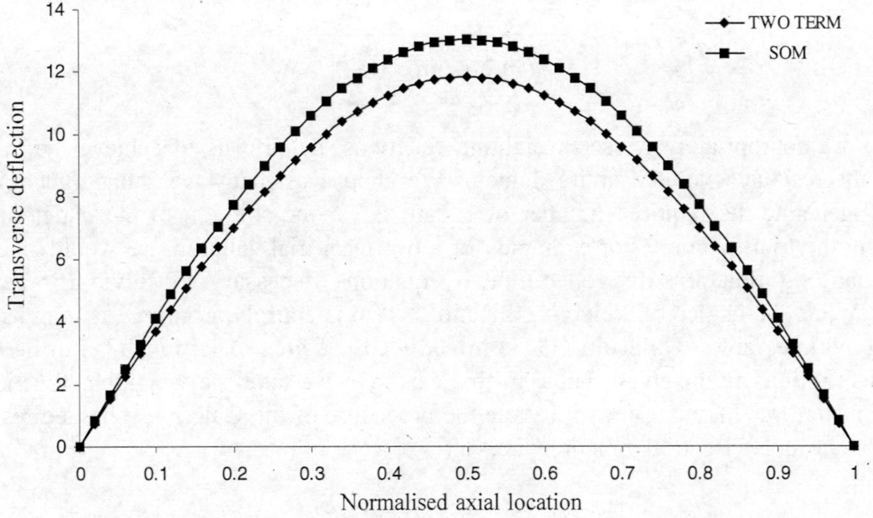

Fig. 2.7   Simply supported beam deflection—Two-parameter solution.

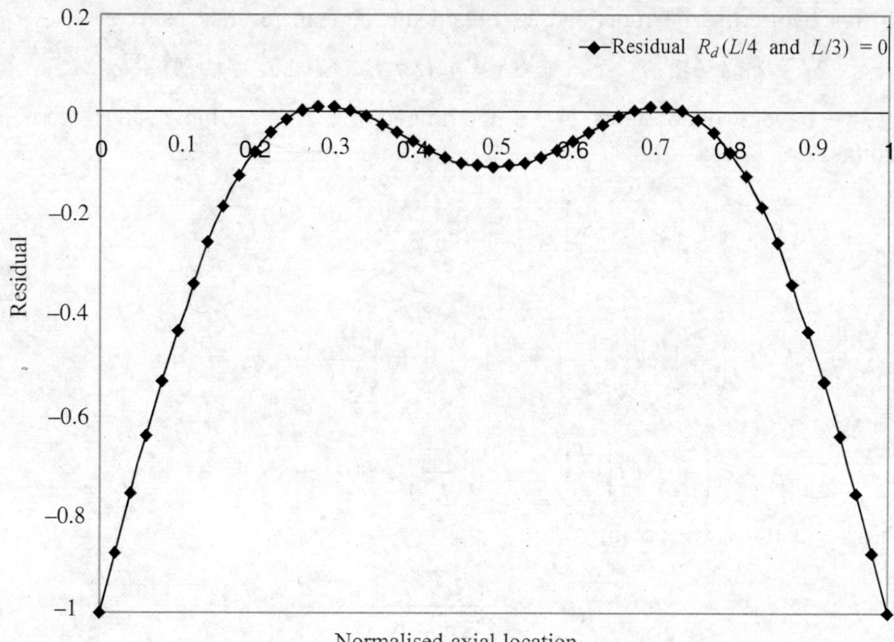

Fig. 2.8   Domain residual—Two-parameter solution $R_d(L/4$ and $L/3) = 0$.

Rather than trying to set the residual exactly zero at a few selected points and having no control over the residual at all the other points in the domain, we will now try to minimise the residual in an overall sense. This technique is called the *weighted residual technique*.

Solution by weighted residual technique.   For the problem at hand, we will formulate this technique as

$$\int_0^L W_i(x)R_d(x)\, dx = 0 \tag{2.35}$$

where $W_i(x)$ are appropriately chosen weighting functions, helping us to achieve the task of minimising the residual over the entire domain. We choose as many weighting functions as necessary to generate the required number of equations for the solution of the undetermined coefficients in the trial function. For example, for a two-term trial solution, we would take two different weighting functions to generate two equations necessary to solve for the two coefficients. While the choice of weighting functions $W(x)$ is entirely arbitrary (as long as they are nonzero and integrable), **Galerkin (1915) introduced the idea of letting $W(x)$ to be same as the trial functions themselves.** Thus, in this example, we take the weighting function to be $\sin(\pi x/L)$, $\sin(3\pi x/L)$, etc. Let us illustrate the procedure of the Galerkin weighted residual technique for the one-term trial function, i.e., let

$$v(x) \approx \hat{v}(x) = c_1 \sin(\pi x/L) \tag{2.36}$$

The domain residual

$$R_d = c_1(\pi/L)^4(EI) \sin (\pi x/L) - q_0 \tag{2.37}$$

By the Galerkin procedure, we set

$$\int_0^L \underbrace{\sin \frac{\pi x}{L}}_{w(x)} \left[\underbrace{c_1\left(\frac{\pi}{L}\right)^4 EI \sin \frac{\pi x}{L} - q_0}_{R_d(x)} \right] dx = 0 \tag{2.38}$$

We have chosen only one weighting function because there is only one constant c_1 to be determined. Thus we have

$$\left(\frac{\pi}{L}\right)^4 EIc_1 \int_0^L \sin^2 \frac{\pi x}{L} dx = \int_0^L q_0 \sin \frac{\pi x}{L} dx \tag{2.39}$$

yielding $c_1 = 0.013071 q_0 L^4/EI$. Therefore,

$$\hat{v}(x) = 0.013071 \frac{q_0 L^4}{EI} \sin \frac{\pi x}{L} \tag{2.40}$$

Figure 2.9 compares the solution $\hat{v}(x)$ obtained in this manner against the classical exact solution. It is observed that the one-term solution is significantly improved by the Galerkin

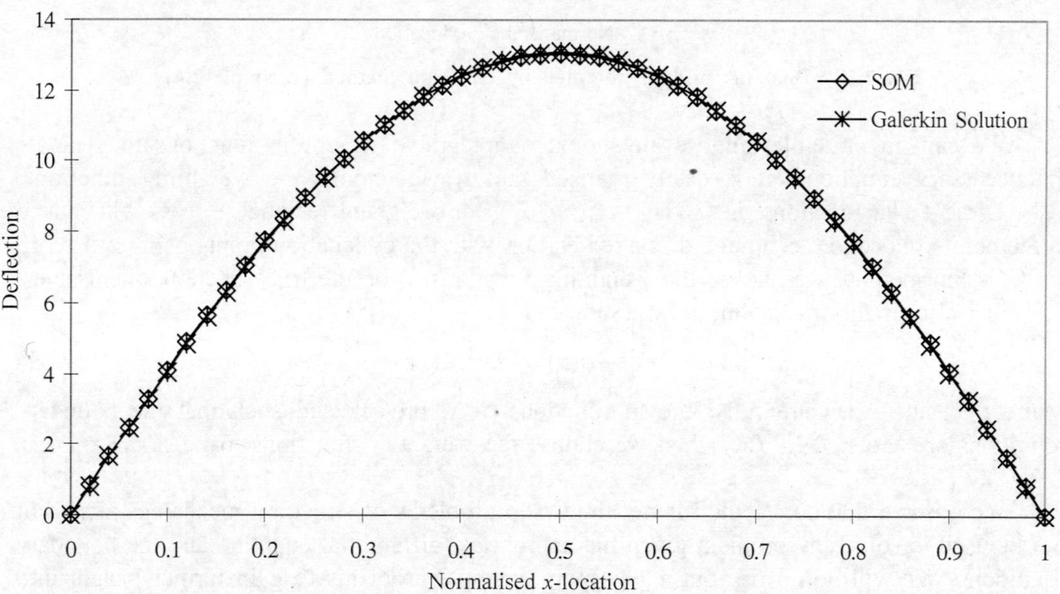

Fig. 2.9   Strength of material (SOM) and Galerkin WR solutions—Simply supported beam (Example 2.4).

weighted residual minimisation process. The error in the deflection at the mid-span is just 0.38%. Figure 2.10 shows a plot of the domain residual obtained by the Galerkin process.

We have thus discussed a very powerful tool for determining approximate solutions to boundary value problems, given the governing differential equation and the boundary conditions. This method requires us to choose trial solutions that satisfy the boundary conditions of the problem. We then generate sufficient number of equations to solve for the undetermined parameters by setting the weighted integral of the residual to zero.

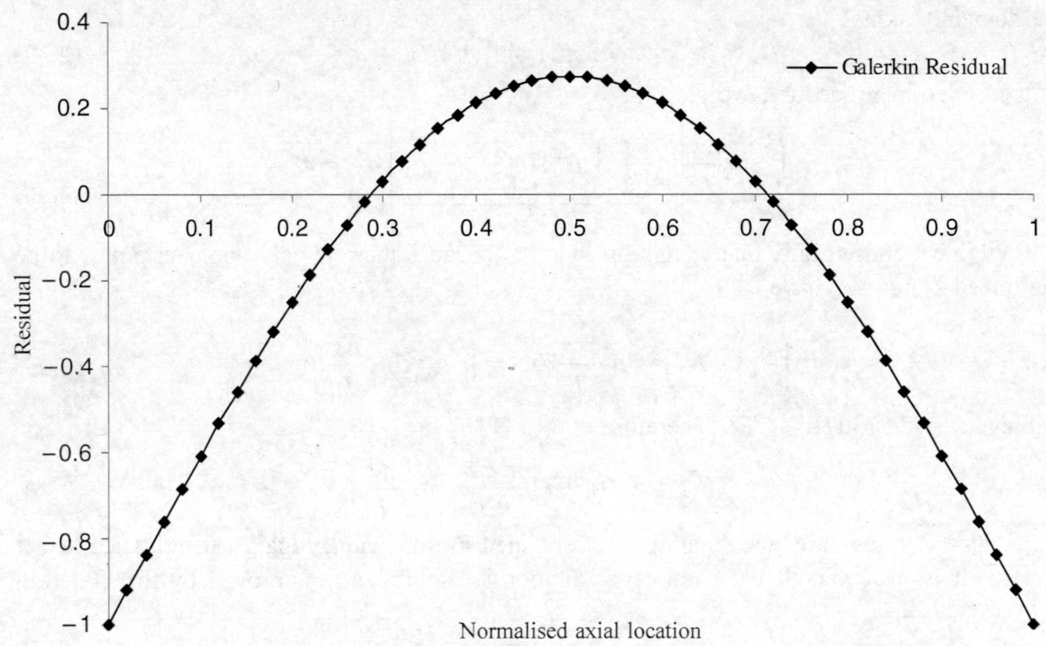

Fig. 2.10   Domain residual obtained by Galerkin method (Example 2.4).

We can, in principle, choose any nonzero, integrable weighting functions to generate the necessary equations. It is easily realised that if we choose the weighting functions, to be Dirac–Delta functions, i.e., $W(x) = \delta(x - x_p)$, the weighted residual process will reduce to the point collocation technique discussed earlier, with the collocation points being $x = x_p$. In the Galerkin method, we choose the weighting functions to be the trial functions themselves. To be precise, if the trial solution is assumed to be

$$f(x) = \phi(x) + \Sigma\, c_i N_i(x) \tag{2.41}$$

where $\phi(x)$ and $N_i(x)$ are fully known functions of x, pre-selected such that the boundary conditions are satisfied by $f(x)$, then we choose the weighting functions to be $W_i(x) = \partial f/\partial c_i = N_i(x)$.

We observe that our method is not limited to problems of structural mechanics alone but can be used to solve any problem given the appropriate differential equation and the boundary conditions. We will now work out a few more example problems to gain further insight into the Galerkin Weighted Residual technique. Our next example illustrates the application of this technique to heat transfer problems.

Example 2.5.   Consider the problem of a long cylinder of radius R with uniformly distributed heat source q_0. The governing differential equation is given by

$$\frac{d^2T}{dr^2} + \frac{1}{r}\frac{dT}{dr} + \frac{q_0}{k} = 0 \tag{2.42}$$

and the boundary conditions are given by

$$T(R) = T_w, \quad \text{the wall temperature}$$

$$q_0 \pi R^2 L = (-k)(2\pi RL)\left.\frac{dT}{dr}\right|_{r=R} \tag{2.43}$$

The second boundary condition implies that heat generated = heat lost.

We wish to determine the temperature distribution T as a function of radial location, r.

Step 1: *Assume a trial solution.* We observe that it may be convenient to choose the trial solution in terms of $(r - R)$ rather than r.

Let us choose the trial solution as

$$T = c_0 + c_1(r - R) + c_2(r - R)^2 \tag{2.44}$$

From the boundary conditions, we find that $c_0 = T_w$ and $c_1 = -(q_0 R/2k)$. Thus we have

$$T = [T_w - (q_0 R/2k)(r - R)] + c_2(r - R)^2 \tag{2.45}$$

In our notation introduced above, we can write $T = \phi(r) + c_2 N(r)$ where $N(r) = (r - R)^2$ will be our weighting function $W(r)$.

Step 2: *Find the domain residual.* The domain residual

$$R_d = 2c_2 + q_0/k + (1/r)[-(q_0 R/2k) + 2c_2(r - R)] \tag{2.46}$$

Step 3: *Minimise the residual.* We observe that the residual is actually minimised over the whole volume of the domain and in radial coordinates, an elemental volume is given by $(r\, dr\, d\theta\, dz)$. Considering the fact that temperature does not vary in the circumferential direction (θ) or the axial direction (z) for this problem, explicit integration in these two directions leads to the constant $(2\pi L)$, and the weighted residual statement is given by

$$2\pi L \int_0^R (r - R)^2 [R_d] r\, dr = 0 \tag{2.47}$$

Solving, we get $c_2 = -\dfrac{q_0}{4k}$. Therefore, our approximation to the temperature distribution is

$$\hat{T}(r) - T_w = \frac{q_0}{4k}(R^2 - r^2) \tag{2.48}$$

Thus we observe that our procedure can be systematically applied to any problem, given the differential equation and the boundary conditions.

We will now illustrate the Galerkin Weighted Residual procedure on a two-term trial function through the following example.

Example 2.6. Consider the uniform bar problem discussed in Example 2.1. Let the bar be subjected to a linearly varying load $q = \alpha x$. The governing differential equation is given by

$$AE\frac{d^2u}{dx^2} + \alpha x = 0 \tag{2.49}$$

with the boundary conditions $u(0) = 0$, $AE\dfrac{du}{dx}\Big|_{x=L} = 0$.

Step 1:   *Assume a trial or guess solution.* In view of the load term, we expect that our trial solution should be a higher degree polynomial than that used in Example 2.1. Let

$$u(x) \approx \hat{u}(x) = c_0 + c_1 x + c_2 x^2 + c_3 x^3 \tag{2.50}$$

From the boundary conditions, we get

$$c_0 = 0, \qquad c_1 = -(2c_2 L + 3c_3 L^2) \tag{2.51}$$

Thus

$$\hat{u}(x) = c_2(x^2 - 2Lx) + c_3(x^3 - 3L^2 x) \tag{2.52}$$

Step 2:   *Find the domain residual*

$$R_d = AE(2c_2 + 6c_3 x) + ax \tag{2.53}$$

Step 3:   *Minimise the residual.* The weighting functions to be used in the Galerkin technique are the trial functions themselves, i.e., $W_1(x) = (x^2 - 2Lx)$ and $W_2(x) = (x^3 - 3L^2 x)$.

Thus we have the weighted residual statements

$$\int_0^L (x^2 - 2Lx)[AE(2c_2 + 6c_3 x) + ax]\, dx = 0 \tag{2.54}$$

$$\int_0^L (x^3 - 3L^2 x)[AE(2c_2 + 6c_3 x) + ax]\, dx = 0 \tag{2.55}$$

Solving for c_2 and c_3, we obtain

$$c_2 = 0, \qquad c_3 = -(a/(6AE)) \tag{2.56}$$

Thus, $\hat{u}(x) = (a/(6AE))(3L^2 x - x^3)$, which is readily verified to tally with the exact solution.

2.2   The General Weighted Residual (WR) Statement

Having understood the basic technique and successfully solved a few problems, we can now attempt to write down the general weighted residual statement.

For the unknown field variable u, we assume an approximate solution of the form

$$\boxed{u \approx \hat{u} = \phi + \sum_{i=1}^{n} c_i N_i} \tag{2.57}$$

where c_i are the independent coefficients to be determined by the WR process and the functions ϕ and N_i are preselected such that $\hat{u}$ satisfies all the prescribed boundary conditions. Let $R_d\,(x, y, z)$ be the domain residual. The WR statement can then be written as

$$\boxed{\int_\Omega W_i R_d\, d\Omega = 0 \quad \text{for } i = 1, 2,..., n} \tag{2.58}$$

where $W_i = N_i$.

It is observed that this criterion is implicitly satisfied by the exact solution *for any and every weighting function* since R_d is trivially zero everywhere. Thus the principal idea of weighted residual approach is that, if we are able to satisfy this criterion for a sufficiently large number of independent weighting functions, then it is likely that the assumed solution will be reasonably close to the exact solution. More strongly, if we have taken a series representation (e.g. polynomial, trigonometric) for the trial function (and, therefore, *W*), we expect that our result will get better as we include more terms in the series. Thus we expect good convergence properties.

Example 2.7.   To show the application of the Galerkin technique to a more complex situation, consider the simply supported rectangular plate subjected to uniform load as shown in Figure 2.11. The governing differential equation is given by

$$\frac{Eh^3}{12(1-v^2)}\left(\frac{\partial^4 w}{\partial x^4} + 2\frac{\partial^4 w}{\partial x^2\,\partial y^2} + \frac{\partial^4 w}{\partial y^4}\right) - q_0 = 0 \tag{2.59}$$

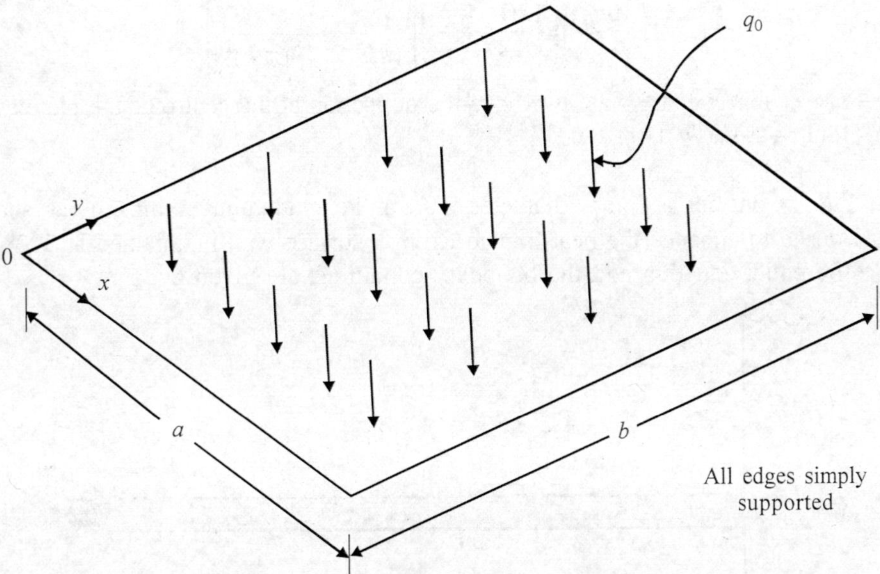

Fig. 2.11   A simply supported plate.

and the boundary conditions by

$$w(0, y) = 0 = w(a, y), \qquad \partial^2 w/\partial x^2 = 0 \text{ for } x = 0 \text{ and } a$$
$$w(x, 0) = 0 = w(x, b) \qquad \partial^2 w/\partial y^2 = 0 \text{ for } y = 0 \text{ and } b$$

Step 1:   *Assume a trial or guess solution.* In view of the boundary conditions for simple supports, we will choose a trigonometric function for trial solution.

Let us choose a one-term trial function given by

$$w(x, y) \approx \hat{w}(x, y) = c_{11} \sin(\pi x/a) \sin(\pi y/b) \tag{2.60}$$

It is easy to see that our trial function satisfies all the boundary conditions of the problem.

Step 2: *Find the domain residual.* The domain residual is obtained by substituting $\hat{w}(x, y)$ in the differential equation as follows:

$$R_d = \frac{Eh^3}{12(1-v^2)} c_{11} \left[\left(\frac{\pi}{a}\right)^2 + \left(\frac{\pi}{b}\right)^2 \right]^2 \sin\frac{\pi x}{a} \sin\frac{\pi y}{b} - q_0 \qquad (2.61)$$

Step 3: *Minimise the residual.* The weighting function is $W(x, y) = \sin(\pi x/a)\sin(\pi y/b)$. The weighted residual statement will now be

$$\int_0^b \int_0^a \left(\sin\frac{\pi x}{a}\sin\frac{\pi y}{b}\right)\left\{ \frac{Eh^3}{12(1-v^2)} c_{11}\left[(\pi/a)^2 + (\pi/b)^2\right]^2 \sin\frac{\pi x}{a}\sin\frac{\pi y}{b} - q_0 \right\} dx\,dy = 0 \quad (2.62)$$

Solving for c_{11}, we get

$$c_{11} = \left(\frac{16q_0}{\pi^2}\right)\left(\frac{12(1-v^2)}{Eh^3}\right)\left[\frac{1}{(\pi/a)^2 + (\pi/b)^2}\right]^2 \qquad (2.63)$$

If $a = b$, i.e., for a square plate, we have the deflection at the centre of the plate as given by $(4/\pi^6)\,[12(1 - v^2)/(Eh^3)]\,(q_0 a^4)$.

Example 2.8. Consider a 1 mm diameter, 50 mm long aluminium pin-fin (as shown in Figure 2.12) used to enhance the heat transfer from a surface wall maintained at 300°C. The governing differential equation and the boundary conditions are given by

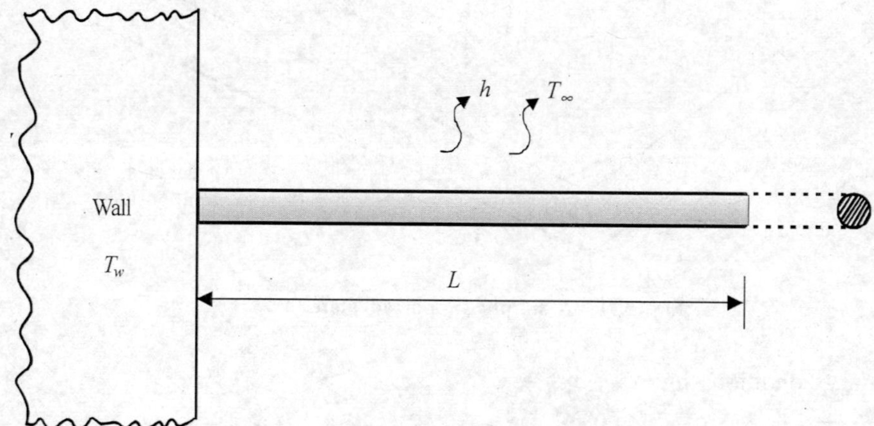

Fig. 2.12   A pin-fin.

$$k\frac{d^2T}{dx^2} = \frac{Ph}{A_c}(T - T_\infty)$$

$$T(0) = T_w = 300°C$$

$$\frac{dT}{dx}(L) = 0 \qquad \text{(insulated tip)} \qquad (2.64)$$

where

    k = coefficient of thermal conductivity,
    P = perimeter,
    A_c = cross-sectional area,
    h = convective heat transfer coefficient,
    T_w = wall temperature,
    T_∞ = ambient temperature.

Let $k = 200$ W/m/°C for aluminium, $h = 20$ W/m²°C, $T_\infty = 30$°C. Estimate the temperature distribution in the fin using the Galerkin weighted residual method.

    With the given numerical values, we have

$$\frac{d^2T}{dx^2} = 400(T - 30) \tag{2.65}$$

with $T(0) = 300$°C, $\dfrac{dT}{dx}(L) = 0$.

Step 1: *Assume a trial solution.* Let

$$T(x) \approx \hat{T}(x) = c_0 + c_1 x + c_2 x^2$$

From the boundary conditions, $c_0 = 300$, $c_1 = -2c_2 L$. Therefore,

$$\hat{T}(x) = 300 + c_2(x^2 - 2Lx) \tag{2.66}$$

Step 2: *Compute the residual.* Substituting $\hat{T}(x)$ in the differential equation, we get

$$R_d(x) = 2c_2 - 400[270 + c_2(x^2 - 2Lx)] \tag{2.67}$$

Step 3: *Minimise the residual.* The weight function $W(x) = x^2 - 2Lx$. Thus we have

$$\int_0^L (x^2 - 2Lx)R_d(x)\,dx = 0 \tag{2.68}$$

Solving for c_2, we get $c_2 = 38{,}751.43$. Hence,

$$\hat{T}(x) = 300 + 38{,}751.43(x^2 - 2Lx) \tag{2.69}$$

We can readily obtain the exact solution as

$$T(x)\Big|_{\text{exact}} = T_\infty + \left(\frac{T_w - T_\infty}{\cosh(mL)}\right)\cosh[m(L - x)]$$

where

$$m^2 = \frac{hP}{kA_c} \tag{2.70}$$

    We observe that the exact solution is an exponential function. The approximate solution obtained just now is quadratic. We can improve our approximation by taking higher degree terms in the polynomial trial function. Let

$$\hat{T}(x) = c_0 + c_1 x + c_2 x^2 + c_3 x^3 \tag{2.71}$$

From the boundary conditions, $c_0 = 300°C$.

$$c_1 = -2c_2L - 3c_3L^2 \tag{2.72}$$

Thus,

$$\hat{T}(x) = 300 + c_2(x^2 - 2Lx) + c_3(x^3 - 3L^2x) \tag{2.73}$$

The weighted residual equations can be developed using the two weighting functions $(x^2 - 2Lx)$ and $(x^3 - 3L^2x)$. We can get the solution as

$$c_2 = 48,860.4, \qquad c_3 = -109,229 \tag{2.74}$$

Thus,

$$\hat{T}(x) = 300 + 48,860.4(x^2 - 2Lx) - 109,229(x^3 - 3L^2x) \tag{2.75}$$

We can further improve upon our solution by taking one more term in the series, i.e.

$$\hat{T}(x) = c_0 + c_1x + c_2x^2 + c_3x^3 + c_4x^4 \tag{2.76}$$

From the boundary conditions we have

$$c_0 = 300, \qquad c_1 = -2c_2L - 3c_3L^2 - 4c_4L^3 \tag{2.77}$$

Thus,

$$\hat{T}(x) = 300 + c_2(x^2 - 2Lx) + c_3(x^3 - 3L^2x) + c_4(x^4 - 4L^3x) \tag{2.78}$$

We can develop the three weighted residual equations using the weighting functions $(x^2 - 2Lx)$, $(x^3 - 3L^2x)$ and $(x^4 - 4L^3x)$. We can get the solution as

$$c_2 = 53,702.8, \qquad c_3 = -255,668, \qquad c_4 = 1.32 \times 10^6$$

Therefore,

$$\hat{T}(x) = 300 + 53,702.8\,(x^2 - 2Lx) - 255,668\,(x^3 - 3L^2x) + 1.32 \times 10^6\,(x^4 - 4L^3x) \tag{2.79}$$

Table 2.1 compares the various approximate solutions with the exact solution. It is observed that the accuracy of the Galerkin approximate solutions can be systematically improved by taking more and more terms in the series solution. However, the mathematical

Table 2.1   Comparison of Various Solutions for Temperature in a Fin

Axial location	Quadratic solution	Cubic solution	Quartic solution	Exact solution
0	300	300	300	300
0.005	281.59	280.87	280.75	280.75
0.01	265.12	264.11	264.00	264.02
0.015	250.59	249.62	249.60	249.62
0.02	238.00	237.33	237.39	237.43
0.025	227.34	227.16	227.27	227.31
0.03	218.62	219.02	219.12	219.16
0.035	211.84	212.83	212.86	212.91
0.04	207.00	208.51	208.43	208.49
0.045	204.09	205.98	205.79	205.85
0.05	203.12	205.16	204.91	204.97

manipulations become very tedious. We will instead try to improve the solution while working with simpler, lower order polynomials. Towards this purpose, we will first introduce the weak form of the weighted residual statement.

2.3 Weak (Variational) Form of the Weighted Residual Statement

Considering the general weighted residual statement (Eq. 2.58), it is possible to carry out integration by parts and recast the general WR statement. This will reduce the continuity requirement on the trial function assumed in the solution. We will thus have a much wider choice of trial functions. Let us illustrate this with the following example.

Example 2.9. Let us reconsider Example 2.6 in which we had the governing differential equation

$$AE\frac{d^2u}{dx^2} + ax = 0 \tag{2.80}$$

with the boundary conditions $u(0) = 0$, $AE\frac{du}{dx}\bigg|_{x=L} = 0$.

Let $\hat{u}$ be the trial solution assumed. Substituting in the differential equation, we find the domain residual as $R_d = AE\,(d^2\hat{u}/dx^2) + ax$.

With $W(x)$ as the weighting function, we have the weighted residual statement

$$\int_0^L W(x)\left[AE\frac{d^2\hat{u}}{dx^2} + ax\right]dx = 0 \tag{2.81}$$

i.e.

$$\int_0^L W(x)AE\frac{d^2\hat{u}}{dx^2}\,dx + \int_0^L W(x)\,ax\,dx = 0 \tag{2.82}$$

or

$$\int_0^L W(x)\,d\left(AE\frac{d\hat{u}}{dx}\right) + \int_0^L W(x)ax\,dx = 0 \tag{2.83}$$

Recall the standard formula for integration by parts (for a definite integral)

$$\int_\alpha^\beta u\,dv = [uv]_\alpha^\beta - \int_\alpha^\beta v\,du = (uv)\big|_\beta - (uv)\big|_\alpha + \int_\alpha^\beta v\,du \tag{2.84}$$

We observe that, in our case, $u = W(x)$ and $v = AE\dfrac{d\hat{u}}{dx}$. Integrating the first term in Eq. (2.83) by parts, we have

$$\left[W(x)AE\frac{d\hat{u}}{dx}\right]_0^L - \int_0^L\left(AE\frac{d\hat{u}}{dx}\right)\frac{dW}{dx}\,dx + \int_0^L W(x)\,ax\,dx = 0 \tag{2.85}$$

Writing $AE(d\hat{u}/dx) = P$, the axial force at the section, we can rewrite the above equation as

$$\boxed{W(L)P_L - W(0)P_0 - \int_0^L AE\frac{d\hat{u}}{dx}\frac{dW}{dx}\,dx + \int_0^L W(x)\,ax\,dx = 0} \tag{2.86}$$

We observe that the continuity demanded on $u(x)$ has gone down. In the original weighted residual statement, Eq. (2.81), we had the term $d^2\hat{u}/dx^2$ while in Eq. (2.86) we have only $d\hat{u}/dx$. Thus our choice of $\hat{u}(x)$ should at least be quadratic for the original weighted residual statement, while with the modified form, even linear trial functions are permissible. Also, the force (natural) boundary conditions have been explicitly brought out in the WR statement itself. We can actually substitute for known natural boundary conditions on P_0 or P_L. (Thus the trial function assumed need only satisfy the "essential" boundary condition at $x = 0$, i.e. $u(0) = 0$.) This form of the WR statement is called the *weak (or variational) form* of the weighted residual statement. It is referred to as the weak form because of the weaker continuity demand on the trial solution. From now on, we will use only this form of the WR technique.

From the prescribed natural boundary condition, $AE\dfrac{du}{dx}\bigg| = P_L = 0$. Also, in view of the prescribed essential boundary condition at $x = 0$ on $u(0)$, we require that the weighting function $W(x)$ be such that $W(0) = 0$. Thus the weak form becomes

$$\boxed{\int_0^L AE\frac{d\hat{u}}{dx}\frac{dW}{dx}\,dx = \int_0^L W(x)\,ax\,dx \qquad \text{subject to } u(0) = 0, \qquad W(0) = 0}$$

(2.87)

Let us clearly state at this stage that we now have three equivalent ways of formulating the problem as shown in Table 2.2.

Table 2.2   Equivalent Representation of a Problem Statement

Differential equation	Weighted residual statement	Weak form
$AE\dfrac{d^2u}{dx^2} + ax = 0$	$\int_0^L W\left(AE\dfrac{d^2\hat{u}}{dx^2} + ax\right)dx = 0$	$\int_0^L AE\dfrac{d\hat{u}}{dx}\dfrac{dW}{dx}\,dx = \int_0^L W\,ax\,dx$
subject to $u(0) = 0$	subject to $\hat{u}(0) = 0$	subject to $\hat{u}(0) = 0$
$AE\dfrac{du}{dx}(L) = 0$	$AE\dfrac{d\hat{u}}{dx}(L) = 0$	$W(0) = 0$

To illustrate the solution of the weak form statement of the above problem, let

$$u(x) \approx \hat{u}(x) = c_1 x + c_2 x^2 \tag{2.88}$$

Then we have

$$d\hat{u}/dx = c_1 + 2c_2 x, \qquad W_1 = x, \qquad W_2 = x^2$$
$$dW_1/dx = 1, \qquad dW_2/dx = 2x$$

We observe that $\hat{u}(0) = 0$, $W_1(0) = 0$ and $W_2(0) = 0$, as required.

Weak form w.r.t. W_1

From Eq. (2.87),

$$\int_0^L (AE)(c_1 + 2c_2 x)(1)\, dx = \int_0^L (x)(ax)\, dx$$

i.e.,

$$AE(c_1 L + c_2 L^2) = aL^3/3 \tag{2.89}$$

Weak form w.r.t, W_2

$$\int_0^L (AE)(c_1 + 2c_2 x)(2x)\, dx = \int_0^L (x^2)(ax)\, dx$$

i.e.,

$$AE\left(\frac{c_1 L^2 + 4c_2 L^3}{3}\right) = \frac{aL^4}{4} \tag{2.90}$$

Rearranging the two equations, we have

$$c_1 + c_2 L = \frac{aL^2}{3\,AE}$$

$$\tag{2.91}$$

$$c_1 + \frac{4}{3} c_2 L = \frac{aL^2}{4\,AE}$$

yielding

$$c_1 = \frac{7aL^2}{12AE}, \qquad c_2 = -\frac{aL}{4\,AE}$$

Thus the deflection $\hat{u}(x)$ is obtained as

$$\hat{u}(x) = \frac{aL}{12\,AE}(7xL - 3x^2) \tag{2.92}$$

It is easily verified that the exact solution is

$$u(x) = \frac{a}{6AE}(3L^2 x - x^3)$$

Thus our approximate solution tallies with the exact solution only at both the ends but is otherwise deficient (being only quadratic). From Example 2.6, it is observed that we needed a cubic trial function to get the exact solution using the Galerkin weighted residual statement. It is left as an exercise for the reader to show that if we were to take the same quadratic trial solution, i.e. $\hat{u}(x) = c_0 + c_1 x + c_2 x^2$, but work with the original Galerkin weighted residual form, we would get $\hat{u}(x) = (5aL/(16AE))(2Lx - x^2)$. Figure 2.13 compares the two quadratic approximations with the exact solution. It is observed that the weak form approxima-tion is much better than the WR approximation.

We will now review the three different mathematical statements of the same physical problem as discussed above and compare their features.

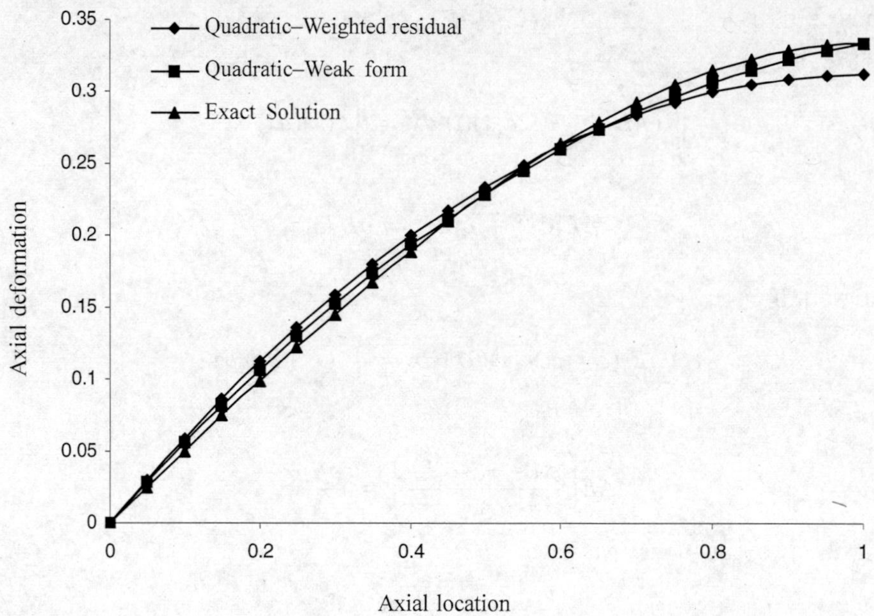

Fig. 2.13   Comparison of approximate solutions for a rod under load (Example 2.9).

2.4   Comparison of Differential Equation, Weighted Residual and Weak Forms

Consider the example of the axial deformation of a rod, described by the differential equation

$$AE\frac{d^2u}{dx^2} + q = 0 \tag{2.93}$$

and the boundary conditions

$$u(0) = u_0$$

$$AE\frac{du}{dx}\bigg|_{x=L} = P_L \tag{2.94}$$

In the context of finding approximate solutions, we have developed the weighted residual statement as follows:

$$\int_0^L W(x)\left(AE\frac{d^2\hat{u}}{dx^2} + q\right)dx = 0 \tag{2.95}$$

subject to the boundary conditions

$$\hat{u}(0) = u_0$$

$$AE\frac{d\hat{u}}{dx}\bigg|_{x=L} = P_L \tag{2.96}$$

We make the following observations:

1. If the function $\hat{u}(x)$ is the exact solution, then the weighted residual form is implicitly satisfied for any and every weighting function $W(x)$ as the domain residual term given within the parentheses is identically zero; otherwise, the error (or residual) is minimised over the entire domain in a weighted integral sense.

2. For any given differential equation (linear/nonlinear, ordinary/partial, etc.), we can write the weighted residual statement.

3. The weighted residual statement is equivalent only to the differential equation and does not account for boundary conditions. The trial solution is expected to satisfy all the boundary conditions and it should be differentiable as many times as required in the original differential equation.

4. The weighting functions $W(x)$ can, in principle, be any nonzero, integrable function. By choosing "n" different weighting functions, we generate the necessary n equations for the determination of coefficients $c_i(i = 1, 2, \ldots, n)$ in the trial solution field assumed.

5. In general, the weighting function $W(x)$ is subject to less stringent continuity requirement than the dependent field variable u. For example, in the above case, the trial solution $\hat{u}(x)$ has to be at least twice differentiable (because of the term $d^2\hat{u}/dx^2$), but no such restriction applies for the weighting function. The objective of performing integration by parts and developing the weak form is to distribute the continuity demand uniformly between the trial solution $\hat{u}(x)$ and the weighting function $W(x)$.

Upon performing integration by parts, we obtain the weak form statement of the problem as

$$\left[W(x)AE\frac{d\hat{u}}{dx} \right]_0^L - \int_0^L AE\frac{d\hat{u}}{dx}\frac{dW}{dx}\,dx + \int_0^L W(x)q\,dx = 0 \tag{2.97}$$

Substituting the prescribed natural (force) boundary condition at $x = L$, i.e.

$AE\left.\dfrac{d\hat{u}}{dx}\right|_{x=L} = P_L$, and requiring that $W(0) = 0$ in view of prescribed essential boundary

condition $u(0) = u_0$, we get the weak form statement as

$$\int_0^L AE\frac{d\hat{u}}{dx}\frac{dw}{dx}\,dx = \int_0^L W(x)q\,dx + W(L)P_L \tag{2.98}$$

subject to $\hat{u}(0) = u_0$, $W(0) = 0$.

Observations on the Weak Form

1. In the weak form obtained by integration by parts, the continuity demand on trial function u has gone down and that on the weighting function W has increased.

2. For even order differential equations (the only type we will be dealing with, in this book), we will be able to distribute the continuity demand (or the differentiation) equally between the trial solution function, and the weighting function by performing integration by parts sufficient number of times.

3. Since the natural boundary condition explicitly arises when we perform integration by parts, we can directly substitute the prescribed natural boundary conditions. Only the remaining essential boundary conditions have to be satisfied by the trial solution of the weak form equation.

4. In view of the above observations, a much wider choice of trial functions (e.g. significantly lower order polynomials than is possible with weighted residual forms) is available.

5. The weak form can be developed for the weighted residual form of any given second- and higher-order differential equation or its corresponding weighted residual form.

6. In view of the prescribed essential boundary conditions, we require that the weighting function be zero at all those points on the boundary where such conditions are specified, i.e., the weighting function should satisfy the homogeneous part of the prescribed essential boundary conditions. Though at this stage this appears to be an arbitrary requirement, it is not really so. The weight function can be interpreted as a virtual displacement (see Section 3.2) and, therefore, to be consistent with the supports, it must vanish at all those points where a prescribed displacement boundary condition exits.

We will now illustrate the intricacies of the weak form solution through the following interesting example.

Example 2.10.   Let us revisit Example 2.4, where we considered the deflection of a simply supported beam under uniformly distributed load. The governing differential equation was given by

$$EI\frac{d^4v}{dx^4} - q = 0 \tag{2.99}$$

and the boundary conditions

$$\text{Essential: } v(0) = 0, \qquad v(L) = 0 \qquad \text{(zero displacements)}$$

$$\text{Natural: } \frac{d^2v}{dx^2}(0) = 0, \qquad \frac{d^2v}{dx^2}(L) = 0 \qquad \text{(zero moments)}$$

Let $\hat{v}(x)$ be the assumed trial solution and let $W(x)$ be the weighting function. The weighted residual statement can be written as

$$\int_0^L W(x)\left(EI\frac{d^4\hat{v}}{dx^4} - q\right)dx = 0 \tag{2.100}$$

i.e.

$$\int_0^L W(x)d\left\{EI\frac{d^3\hat{v}}{dx^3}\right\} = \int_0^L Wq\,dx \qquad (2.101)$$

On integration by parts, we get

$$\left[W(x)EI\frac{d^3\hat{v}}{dx^3}\right]_0^L - \int_0^L\left(EI\frac{d^3\hat{v}}{dx^3}\right)\frac{dW}{dx}\,dx = \int_0^L Wq\,dx \qquad (2.102)$$

Integrating by parts once again, we obtain

$$\left[W(x)EI\frac{d^3\hat{v}}{dx^3}\right]_0^L - \left\{\left[\frac{dW(x)}{dx}EI\frac{d^2\hat{v}}{dx^2}\right]_0^L - \int_0^L EI\frac{d^2\hat{v}}{dx^2}\frac{d^2W(x)}{dx^2}dx\right\} = \int_0^L Wq\,dx \qquad (2.103)$$

Thus we now have equal order of differentiation on both the dependent variable $\hat{v}(x)$ and the weighting function $W(x)$. Substituting for the prescribed natural boundary conditions at either end that $d^2\hat{v}/dx^2$ be zero, we have the resulting weak form:

$$\int_0^L EI\frac{d^2\hat{v}}{dx^2}\frac{d^2W(x)}{dx^2}dx + \left(W(x)EI\frac{d^3\hat{v}}{dx^3}\right)_0^L = \int_0^L Wq\,dx \qquad (2.104)$$

subject to the essential boundary condition that $\hat{v}(0) = 0$, $\hat{v}(L) = 0$. In view of these prescribed essential boundary conditions, we require that the weighting function be such that $W(0) = 0$, $W(L) = 0$. Thus the weak form becomes

$$\int_0^L EI\frac{d^2\hat{v}}{dx^2}\frac{d^2W(x)}{dx^2}\,dx = \int_0^L Wq\,dx \qquad (2.105)$$

subject to

$$\hat{v}(0) = 0 = \hat{v}(L)$$
$$W(0) = 0 = W(L)$$

We observe that our trial solution $\hat{v}(x)$ for Eq. (2.105) need only be twice differentiable while a trial solution for Eq. (2.100) should be differentiable at least four times. Thus we can now take a simple quadratic trial function.

Let $\hat{v}(x) = c_0(x)(L - x)$ which satisfies the essential boundary condition that the displacement be zero at either end. Also, following the Galerkin procedure, we choose the weighting function to be the trial function itself and thus we have $W(x) = x(L - x)$, which also satisfies the required conditions on $W(x)$. For simplicity, let $q(x) = q_0$, a uniformly distributed load. Substituting these in the weak form, we obtain

$$\int_0^L EI(c_0)(-2)(-2)\,dx = \int_0^L (x)(L - x)q_0\,dx \qquad (2.106)$$

i.e.

$$(4EIL)c_0 = \frac{q_0 L^3}{6}$$

or

$$c_0 = \frac{q_0 L^2}{24\,EI}$$

Thus our approximate solution to the given problem is

$$\hat{v}(x) = \frac{q_0 L^2}{24EI}(x)(L - x) \tag{2.107}$$

The exact solution can be readily obtained as

$$v(x) = \frac{q_0 x}{24EI}(L^3 - 2Lx^2 + x^3) \tag{2.108}$$

Our approximation to the maximum deflection of the beam at the mid-span is seen to be 20% in error. However, it is interesting to see that the slopes at either end tally with the exact solution! The reader is urged to ponder over this result.

Observations

1. We observe that in the form of the WR statement of Example 2.4, we could not have considered this trial function, whereas with the weaker continuity demanded on $\hat{v}(x)$ now, we are able to consider lower order polynomials.

2. While we obtained a not-so-bad estimate to the field variable with such a lower order polynomial trial function, it would be interesting to study the behaviour of the derivatives of the solution obtained in Example 2.4 and the present one. Figure 2.14 shows a comparison of the nature of the function (i.e. deflection here) in both the

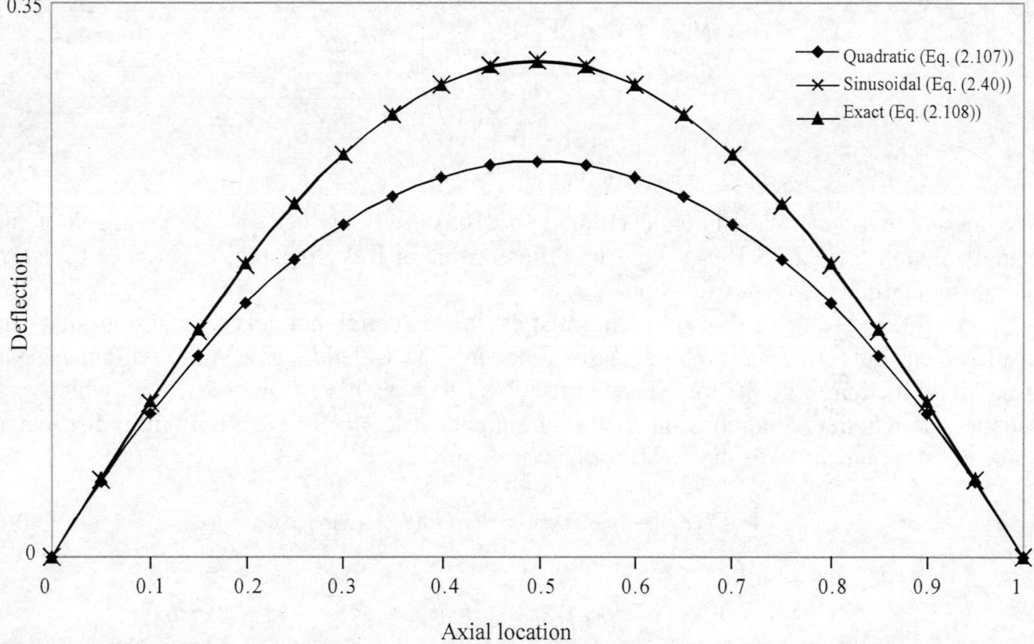

Fig. 2.14   Comparison of approximate solutions for deflection of a simply supported beam (Example 2.10).

cases. The first and second derivatives are compared in Figures 2.15(a) and 2.15(b). Thus we see that a trial function may approximate the field variable itself reasonably well but the derivative quantities may be very poorly approximated. The derivatives yield us information regarding physical variables of interest in the problem, e.g. strain/ stress, heat flux, etc. **It is important to predict not only the dependent variable (e.g. displacements, temperatures, etc.) but also its derivatives fairly accurately.**

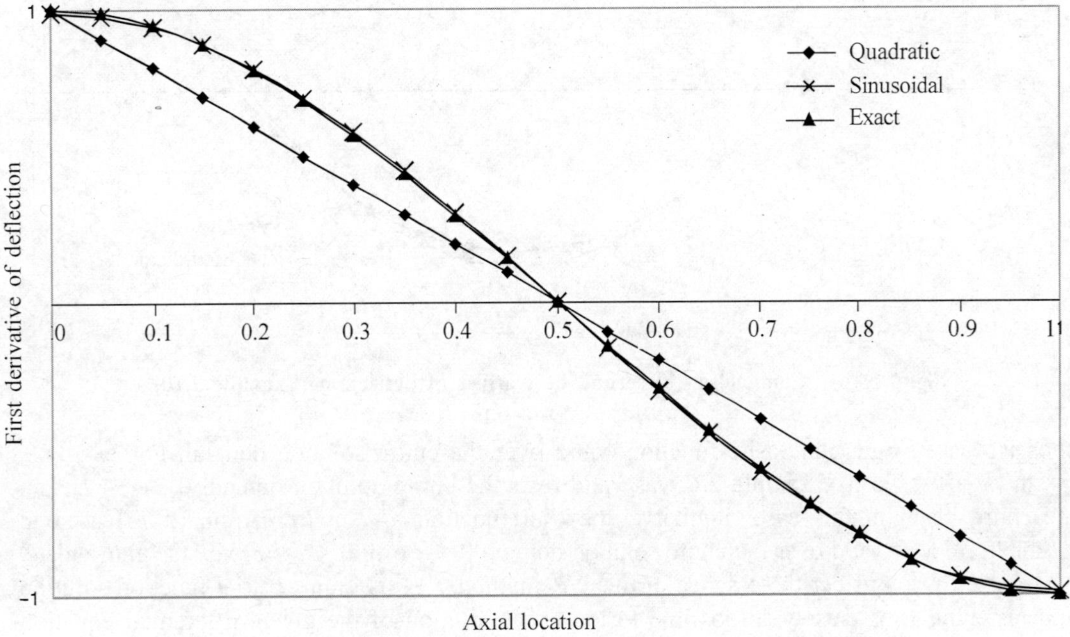

Fig. 2.15(a)   Comparison of first derivatives of deflection (Example 2.10).

3. We can use a higher order polynomial to improve our approximation. However, one of the primary purposes in converting the weighted residual form into the weak form is to be able to work with lower order polynomials. Thus we would like to improve our approximation by retaining the lower order polynomials. We propose to achieve this by using many such polynomials over the domain. We now describe the technique of approximating the solution in a piece-wise manner, rather than using a single continuous trial function for the whole domain.

2.5   Piece-wise Continuous Trial Function Solution of the Weak Form

We have seen that the general method of weighted residual technique consisted in assuming a trial function solution and minimising the residual in an overall sense. The Galerkin method gave us a way of choosing the appropriate weighting functions. We have so far used different trial functions such as polynomial and trigonometric series. However, in each case the trial function

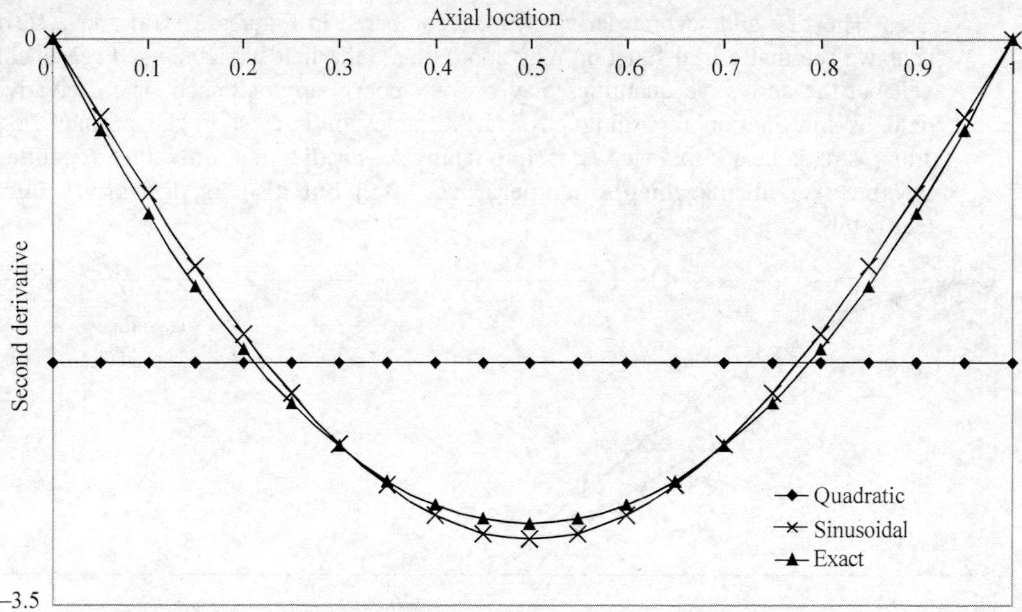

Fig. 2.15(b)   Comparison of second derivatives of deflection (Example 2.10).

chosen was a single composite function, valid over the entire solution domain. For example, $c_1 \sin(\pi x/L)$ used in Example 2.4 was valid over the entire solution domain $0 < x < L$, i.e., the entire beam in this case. Similarly, the trial function $c_{11} \sin(\pi x/a) \sin(\pi y/b)$ used in Example 2.7 was valid over the entire solution domain $0 < x < a$, $0 < y < b$, i.e., the entire plate.

When we consider the essence of the WR method, i.e., assuming a trial function solution and matching it as closely as possible to the exact solution of the given differential equation and the boundary conditions, we realise that this is essentially a process of "curve fitting". It is well known that curve fitting is best done "piece-wise"; the more the number of pieces, the better the fit. Figure 2.16(a) illustrates this basic idea on a function $f(x) = \sin(\pi x/L)$ being approximated using straight line segments. Since a straight line can be drawn through any two points, we can generate one such approximation by drawing a line through the function values at $x = 0$ and $x = L/2$. Clearly, this is a poor approximation of the function for $x > L/2$. If we were to use two straight line segments instead of only one, we could draw two line segments— one through the function values at $x = 0$ and $x = L/2$; and the other through the function values at $x = L/2$ and $x = L$. Each of these is our approximation to the function within that piece of the solution domain—the former segment in the sub-domain $0 < x < L/2$, and the latter in the sub-domain $L/2 < x < L$. Clearly, this is a better approximation to the function. It can be further improved by taking four-line segment approximation as shown in Figure 2.16(b).

This is the essential idea of piece-wise continuous trial function approximation for the weak form. Of course, when we consider piece-wise trial functions, we need to ensure continuity of the field variable and its derivatives at the junctions. While the exact solution will ensure continuity of the field variable and ALL the derivatives, we expect to be able to satisfy this to derivatives of only desired degree with our approximate solution. This leads to an

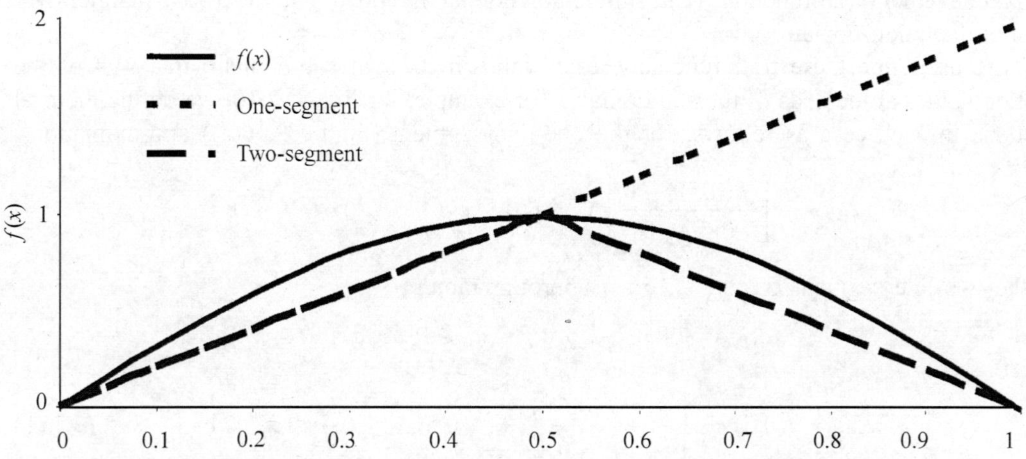

Fig. 2.16(a)   One and two-line segment approximation of a function.

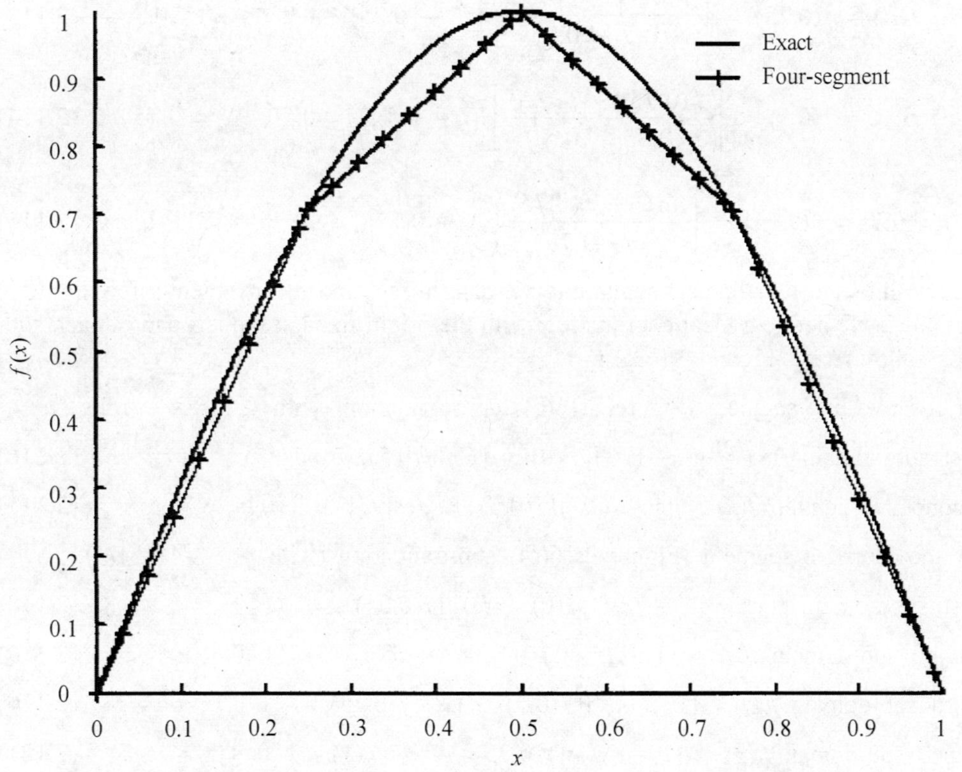

Fig. 2.16(b)   Four-line segment approximation of a function.

inherent error in our solution. We aim, however, to obtain a reasonably accurate solution using several convenient trial functions (e.g. simple polynomials) defined in a piece-wise manner over the entire solution domain.

Let us define these trial functions, each valid in its own sub-domain, in terms of the function values at the ends of the sub-domain. For example, for Figure 2.16, we define the trial functions in each case as follows: for the one-line segment (Figure 2.16(a)) approximation,

$$f(X) \approx \left[\frac{f(0.5) - 0}{0.5 - 0}\right] X = f(0.5) * 2X, \qquad 0 < X < 1 \tag{2.109}$$

for the two-line segment (Figure 2.16(a)) approximation,

$$f(X) \approx f(0.5) * 2X, \qquad 0 < X < 0.5 \tag{2.110}$$

$$f(X) \approx f(0.5) + \left[\frac{0 - f(0.5)}{1 - 0.5}\right](X - 0.5), \qquad 0.5 < X < 1 \tag{2.111}$$

for the four-line segment (Figure 2.16(b)) approximation,

$$f(X) \approx f(0.25) * 4X, \qquad 0 < X < 0.25 \tag{2.112}$$

$$f(X) \approx f(0.25) + \left[\frac{f(0.5) - f(0.25)}{0.5 - 0.25}\right](X - 0.25), \quad 0.25 < X < 0.5 \tag{2.113}$$

$$f(X) \approx f(0.5) + \left[\frac{f(0.75) - f(0.5)}{0.75 - 0.5}\right](X - 0.5), \qquad 0.5 < X < 0.75 \tag{2.114}$$

$$f(X) \approx f(0.75) + \left[\frac{f(1) - f(0.75)}{1 - 0.75}\right](X - 0.75), \qquad 0.75 < X < 1 \tag{2.115}$$

For two-line and four-line segmentation, we can, in fact, more conveniently rewrite these trial functions if we define a local coordinate x with the origin fixed at the left end of each sub-domain as follows (ref. Figure 2.16(c)):

For the two-line segment (Figure 2.16(c)) approximation (with $\ell = L/2 = 1/2$),

First sub-domain $f(x) \approx [1 - (x/\ell)]\, f(0) + [x/\ell]\, f(0.5), \qquad 0 < x < \ell \tag{2.116}$

Second sub-domain $f(x) \approx [1 - (x/\ell)]\, f(0.5) + [x/\ell]\, f(1), \qquad 0 < x < \ell \tag{2.117}$

For the four-line segment (Figure 2.16(c)) approximation (with $\ell = L/4 = 1/4$),

First sub-domain $f(x) \approx [1 - (x/\ell)]\, f(0) + [x/\ell]\, f(0.25), \qquad 0 < x < \ell \tag{2.118}$

Second sub-domain $f(x) \approx [1 - (x/\ell)]\, f(0.25) + [x/\ell]\, f(0.5), \qquad 0 < x < \ell \tag{2.119}$

Third sub-domain $f(x) \approx [1 - (x/\ell)]\, f(0.5) + [x/\ell]\, f(0.75), \qquad 0 < x < \ell \tag{2.120}$

Fourth sub-domain $f(x) \approx [1 - (x/\ell)]\, f(0.75) + [x/\ell]\, f(1), \qquad 0 < x < \ell \;. \tag{2.121}$

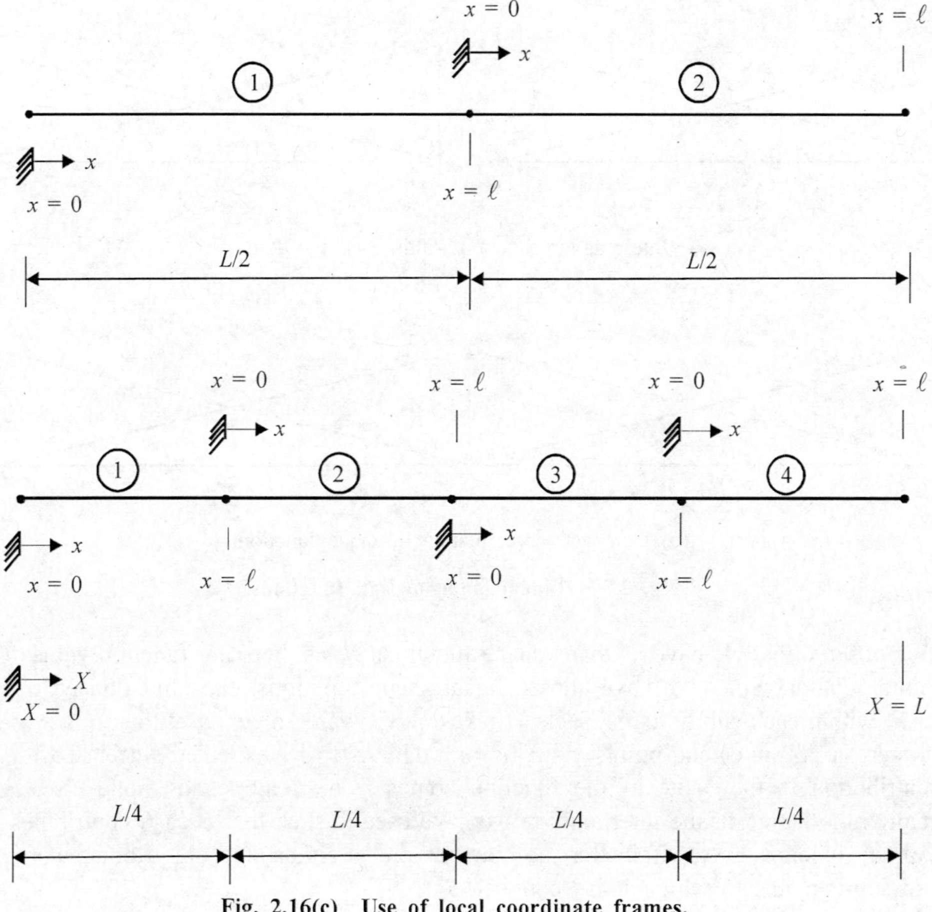

Fig. 2.16(c)   Use of local coordinate frames.

We observe that we have used the characteristic functions $[1 - (x/\ell)]$ and $[x/\ell]$ to interpolate the value of the function at any point within the sub-domain from the values of the function at the ends of the sub-domain. We can thus generate very conveniently any desired number of such line segments (say, n) to approximate the function. For the kth line segment we have

$$\boxed{f(x) \approx [1 - (x/\ell)]\, f_{k-1} + [x/\ell]\, f_k, \qquad 0 < x < \ell = L/n} \tag{2.122}$$

These characteristic functions $[1 - (x/\ell)]$ and $[x/\ell]$ used in our interpolation are called *interpolation functions* or *shape functions* since they determine the shape of variation of the function within each sub-domain. We will use the notation $N_i (i = 1, 2$, in this case) to designate the shape functions.

These characteristic functions dictate the contribution of a given f_k to the value of the function at any point P within the domain $0 < X < L$. Figure 2.17 shows these interpolation functions for the two-line segment case discussed earlier. Figure 2.17(a) shows the interpolation functions separately for the two sub-domains while the Figure 2.17(b) shows them more compactly for the whole domain $0 < X < L$. Figure 2.18 shows these shape functions for the four-line segment case.

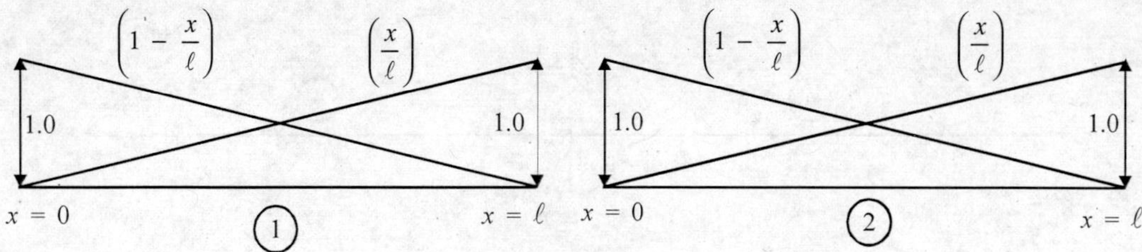

(a) Interpolation function within each sub-domain

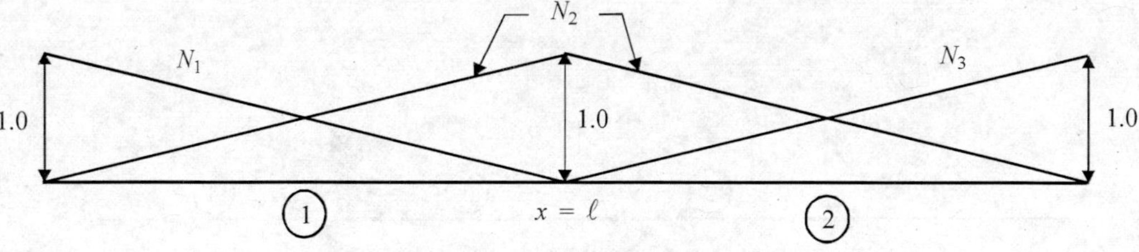

(b) Compact representation of shape function

Fig. 2.17   Linear interpolation functions.

We observe that we have as many shape functions N_k as there are function values f_k used in the interpolation. Since we have chosen linear shape functions, each function N_k ramps up and down within each sub-domain. Each N_k takes on the value of unity at the point $x = x_k$ and goes to zero at all other end-points $x = x_i (i \neq k)$. This is to be expected since, at $x = x_k$, the full contribution to the value of the function comes from f_k alone and none else. Equally importantly, the nature of the interpolation we have used is such that each f_k contributes to the value of the function only within the sub-domains on its either side or, in other words, only in those sub-domains to which it is "connected".

With respect to the solution of differential equations, the solution to be determined can itself be considered to be the function $f(x)$ which is being approximated in a piece-wise manner as discussed above. Thus, all the values of the function f_k, $k = 1, 2, 3, \ldots, n$, replace the coefficients c_1, c_2, $\ldots$, which were earlier determined satisfying the weak form. Considering the computation of the integrals over the entire domain, when we used a single trial function we achieved this task by evaluating the integral $\int W_i(x) R_d(x)\, dx$ over the domain $(0 < X < L)$. With the use of interpolation functions as explained above, we need to rewrite this as $\Sigma \int W_i(x) R_d(x)\, dx$, where the integration is carried out over each sub-domain and the summation is carried out over all the sub-domains. **By interpolating the value of the function at an interior point from the function values at the ends of the sub-domain, we gain two advantages—first, continuity of the function at the junction of two sub-domains is assured and, secondly, the prescribed essential boundary conditions can be readily imposed.** Since the weak form is equivalent to the differential equation and the natural boundary condition, the essential boundary condition has to be satisfied by the trial function and thus the second advantage comes in handy.

We reiterate the fact that the unknown coefficients to be determined by the weak form are now the values of the field variable f_k at the ends of the sub-domains. If we discretise the physical domain $0 < X < L$ into 'n' sub-domains, we thus have $n + 1$ unknown values of the function to be determined through satisfying the weak form. We expect to generate '$n + 1$' algebraic equations, compactly written in matrix form as $[A]\{f\} = \{b\}$. The '$n + 1$' weighting functions to generate the necessary equations are the interpolation functions themselves, following the Galerkin procedure. These interpolation or weighting functions are the same as those shown in Figures 2.17 and 2.18. It is observed that these functions simply repeat themselves from one sub-domain to another. Thus, it is convenient for us to do our

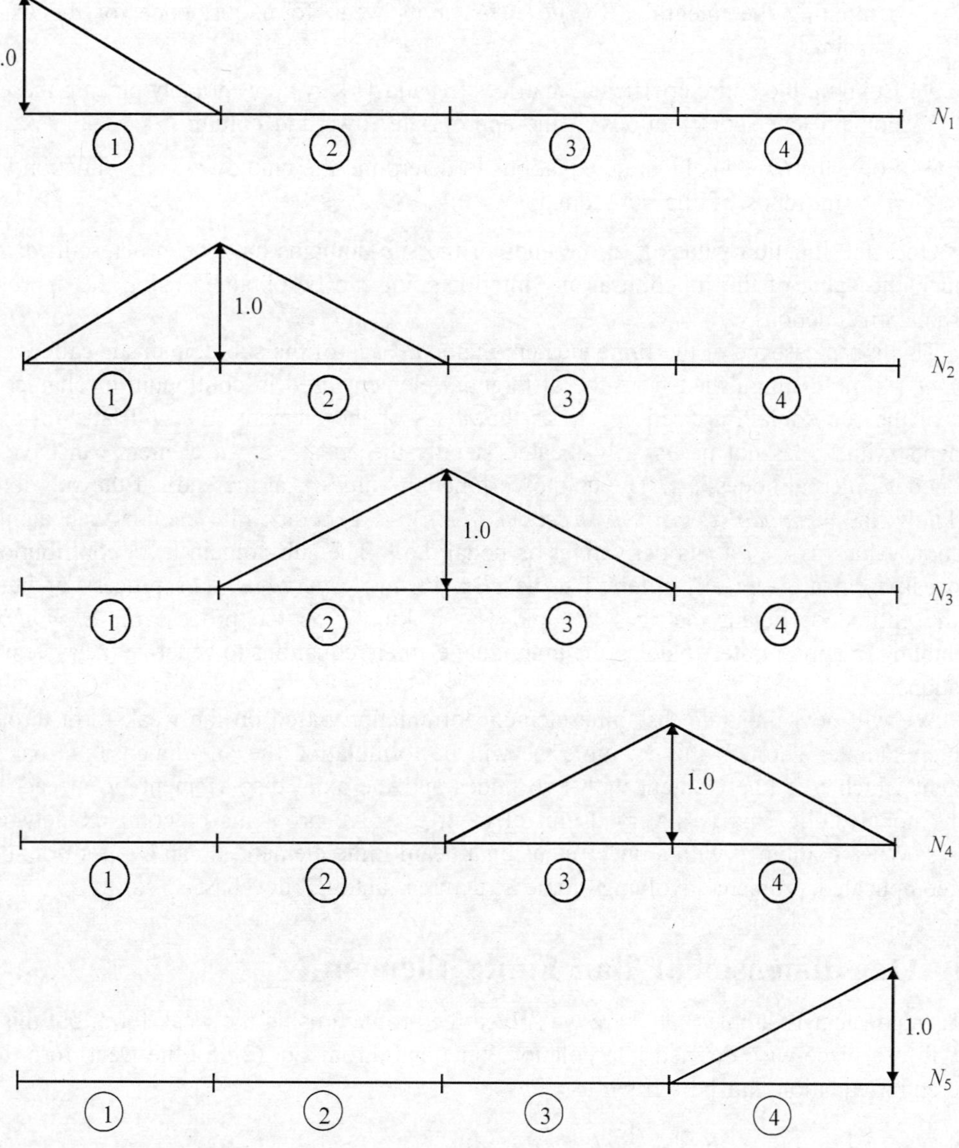

Fig. 2.18 **Interpolation functions for four-segment approximation.**

computations such as evaluation of the integral $\int W_i(x)R_d(x)\,dx$ or its weak form, just once for the kth sub-domain, and merely reuse the same result for all other sub-domains when we perform the summation over all the sub-domains. We note that the contributions from each sub-domain to the elements of the matrix $[A]$ above will essentially be the same except that the corresponding coefficients they multiply will keep shifting from one sub-domain to another as given by f_k and f_{k+1}. Thus the process of evaluating the total weighted residual or its weak form over the entire domain is greatly simplified and is summarised now.

> ➤ Evaluate the sub-domain level contributions to the weighted residual by merely computing the integral $\int W_i(x)R_d(x)\,dx$ or its weak form, just once for the kth sub-domain.

> ➤ Build up the entire coefficient matrices $[A]$ and $\{b\}$ by appropriately placing these sub-domain level contributions in the appropriate rows and columns.

> ➤ Solve the $(n + 1)$ algebraic equations to determine the unknowns, viz., function values f_k at the ends of the sub-domains.

Once the function values f_k at the ends of the sub-domains have been determined in this manner, the value of the function at any interior point can be obtained using the appropriate interpolation functions.

This is the essence of the finite element method. Each of the sub-domains is called a *finite element*—to be distinguished from the differential element used in continuum mechanics. The ends of the sub-domain are referred to as the *nodes* of the element. We will later on discuss elements with nodes not necessarily located at only the ends, e.g. an element can have mid-side nodes, internal nodes, etc. The unknown function values f_k at the ends of the sub-domains are known as the *nodal degrees of freedom (d.o.f.)*. A general finite element can admit the function values as well as its derivatives as nodal d.o.f. The sub-domain level contributions to the weak form are typically referred to as *element level equations*. The process of building up the entire coefficient matrices $[A]$ and $\{b\}$ is known as the process of *assembly,* i.e., assembling or appropriately placing the individual element equations to generate the system level equations.

We will now illustrate the finite element formulation based on the weak form through a simple example. Through this example we will be formulating the one-dimensional bar finite element which is a line element with two nodes and the axial displacement 'u' at each node as the nodal d.o.f. This is a very useful element to solve problems of complex network of trusses. When combined with a one-dimensional beam finite element, it can be used for solving any complicated problem involving frame structures, automobile chassis, etc.

2.6   One-dimensional Bar Finite Element

Let us reconsider Example 2.9 where we solved the problem using the weak form, but this time we will use piece-wise defined interpolation functions. From Eq. (2.85), the weak form of the differential equation can be written as

$$\left[W(X)AE\frac{d\hat{u}}{dX}\right]_0^L - \int_0^L AE\frac{d\hat{u}}{dX}\frac{dW}{dX}\,dX + \int_0^L W(X)q(X)\,dX = 0 \qquad (2.123)$$

We use the symbol "X" for "global" coordinate running along the length of the entire bar (0 to L), and the symbol "x" for representing the "local" coordinate within each element (0 to ℓ). Rewriting Eq. (2.123), we have

$$\int_0^L \left(AE\frac{d\hat{u}}{dX} \right)\left(\frac{dW}{dX} \right) dX = \int_0^L W(X)q(X)\,dX + \left[W(X)AE\frac{d\hat{u}}{dX} \right]_0^L \tag{2.124}$$

Let us "discretise" the bar into "n" finite elements, not necessarily of the same length, as shown in Figure 2.19. Figure 2.20 shows a typical bar element of length ℓ, two nodes and axial displacement u as the nodal d.o.f.

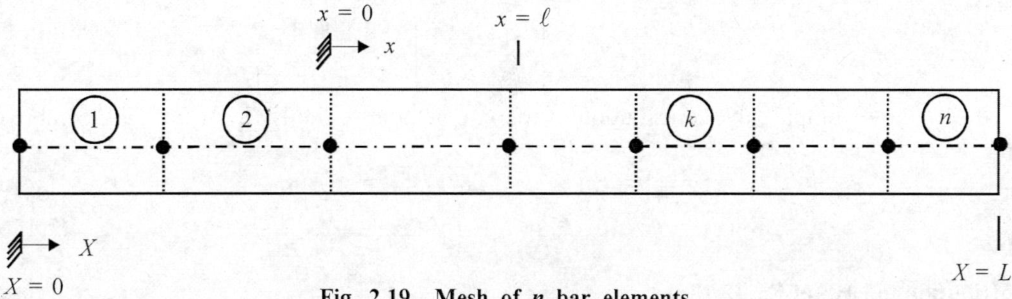

Fig. 2.19   Mesh of n bar elements.

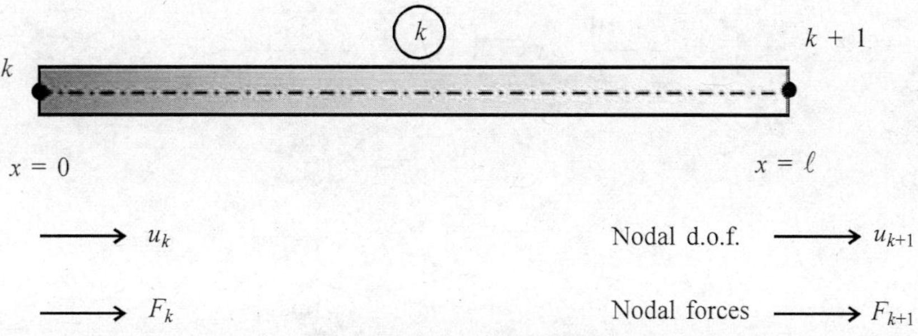

Fig. 2.20   Typical bar element.

For this n-element mesh, let us rewrite the above equation as

$$\sum_{k=1}^{n} \int_0^\ell AE\frac{d\hat{u}}{dx}\frac{dW}{dx}\,dx = \sum_{k=1}^{n} \int_0^\ell W(x)q(x)\,dx + \sum_{k=1}^{n} \left[W(x)AE\frac{d\hat{u}}{dx} \right]_0^\ell \tag{2.125}$$

where we have replaced the integral over 0 to L, by summation over all the finite elements. Let $\hat{u}(x)$ within each element be given by the interpolation functions introduced earlier, i.e., for a kth element,

$$\hat{u}(x) = \left(1 - \frac{x}{\ell} \right)\hat{u}_k + \left(\frac{x}{\ell} \right)\hat{u}_{k+1} = [N]\,\{\hat{u}\}^k \tag{2.126}$$

where we have introduced the generic representation of interpolation for the field variable within an element—valid for any element, with any number of nodes or nodal d.o.f.—for generalising our method of formulation. It is observed that the shape function array $[N]$ and the nodal d.o.f. array $\{\hat{u}\}^k$ are of appropriate size depending on the element under consideration. For the present case,

$$[N] = \left[\left(1 - \frac{x}{\ell}\right) \quad \left(\frac{x}{\ell}\right)\right], \quad \{\hat{u}\}^k = \begin{Bmatrix} \hat{u}_k \\ \hat{u}_{k+1} \end{Bmatrix} \tag{2.127}$$

$$\frac{d\hat{u}}{dx} = \frac{-1}{\ell}\hat{u}_k + \frac{1}{\ell}\hat{u}_{k+1} = \frac{\hat{u}_{k+1} - \hat{u}_k}{\ell} = [B]\{\hat{u}\}^k \tag{2.128}$$

where

$$[B] = \left[\frac{-1}{\ell} \quad \frac{1}{\ell}\right]$$

Let us now compute the contribution of the kth element to the LHS of Eq. (2.125) above, using

$$W_1(x) = (1 - x/\ell), \qquad dW_1/dx = -1/\ell$$
$$W_2(x) = (x/\ell), \qquad dW_2/dx = 1/\ell \tag{2.129}$$

Contribution to LHS of Eq. (2.125)

Now,

$$\int_0^\ell AE \frac{d\hat{u}}{dx} \frac{dW_1}{dx} dx = \int_0^\ell AE \frac{\hat{u}_{k+1} - \hat{u}_k}{\ell}\left(\frac{-1}{\ell}\right) dx$$

$$= \frac{AE}{\ell}(\hat{u}_k - \hat{u}_{k+1})$$

$$= \frac{AE}{\ell}[1 \quad -1]\begin{Bmatrix} \hat{u}_k \\ \hat{u}_{k+1} \end{Bmatrix} \tag{2.130}$$

Similarly,

$$\int_0^\ell AE \frac{d\hat{u}}{dx} \frac{dW_2}{dx} dx = \int_0^\ell AE \frac{\hat{u}_{k+1} - \hat{u}_k}{\ell}\left(\frac{1}{\ell}\right) dx$$

$$= \frac{AE}{\ell}(\hat{u}_{k+1} - \hat{u}_k)$$

$$= \frac{AE}{\ell}[-1 \quad 1]\begin{Bmatrix} \hat{u}_k \\ \hat{u}_{k+1} \end{Bmatrix} \tag{2.131}$$

Contribution to RHS of Eq. (2.125)

We observe that the first term on the RHS of Eq. (2.125) requires specifying the nature of externally applied force. Assuming $q(x) = q_0$ to be a constant, we can compute the RHS terms as

$$\int_0^\ell W_1(x)q(x)\,dx = \int_0^\ell \left(1 - \frac{x}{\ell}\right)q_0\,dx = \frac{q_0\ell}{2}$$

$$\int_0^\ell W_2(x)q(x)\,dx = \int_0^\ell \left(\frac{x}{\ell}\right)q_0\,dx = \frac{q_0\ell}{2}$$

(2.132)

For any other specified $q(x)$, similar computation can be readily performed. Contributions to the second term of RHS can be computed as follows:

$$\left[W_1(x)AE\frac{d\hat{u}}{dx}\right]_0^\ell = \left[\left(1 - \frac{x}{\ell}\right)P\right]_0^\ell = -P_0$$

$$\left[W_2(x)AE\frac{d\hat{u}}{dx}\right]_0^\ell = \left[\left(\frac{x}{\ell}P\right)\right]_0^\ell = P_\ell$$

(2.133)

where we have used the fact that the term $AE(d\hat{u}/dx)$ stands for the axial force P in the bar at that section. Thus, P_0 and P_ℓ stand for the forces at either end of the element. It is observed that these designate the algebraic sum of the externally applied forces F at these nodes and the internal reaction forces R (i.e., those forces that are exerted on this element by the adjoining elements), that get exposed when we consider this element alone as a free body.

Combining Eqs. (2.130)–(2.134), we can write the contributions of the kth element to the overall system equations given in Eq. (2.125) as

$$\begin{array}{cc} \text{LHS} & \text{RHS} \end{array}$$

$$\frac{AE}{\ell}\begin{bmatrix} 1 & -1 \\ -1 & 1 \end{bmatrix}\begin{Bmatrix} u_k \\ u_{k+1} \end{Bmatrix} \qquad \begin{Bmatrix} q_0\ell/2 \\ q_0\ell/2 \end{Bmatrix} + \begin{Bmatrix} -P_0 \\ P_\ell \end{Bmatrix}$$

(2.134)

These two sets of terms constitute the "characteristic matrices" or "characteristic equations" for the bar element. We observe that (AE/ℓ) represents the axial stiffness of a rod; the first force vector represents the nodal forces equivalent to the distributed force q_0, and the element equations are simply the nodal force equilibrium equations. In view of the fact that these equations represent the force-deflection relations for the bar element, the LHS matrix is termed as the *element stiffness matrix* and the RHS vectors as the *element nodal force vectors*:

Now we need to sum up all the element level contributions to generate the system level equations. As already mentioned, this process would simply involve repetitive use of the generic element matrices given above. To illustrate how the element matrices get simply "placed appropriately" in system level equations, we show the detailed computations for a two-element mesh. Based on this illustration, we provide a general method of "assembling element equations" for later use.

Example 2.11. *Illustration of assembly.* Consider a rod of total length $L = 2\ell$, discretised into two elements, as shown in Figure 2.21. The weak form of the differential equation is given by

$$\int_0^L AE\frac{d\hat{u}}{dx}\,\frac{dW}{dx}\,dx = \int_0^L Wq\,dx + \left[WAE\frac{d\hat{u}}{dx}\right]_0^L \tag{2.135}$$

We observe that the weighting functions can be depicted as shown in Figure 2.22, similar

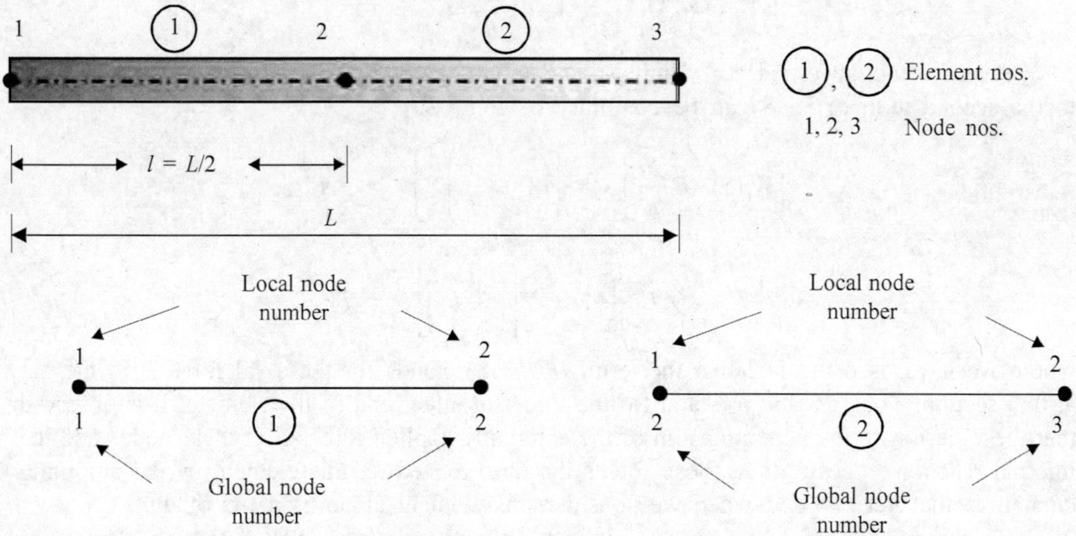

Fig. 2.21   Two-element discretisation of a rod.

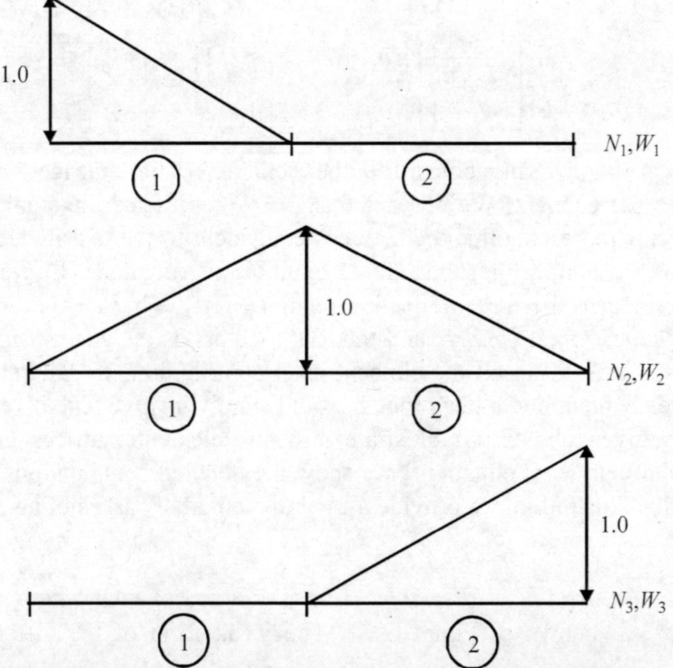

Fig. 2.22   Weight (shape) functions for two linear bar elements.

to Figures 2.17–2.18 given earlier. In other words, we can write the weighting function as follows:

$$
\begin{aligned}
W_1(x) &= 1 - x/\ell, & 0 < x < \ell \ \text{in the 1st element} \\
&= 0 & \text{in the 2nd element} \\
W_2(x) &= x/\ell, & 0 < x < \ell \ \text{in the 1st element} \\
&= 1 - x/\ell, & 0 < x < \ell \ \text{in the 2nd element} \\
W_3(x) &= 0 & \text{in the 1st element} \\
&= x/\ell, & 0 < x < \ell \ \text{in the 2nd element}
\end{aligned}
\tag{2.136}
$$

We have, in all, three field variables u_1, u_2 and u_3, and we have three weighting functions required to generate the three equations necessary to solve for the three field variables. Substituting these in the weak form, we get the necessary equations as follows:

LHS of Eq. (2.124)

$$
\begin{aligned}
\int_0^L AE \frac{d\hat{u}}{dX} \frac{dW_1}{dX} \, dX &= \sum_1^2 \int_0^\ell AE \frac{d\hat{u}}{dx} \frac{dW_1}{dx} \, dx \\
&= \int_0^\ell AE \left(\frac{u_2 - u_1}{\ell} \right) \left(\frac{-1}{\ell} \right) dx + 0 \\
&= \frac{AE}{\ell} (u_1 - u_2)
\end{aligned}
\tag{2.137}
$$

$$
\begin{aligned}
\int_0^L AE \frac{d\hat{u}}{dX} \frac{dW_2}{dX} \, dX &= \sum_1^2 \int_0^\ell AE \frac{d\hat{u}}{dx} \frac{dW_2}{dx} \, dx \\
&= \int_0^\ell AE \left(\frac{u_2 - u_1}{\ell} \right) \left(\frac{1}{\ell} \right) dx + \int_0^\ell AE \left(\frac{u_3 - u_2}{\ell} \right) \left(\frac{-1}{\ell} \right) dx \\
&= \frac{AE}{\ell} [(-u_1 + u_2) + (u_2 - u_3)]
\end{aligned}
\tag{2.138}
$$

$$
\begin{aligned}
\int_0^L AE \frac{d\hat{u}}{dX} \frac{dW_3}{dX} \, dX &= \sum_1^2 \int_0^\ell AE \frac{d\hat{u}}{dx} \frac{dW_3}{dx} \, dx \\
&= 0 + \int_0^\ell AE \left(\frac{u_3 - u_2}{\ell} \right) \left(\frac{1}{\ell} \right) dx \\
&= \frac{AE}{\ell} [(u_3 - u_2)
\end{aligned}
\tag{2.139}
$$

RHS first term of Eq. (2.124)

$$
\int_0^L W_1 q_0 \, dX = \sum_1^2 \int_0^\ell W_1 q_0 \, dx = \int_0^\ell \left(1 - \frac{x}{\ell} \right) q_0 \, dx + 0 = \frac{q_0 \ell}{2}
\tag{2.140}
$$

$$\int_0^L W_2 q_0 \, dX = \sum_1^2 \int_0^\ell W_2 q_0 \, dx = \int_0^\ell \left(\frac{x}{\ell}\right) q_0 \, dx + \int_0^\ell \left(1 - \frac{x}{\ell}\right) q_0 \, dx = \frac{q_0 \ell}{2} + \frac{q_0 \ell}{2} \qquad (2.141)$$

$$\int_0^L W_3 q_0 \, dX = \sum_1^2 \int_0^\ell W_3 q_0 \, dx = 0 + \int_0^\ell \left(\frac{x}{\ell}\right) q_0 \, dx = \frac{q_0 \ell}{2} \qquad (2.142)$$

RHS second term of Eq. (2.124)

$$\left[W_1(X) AE \frac{d\hat{u}}{dX} \right]_0^L = \sum_1^2 \left[W_1(x) AE \frac{d\hat{u}}{dx} \right]_0^\ell = \left[\left(1 - \frac{x}{\ell}\right) P \right]_0^\ell + 0 = -P_0^{(1)} \qquad (2.143)$$

Note: The superscript on the force term (e.g. $P_0^{(1)}$) represents element number and the subscript refers to $x = 0$ or ℓ.

$$\left[W_2(X) AE \frac{d\hat{u}}{dX} \right]_0^L = \sum_1^2 \left[W_2(x) AE \frac{d\hat{u}}{dx} \right]_0^\ell$$

$$= \left[\frac{x}{\ell} P \right]_0^\ell + \left[\left(1 - \frac{x}{\ell}\right) P \right]_0^\ell$$

$$= P_\ell^{(1)} - P_0^{(2)} \qquad (2.144)$$

$$\left[W_3(X) AE \frac{d\hat{u}}{dX} \right]_0^L = \sum_1^2 \left[W_3(x) AE \frac{d\hat{u}}{dx} \right]_0^\ell = 0 + \left[\frac{x}{\ell} P \right]_0^\ell = P_\ell^{(2)} \qquad (2.145)$$

Thus the system equations are (from Eqs. (2.137)–(2.145))

$$\frac{AE}{\ell} \begin{bmatrix} 1 & -1 & 0 \\ -1 & 1+1 & -1 \\ 0 & -1 & 1 \end{bmatrix} \begin{Bmatrix} u_1 \\ u_2 \\ u_3 \end{Bmatrix} = \begin{Bmatrix} q_0 \ell/2 \\ q_0 \ell/2 \\ \end{Bmatrix} + \begin{Bmatrix} q_0 \ell/2 \\ q_0 \ell/2 \end{Bmatrix} + \begin{Bmatrix} -P_0^{(1)} \\ P_\ell^{(1)} \\ \end{Bmatrix} - \begin{Bmatrix} P_0^{(2)} \\ P_\ell^{(2)} \end{Bmatrix}$$

$$= \begin{Bmatrix} q_0 \ell/2 \\ q_0 \ell/2 \\ \end{Bmatrix} + \begin{Bmatrix} q_0 \ell/2 \\ q_0 \ell/2 \end{Bmatrix} + \begin{Bmatrix} F_1 \\ F_2 \\ F_3 \end{Bmatrix} \qquad (2.146)$$

With respect to the system of equations given here, we can make the following observations:

1. The system level equations can be readily built up from the element level equations by appropriately "placing" them in the system matrices.

2. Since each bar element can be viewed as a spring of stiffness AE/ℓ, the bar element mesh can be viewed as an assemblage of springs connected together at the junction nodes. Since they share the same nodal displacement, they act as springs in parallel at the junction node, and hence their individual stiffness gets added up (for example, at junction node 2 here).

3. The first term on the RHS represents the equivalent nodal force for the distributed force and, at the junction nodes, the forces too get added up.

4. We may recall that the second term on RHS represents net axial force at the section, consisting of any concentrated external forces (F) applied at that section as well as the internal reaction forces (R) that get exposed when we view a single element as a free body. However, when we assemble the elements together to get the system level equations, the internal reaction forces annul each other (based on Newton's third law), leaving out the externally applied concentrated nodal forces.

5. Each of the equations represents the nodal force equilibrium equation. Thus, by viewing the structure as being made up of a set of discrete finite elements connected together at only their nodes, we have effectively replaced the differential equations of equilibrium of continuum mechanics by the present set of algebraic equilibrium equations applicable only at the nodes. This leads us to the interesting question of whether or not equilibrium equations are satisfied at interior points in a finite element. It can be shown that we cannot guarantee that the equilibrium equations are satisfied at interior points in a general finite element.

6. We may recollect that the weak form was equivalent to only the differential equation and the natural boundary conditions. Thus the finite element equations given above, derived from the weak form, cannot be solved until we substitute for the essential boundary conditions also. In the present context, the essential boundary conditions refer to prescribed values of the displacement "u".

We illustrate the solution of finite element equations using the following two examples.

Example 2.12. *A tip loaded rod.* Consider a 1 m long steel rod (3×10 mm cross-section) held fixed at its left end and subjected to a concentrated force of 100 N at its right end. Let us use the two-element mesh of Example 2.11 to model this problem. Setting

$$A = 30 \text{ mm}^2 = 30 \times 10^{-6} \text{ m}^2, \qquad E = 2 \times 10^{11} \text{ N/m}^2,$$

$$q_0 = 0, \qquad\qquad F_1 = R_1 \text{ (the support reaction force)}$$

$$F_2 = 0, \qquad\qquad F_3 = 100 \text{ N}$$

We have the finite element equations as follows:

$$\frac{(30 \times 10^{-6})(2 \times 10^{11})}{(0.5)}\begin{bmatrix} 1 & -1 & 0 \\ -1 & 2 & -1 \\ 0 & -1 & 1 \end{bmatrix}\begin{Bmatrix} u_1 \\ u_2 \\ u_3 \end{Bmatrix} = \begin{Bmatrix} R_1 \\ 0 \\ 100 \end{Bmatrix} \tag{2.147}$$

At the left end, the rod is clamped and thus the essential boundary condition is that $u_1 = 0$. It is observed that the unknowns in these equations are therefore the nodal displacements u_2, u_3 and the ground reaction force R_1. We therefore make a mental picture of partitioning the system matrices given in Eq. (2.147) above and use the last two equations for determining u_2, u_3 and the first equation for obtaining the reaction force R_1. Solving, we get

$$u_2 = 0.00834 \text{ mm}, \qquad u_3 = 0.01667 \text{ mm}$$

$$R_1 = -100 \text{ N} \tag{2.148}$$

We observe that the tip deflection tallies with the exact value $\left(= \dfrac{F_3 L}{AE} \right)$.

Example 2.13. *Rod under distributed and concentrated forces.* Consider the same rod as in Example 2.12 but subjected to forces as shown in Figure 2.23.

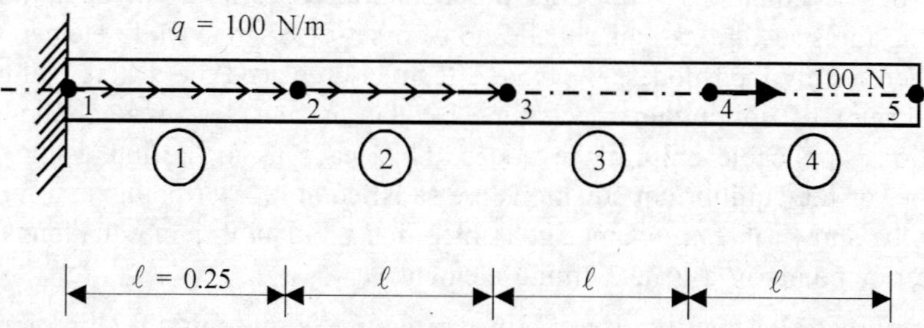

Fig. 2.23   **Rod under distributed and concentrated forces (Example 2.13).**

We will use a four-element mesh depicted in the figure. We have the element level equations as

$$\frac{(30 \times 10^{-6})(2 \times 10^{11})}{(0.25)} \begin{bmatrix} 1 & -1 \\ -1 & 1 \end{bmatrix} \begin{Bmatrix} u_1 \\ u_2 \end{Bmatrix} = \begin{Bmatrix} 12.5 \\ 12.5 \end{Bmatrix} + \begin{Bmatrix} -P_0^{(1)} \\ P_\ell^{(1)} \end{Bmatrix}$$

$$\frac{(30 \times 10^{-6})(2 \times 10^{11})}{(0.25)} \begin{bmatrix} 1 & -1 \\ -1 & 1 \end{bmatrix} \begin{Bmatrix} u_2 \\ u_3 \end{Bmatrix} = \begin{Bmatrix} 12.5 \\ 12.5 \end{Bmatrix} + \begin{Bmatrix} -P_0^{(2)} \\ P_\ell^{(2)} \end{Bmatrix}$$

$$\frac{(30 \times 10^{-6})(2 \times 10^{11})}{(0.25)} \begin{bmatrix} 1 & -1 \\ -1 & 1 \end{bmatrix} \begin{Bmatrix} u_3 \\ u_4 \end{Bmatrix} = \begin{Bmatrix} 0 \\ 0 \end{Bmatrix} + \begin{Bmatrix} -P_0^{(3)} \\ P_\ell^{(3)} \end{Bmatrix}$$

$$\frac{(30 \times 10^{-6})(2 \times 10^{11})}{(0.25)} \begin{bmatrix} 1 & -1 \\ -1 & 1 \end{bmatrix} \begin{Bmatrix} u_4 \\ u_5 \end{Bmatrix} = \begin{Bmatrix} 0 \\ 0 \end{Bmatrix} + \begin{Bmatrix} -P_0^{(4)} \\ P_\ell^{(4)} \end{Bmatrix} \tag{2.149}$$

Assembling together all the element level equations, we get the system level equations as

$$\frac{(30 \times 10^{-6})(2 \times 10^{11})}{(0.25)} \begin{bmatrix} 1 & -1 & 0 & 0 & 0 \\ -1 & 2 & -1 & 0 & 0 \\ 0 & -1 & 2 & -1 & 0 \\ 0 & 0 & -1 & 2 & -1 \\ 0 & 0 & 0 & -1 & 1 \end{bmatrix} \begin{Bmatrix} u_1 \\ u_2 \\ u_3 \\ u_4 \\ u_5 \end{Bmatrix} = \begin{Bmatrix} 12.5 \\ 25 \\ 12.5 \\ 0 \\ 0 \end{Bmatrix} + \begin{Bmatrix} R_1 \\ 0 \\ 0 \\ 100 \\ 0 \end{Bmatrix} \qquad (2.150)$$

Here, $R_1 = -P_0^{(1)}$ represents the reaction force to be exerted by the support. Since there is no externally applied concentrated force at node 2 or 3, we have $P_\ell^{(1)} = P_0^{(2)}$ and $P_\ell^{(2)} = P_0^{(3)}$. At node 4 we have $P_\ell^{(3)} - P_0^{(4)} = 100$, and node 5 being free edge, $P_\ell^{(4)} = 0$.

Substituting for the essential boundary condition that $u_1 = 0$, we can solve the equations to get the unknown nodal deflections and the support reaction force as

$$u_2 = 0.00573 \text{ mm}, \quad u_3 = 0.0104 \text{ mm}, \quad u_4 = 0.0146 \text{ mm}, \quad u_5 = 0.0146 \text{ mm}$$

$$R_1 = -150 \text{ N} \qquad (2.151)$$

We observe that the axial deflection of nodes 4 and 5 is the same as expected (rigid body motion). The reaction force at support equals the total external force applied on the structure.

We have demonstrated how to develop the finite element equations, starting from the weak form of the differential equation, for the axial deformation of a bar. We will now show that our method is quite generic and can be applied to any given differential equation. We demonstrate that exactly the same procedure can be used to formulate, for example, the heat transfer element.

2.7 One-dimensional Heat Transfer Element

Consider a plane wall with a uniformly distributed heat source. The governing differential equation for the steady state one-dimensional conduction heat transfer is given by

$$k\frac{d^2T}{dX^2} + q_0 = 0 \qquad (2.152)$$

where k is the coefficient of thermal conductivity of the material, T represents the temperature, and q_0 the internal heat source per unit volume. We assume that k is not a function of temperature.

The weighted residual statement can be written as

$$\int_0^L W\left(k\frac{d^2T}{dX^2} + q_0\right)dX = 0 \qquad (2.153)$$

Integrating by parts, the weak form of the differential equation can be obtained as

$$\left[Wk\frac{dT}{dX} \right]_0^L - \int_0^L k\frac{dW}{dX}\frac{dT}{dX}dX + \int_0^L Wq_0\,dX = 0 \tag{2.154}$$

i.e.

$$\int_0^L k\frac{dW}{dX}\frac{dT}{dX}dX = \int_0^L Wq_0\,dX + \left[Wk\frac{dT}{dX} \right]_0^L \tag{2.155}$$

The weak form for a typical mesh of n finite elements can be written as

$$\sum_{k=1}^n \left[\int_0^\ell k\frac{dW}{dx}\frac{dT}{dx}dx \right] = \sum_{k=1}^n \left[\int_0^\ell Wq_0\,dx + \left[Wk\frac{dT}{dx} \right]_0^\ell \right] \tag{2.156}$$

The attention of the reader is drawn to the similarity of the weak form in the above equation and that in Eq. (2.125) for the bar problem. We now propose to develop the one-dimensional heat transfer element following the procedure we used while developing the equations for the bar element. Figure 2.24 shows a typical element, with its nodes and nodal degree of freedom (viz., temperature). Following the similarity of the bar and heat transfer

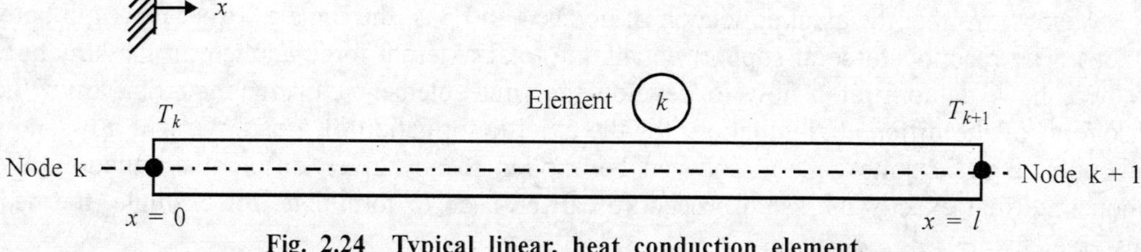

Fig. 2.24   **Typical linear, heat conduction element.**

element, we simply use the same shape functions for interpolating the temperature within the element from the nodal temperatures. Thus,

$$T(x) = \left(1 - \frac{x}{\ell} \right)T_k + \left(\frac{x}{\ell} \right)T_{k+1} \tag{2.157}$$

$$\frac{dT}{dx} = \frac{T_{k+1} - T_k}{\ell} \tag{2.158}$$

$$W_1 = 1 - \frac{x}{\ell}, \qquad \frac{dW_1}{dx} = -\frac{1}{\ell}, \qquad W_2 = \frac{x}{\ell}, \qquad \frac{dW_2}{dx} = \frac{1}{\ell} \tag{2.159}$$

We now compute the element level contributions to the weak form (Eq. (2.156)) above.

LHS first term of Eq. (2.156). Similar to the LHS of the bar element equation, k replaces AE and T replaces u. Thus we have

$$\frac{k}{\ell}\begin{bmatrix} 1 & -1 \\ -1 & 1 \end{bmatrix}\begin{Bmatrix} T_k \\ T_{k+1} \end{Bmatrix} \tag{2.160}$$

RHS first term of Eq. (2.156). Since it is similar to the RHS first term of the bar element equation, we can write

$$(q_0)\begin{Bmatrix} \ell/2 \\ \ell/2 \end{Bmatrix} \tag{2.161}$$

RHS second term of Eq. (2.156). Since it is similar to the RHS second term of bar element, we have

$$\begin{Bmatrix} -Q_0 \\ Q_\ell \end{Bmatrix} \tag{2.162}$$

where Q_0, Q_l represent the net heat flux at the ends of the element.

Thus the element level equations can be written as

$$\left(\frac{k}{\ell} \begin{bmatrix} 1 & -1 \\ -1 & 1 \end{bmatrix} \right) \begin{Bmatrix} T_k \\ T_{k+1} \end{Bmatrix} = (q_0) \begin{Bmatrix} \ell/2 \\ \ell/2 \end{Bmatrix} + \begin{Bmatrix} -Q_0 \\ Q_\ell \end{Bmatrix} \tag{2.163}$$

The LHS may be considered to be the "element conductance matrix" and the first term on the RHS is the "equivalent nodal heat flux" corresponding to the distributed heat source/sink. The second vector on the RHS refers to the net heat flux at the nodes and permits us to incorporate prescribed natural boundary conditions. When the individual elements are assembled together, these Q's algebraically sum up to the resultant externally applied heat flux at the section.

The attention of the reader is drawn to the similarity of finite element equations of a bar and the present heat transfer element, and we note that this is due to the similarity of the governing differential equations and the shape functions used. Thus, essentially the same equations can be used for bar and heat transfer problems except that the variables have different physical meaning in their respective context. It is due to this fact that a single commercial finite element software package is able to solve a wide variety of physical problems.

We will illustrate the use of this heat transfer element with the following example.

Example 2.14. *Plane wall with uniformly distributed heat source.* Consider a four-element mesh for a plane wall of thickness $2L$, as shown in Figure 2.25. The length of each element is given as $\ell = 2L/4 = L/2$. The temperatures at the left and the right ends of the wall are prescribed as T_w. The assembled equations are

$$\frac{k}{\ell} \begin{bmatrix} 1 & -1 & 0 & 0 & 0 \\ -1 & 2 & -1 & 0 & 0 \\ 0 & -1 & 2 & -1 & 0 \\ 0 & 0 & -1 & 2 & -1 \\ 0 & 0 & 0 & -1 & 1 \end{bmatrix} \begin{Bmatrix} T_w \\ T_2 \\ T_3 \\ T_4 \\ T_w \end{Bmatrix} = \begin{Bmatrix} \dfrac{q_0\ell}{2} \\ q_0\ell \\ q_0\ell \\ q_0\ell \\ \dfrac{q_0\ell}{2} \end{Bmatrix} + \begin{Bmatrix} -Q_{\text{left}} \\ 0 \\ 0 \\ 0 \\ Q_{\text{right}} \end{Bmatrix} \tag{2.164}$$

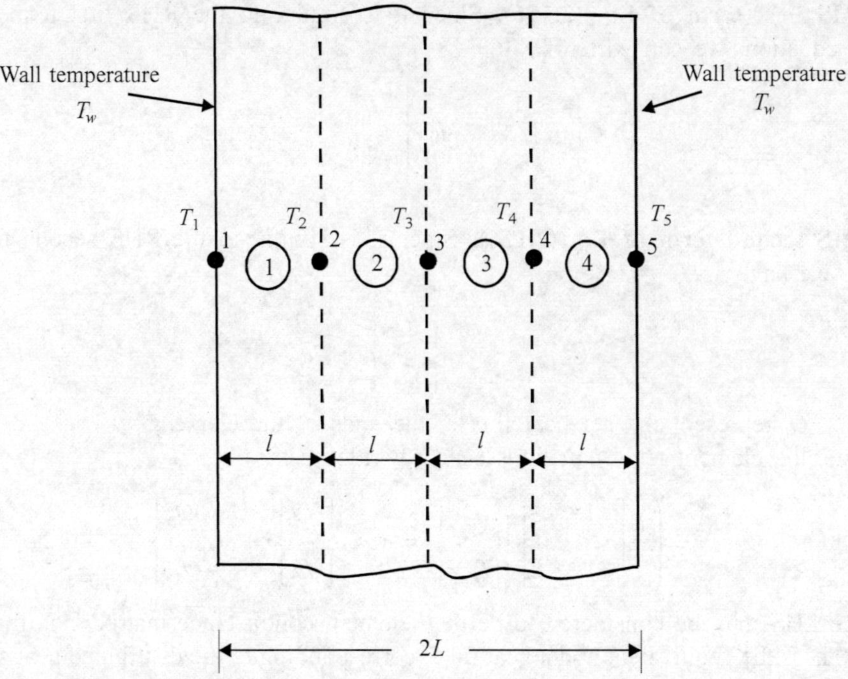

Fig. 2.25 Plane wall with uniform heat source (Example 2.14).

The unknowns are the temperatures at the interior nodes and the heat fluxes at the left and right ends of the wall. We visualise an imaginary partition of the above equation, using the first and the last rows for determination of heat fluxes and the middle three rows for the determination of temperatures. Rearranging the three equations for the three temperature variables, we have

$$\frac{k}{\ell}\begin{bmatrix} 2 & -1 & 0 \\ -1 & 2 & -1 \\ 0 & -1 & 2 \end{bmatrix}\begin{Bmatrix} T_2 \\ T_3 \\ T_4 \end{Bmatrix} = \begin{Bmatrix} q_0\ell \\ q_0\ell \\ q_0\ell \end{Bmatrix} + \begin{Bmatrix} T_w \\ 0 \\ T_w \end{Bmatrix} \tag{2.165}$$

Solving, we get (using $\ell = L/2$)

$$T_2 = T_4 = \frac{3q_0L^2}{8k} + T_w$$

$$\tag{2.166}$$

$$T_3 = \frac{q_0L^2}{2k} + T_w$$

It is easily verified that these results tally with the exact solution.

SUMMARY

We have described the formulation of the finite element equations, assuming the weak form of any differential equation. We have also illustrated the procedure through two typical one-dimensional examples—axial deformation of a rod and the temperature distribution in a wall. We have also demonstrated how to incorporate the boundary conditions and obtain a solution with a given finite element mesh. We will now summarise the procedure for finite element analysis given any differential equation. Our methodology will not be restricted to problems of structural mechanics alone but can equally be applied to other fields.

Procedure for finite element analysis starting from a given differential equation

1. Write down the Weighted Residual statement.

2. Perform integration by parts and develop the weak form of the WR statement. The integration by parts is to be carried sufficient number of times to evenly distribute the differentiation between the field variable and the weighting function. It is observed that the weak form is equivalent to the differential equation and the natural boundary conditions.

3. Rewrite the weak form as a summation over n elements.

4. Define the finite element i.e. the geometry of the element, its nodes, nodal d.o.f.

5. Derive the shape or interpolation functions. Use these as the weighting functions also.

6. Compute the element level equations by substituting these in the weak form.

7. For a given topology of finite element mesh, build up the system equations by assembling together element level equations.

8. Substitute the prescribed boundary conditions and solve for the unknowns.

PROBLEMS

2.1 Solve the following equation using a two-parameter trial solution by (a) the point-collocation method ($R_d = 0$ at $x = 1/3$ and $x = 2/3$); (b) the Galerkin method. Then, compare the two solutions with the exact solution.

$$\frac{dy}{dx} + y = 0, \qquad 0 \leq x \leq 1$$

$$y(0) = 1$$

2.2 Solve the following equation using the Galerkin's method (Use at least a two-parameter solution):

$$\frac{dy}{dx} = 50(1 + \cos x) - 0.05y, \qquad 0 \leq x \leq 2\pi$$

$$y(0) = 200$$

2.3 Determine a three-parameter solution of the following using the Galerkin's method and compare it with the exact solution:

$$\frac{d^2 y}{dx^2} = -\cos \pi x, \qquad 0 \le x \le 1$$

$$u(0) = 0, \qquad u(1) = 0$$

2.4   Determine a two-parameter Galerkin approximation solution of the differential equation:

$$\frac{d^2 u}{dx^2} = f_0, \qquad 0 < x < L$$

$$\frac{du}{dx} = 0 \quad \text{at } x = 0 \text{ and } L$$

2.5   Give a one-parameter Galerkin solution of the following equation, for the two domains shown in Fig. P2.5:

$$\left(\frac{\partial^2 u}{\partial x^2} + \frac{\partial^2 u}{\partial y^2} \right) = 1$$

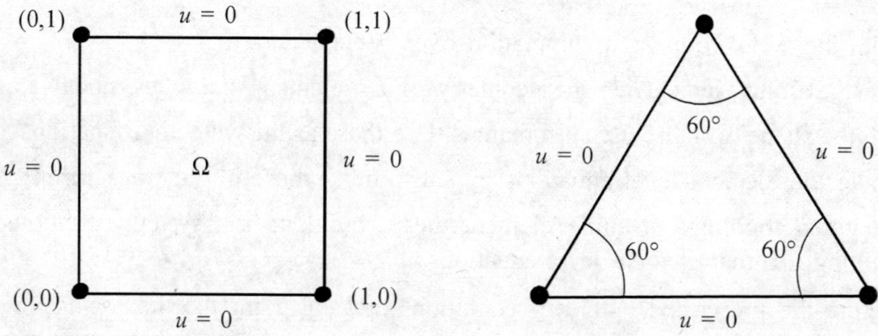

Fig. P2.5   (a) Square domain and   (b) Triangular domain.

2.6   Develop the weak form and find the solution (Fig. P2.6).

$$EI \frac{d^4 v}{dx^4} - q_0 = 0$$

$$v(0) = 0; \quad \frac{dv}{dx}(0) = 0,$$

$$EI \frac{d^2 v}{dx^2}(I) = M,$$

$$EI \frac{d^3 v}{dx^3}(I) = P$$

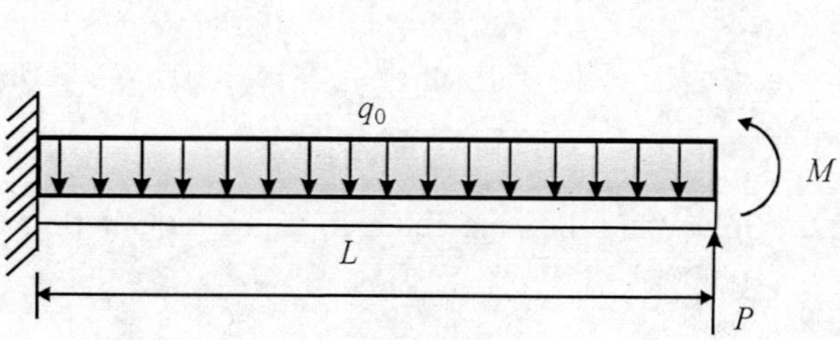

Fig. P2.6   Cantilever beam.

2.7   For fully developed fluid flow in an annular pipe (Fig. P2.7), the governing equation is

$$\frac{1}{r}\frac{d}{dr}\left(r\mu\frac{du}{dr}\right) = k$$

$$u_{r=a} = 0, \qquad u_{r=b} = 0$$

Find the solution using the weak form.

Fig. P2.7   Flow through annular pipe.

2.8   Solve the following by making use of weak formulation (Fig. P2.8):

$$EI\frac{d^4v}{dx^4} - q_0 = 0$$

$$v(0) = 0, \qquad \frac{dv}{dx}(0) = 0$$

$$v(L) = 0, \qquad \frac{dv}{dx}(L) = 0$$

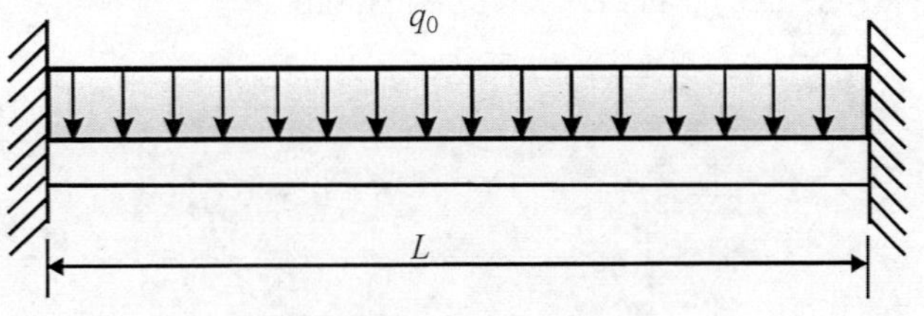

Fig. P2.8   Fixed-fixed beam.

2.9   For heat conduction through an annulus region (Fig. P2.9), the governing equation is given below along with the boundary conditions. Derive the weak form and find the solution

$$\frac{d^2T}{dr^2} + \frac{1}{r}\frac{dT}{dr} + \frac{q_0}{k} = 0$$

$$T(r_i) = T_i, \qquad T(r_o) = T_o$$

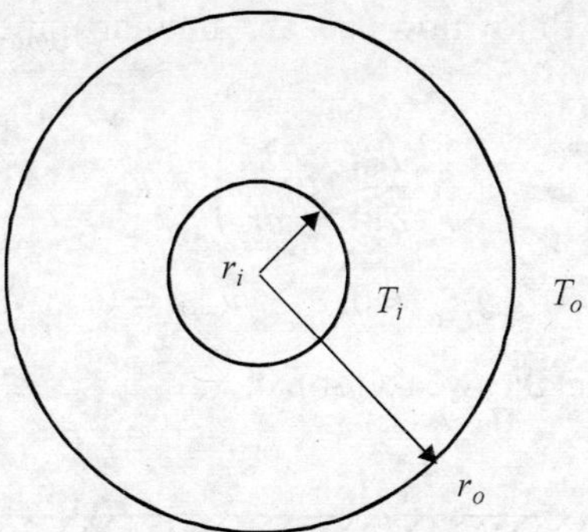

Fig. P2.9   Heat conduction in a disc.

2.10   For the problem given below, find a two-parameter Galerkin approximate solution that contains trigonometric trial functions.

$$\frac{d^2 y}{dx^2} - \cos\frac{\pi x}{\ell} - 10 = 0$$

$$\frac{dy}{dx} = 0 \quad \text{at } x = 0 \text{ and } L$$

2.11   Determine the Galerkin approximation solution of the differential equation

$$A\frac{d^2 u}{dx^2} + B\frac{du}{dx} + C = 0$$

$$u(0) = u(L) = 0$$

2.12   A certain problem of one-dimensional steady heat transfer with a distributed heat source is governed by equation

$$\frac{d^2\phi}{dx^2} + \phi + 1 = 0$$

$$\phi = 0 \quad \text{at } x = 0$$

$$\frac{d\phi}{dx} = -\phi \quad \text{at } x = 1$$

Find the Galerkin approximation solution of the above differential equation.

2.13   Determine a two-parameter Galerkin approximation solution of the differential equation

$$\frac{d^2 u}{dx^2} = f_c, \qquad 0 < x < L$$

$$\frac{du}{dx} = 0 \quad \text{at } x = 0; \ x = L$$

Problem to Investigate

2.14   The coupled, nonlinear governing equations for large deflections of a beam are as follows:

$$AE\frac{d}{dx}\left(\frac{du}{dx} + \frac{1}{2}\left(\frac{dv}{dx}\right)^2\right) + q_a = 0$$

$$EI\frac{d^4v}{dx^4} - AE\frac{d}{dx}\left\{\frac{dv}{dx}\left(\frac{du}{dx} + \frac{1}{2}\left(\frac{dv}{dx}\right)^2\right)\right\} - q_t = 0$$

where A, E, q_a, q_t are known constants; u and v refer to axial and transverse deflections. Develop the weak form statement for this nonlinear problem.

CHAPTER 3

Finite Element Formulation Based on Stationarity of a Functional

3.1 Introduction

In Chapter 2, we discussed the formulation of the finite element equations from the weak form of the Galerkin weighted residual statement. Our discussion typically started with the specification of the governing differential equation and the boundary conditions for the problem at hand. While the formulation of the finite element method starting from the governing differential equation is quite general, in many situations of practical importance, the governing equations can also be given, perhaps more conveniently, in the form of "stationarity of a functional". The term 'functional' is used to represent a function of a function, e.g. an expression such as

$$I = \int \left\{ \left(\frac{1}{2}\right) AE \left(\frac{du}{dx}\right)^2 - (q)(u) \right\} dx \tag{3.1}$$

is a function of u, which itself is a function of x. **We state that the problem of finding $u(x)$ that makes I stationary with respect to small, admissible variations in $u(x)$ is equivalent to the problem of finding $u(x)$ that satisfies the governing differential equation for this problem.** For example, for most problems of structural mechanics, such a functional expression represents the "total potential of the system", a quantity with direct physical significance for the structural analyst. The first term in the above expression for I signifies the strain energy stored in the structure (in this case, just a rod undergoing axial deformation) and the second term represents the potential of external forces. The Principle of Stationary Total Potential (PSTP) characterising the equilibrium configuration of a structure is a well-known principle of structural mechanics. For conservative systems (i.e. no dissipative forces), it can be stated as

Among all admissible displacement fields, the equilibrium configuration of the system is that, which makes the total potential of the system stationary with respect to small, admissible variations of displacement.

Here we use the term 'admissible' to describe the displacement field that satisfies the essential boundary conditions and does not lead to any voids/openings/kinks, etc. in the structure. For long, it has been common practice to use energy methods in structural mechanics and thus for most problems of structural mechanics, it may be convenient to use the PSTP to formulate the governing equations. The finite element equations can be developed starting from such a principle governing the stationarity of a functional. In fact, the finite element method had its origins in the late 50s and early 60s in structural mechanics where it was purely formulated as a technique to solve problems of structural analysis. Subsequently, it was realised that the method belonged to the general class of weighted residual methods and its capability to model a variety of problems in engineering science was fully appreciated. At present it is used in a wide variety of applications, viz., structural mechanics, heat transfer, fluid mechanics, biomechanics, electromagnetic field problems, etc.

In what follows, we will first discuss some basic ideas in variational calculus and the extrema of a functional and show that the governing differential equations (Euler–Lagrange equations) can be derived from this. Thus we establish the equivalence of differential equation representation and the functional representation. We will then discuss the method of finding approximate solutions to the problem of minimisation of a functional, e.g., $u(x)$ that minimises I in the example cited above, using a single composite trial function valid over the entire solution domain. We will then introduce piece-wise continuous trial functions leading to the formulation of the finite element equations starting from a given functional. We will show that the resulting finite element formulation is identical to that obtained, based on the weak form of the governing differential equation.

It must be appreciated that the finite element equations can be derived systematically, with equal ease, starting from either form of representation of governing equations—differential or functional. Given a functional, we can always find the corresponding differential equation for the problem, but some differential equation forms may not have a corresponding functional form. Unlike the total potential of a structure, in many cases the functional (even when it exists) may not have a direct physical meaning. While traditionally the functional form (via PSTP for structures) is used to introduce the finite element method, it must be appreciated that the finite element formulation itself does not require the existence of a functional. In the subsequent chapters, we will use both these forms interchangeably for developing the necessary finite elements.

3.2 Functional and Differential Equation Forms

It is beyond the scope of the present discussion to provide the general theory of variational calculus. The interested reader is advised to refer to the standard texts on the subject. We will present a brief introduction to the relevant concepts of variational calculus.

It is well known that for a function of single variable, viz., $f(x)$, the condition for an extremum is given by $df/dx = 0$. A function of several variables $f(x_1, x_2, \ldots, x_n)$ will have an extremum at a given $(x_1, x_2, \ldots, x_n)$ if

$$\frac{\partial f}{\partial x_1} = \frac{\partial f}{\partial x_2} = \cdots = \frac{\partial f}{\partial x_n} = 0 \tag{3.2}$$

Now consider a "functional" (i.e. function of functions) such as

$$I(u) = \int_a^b F\left(u, \frac{du}{dx}, x\right) dx \tag{3.3}$$

We wish to find $u(x)$ which makes the functional stationary, subject to the end conditions $u(a) = u_a$ and $u(b) = u_b$. As an example, if $F = (1/2)AE(du/dx)^2 - (q)(u)$, and the prescribed end conditions are $u(0) = 0$; $u(L) = 0$, then this represents the problem of a uniform bar clamped at both ends and subjected to a distributed load $q(x)$. In Figure 3.1, we represent the possible set of $\overline{u}(x)$ as also the as yet unknown solution $u(x)$. We observe that all $\overline{u}(x)$ satisfy

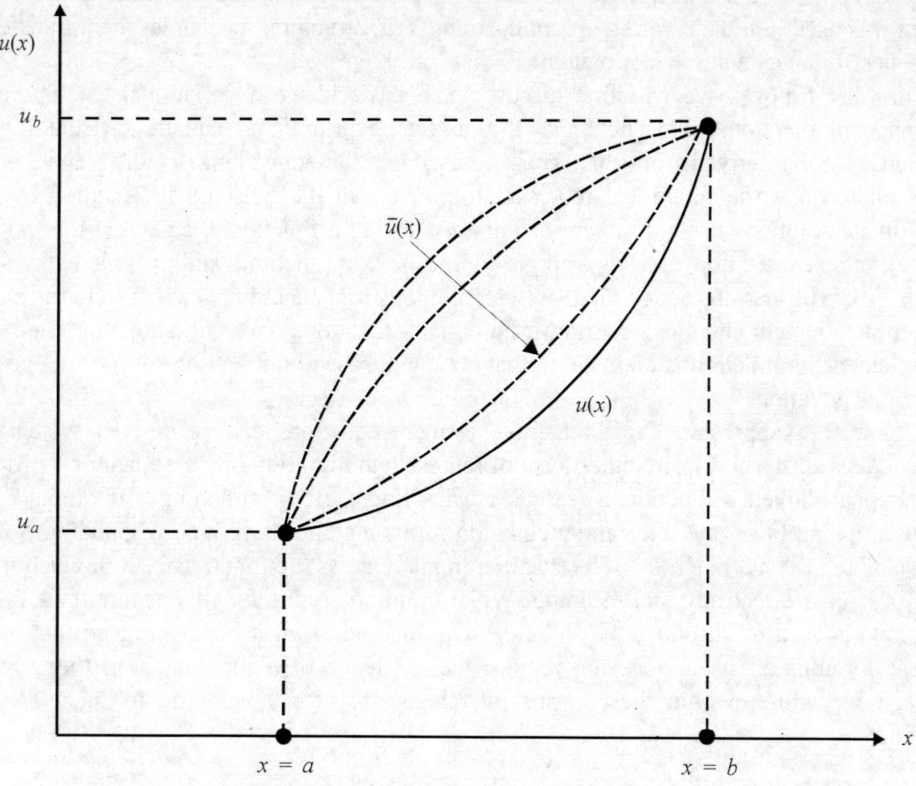

Fig. 3.1   Admissible solutions $u(x)$.

the prescribed end conditions, and thus we are dealing with only fixed end variations. We wish to study what happens to I, if $u(x)$ is slightly changed to $\overline{u}(x)$, i.e.,

$$\overline{u}(x) = u(x) + \varepsilon v(x) \tag{3.4}$$

where ε is a small parameter and $v(a) = 0 = v(b)$.

The difference between $\overline{u}(x)$ and $u(x)$ is termed the "variation" in $u(x)$, and we denote this by $\delta u(x)$. Thus,

$$\delta u(x) = \overline{u}(x) - u(x) = \varepsilon v(x) \tag{3.5}$$

Figure 3.2 shows the distinction between du and δu at a given position x, the variation δu refers to the difference between $\bar{u}(x)$ and $u(x)$ while du refers to the differential change in $u(x)$ as x is changed to $x + dx$.

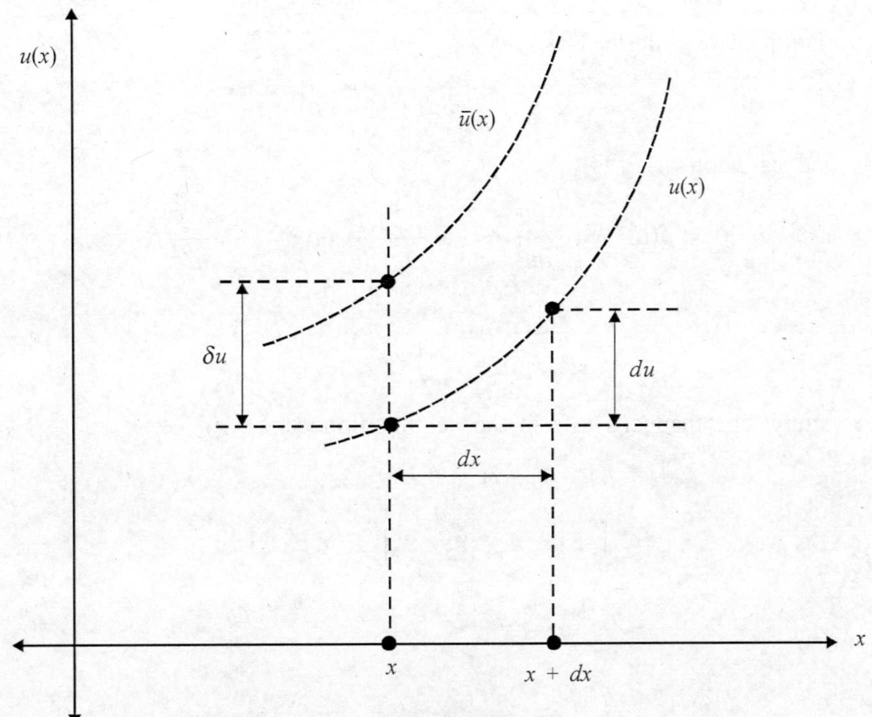

Fig. 3.2　Distinction between δu, du.

We define

$$\delta(u') = \text{difference in slope of } \bar{u}(x) \text{ and } u(x)$$
$$= \bar{u}'(x) - u'(x) = u'(x) + \varepsilon v'(x) - u'(x)$$
$$= \varepsilon v'(x) = [\delta u]' \tag{3.6}$$

where $(\ )'$ designates differentiation with respect to x. Thus we observe that the process of variation and differentiation are permutable. Now, for a given x, as we move from $u(x)$ to $\bar{u}(x)$, we have

$$\Delta F = F(\bar{u},\ \bar{u}',\ x) - F(u,\ u',\ x) = F(u + \delta u,\ u' + \delta u',\ x) - F(u,\ u',\ x) \tag{3.7}$$

Expanding the first term in Taylor's series, we have

$$F(u + \delta u,\ u' + \delta u',\ x) = F(u,\ u',\ x) + \frac{\partial F}{\partial u}\delta u + \frac{\partial F}{\partial u'}\delta u'$$

$$+ \frac{1}{2!}\left[\frac{\partial^2 F}{\partial u^2}\delta u^2 + \frac{2\partial^2 F}{\partial u\,\partial u'}\delta u\,\delta u' + \frac{\partial^2 F}{\partial u'^2}\delta u'^2\right] + \cdots \tag{3.8}$$

Thus,

$$\Delta F = \left(\frac{\partial F}{\partial u} \delta u + \frac{\partial F}{\partial u'} \delta u' \right) + \frac{1}{2!} \left(\frac{\partial^2 F}{\partial u^2} \delta u^2 + \frac{2 \partial^2 F}{\partial u\, \partial u'} \delta u\, \delta u' + \frac{\partial^2 F}{\delta u'^2} \delta u'^2 \right) + \cdots \quad (3.9)$$

The first variation of F is defined as

$$\delta F = \frac{\partial F}{\partial u} \delta u + \frac{\partial F}{\partial u'} \delta u' \tag{3.10}$$

and the second variation of F as

$$\delta^2 F = \delta(\delta F) = \frac{\partial^2 F}{\partial u^2} \delta u^2 + \frac{2 \partial^2 F}{\partial u\, \partial u'} \delta u\, \delta u' + \frac{\partial^2 F}{\partial u'^2} \delta u'^2 \tag{3.11}$$

Therefore,

$$\Delta F = \delta F + \frac{1}{2!} \delta^2 F + \cdots \tag{3.12}$$

Now let us study what happens to I in the neighbourhood of $u(x)$, i.e.

$$\Delta I = I(\overline{u}, \overline{u}', x) - I(u, u', x)$$

$$= \int_a^b F(\overline{u}, \overline{u}', x)\, dx - \int_a^b F(u, u', x)\, dx$$

$$= \int_a^b \Delta F\, dx$$

$$= \int_a^b \left(\delta F + \frac{1}{2!} \delta^2 F + \cdots \right) dx \tag{3.13}$$

The first variation of I is defined as

$$\delta I = \int_a^b \delta F\, dx \tag{3.14}$$

and the second variation of I as

$$\delta^2 I = \int_a^b \delta^2 F\, dx \tag{3.15}$$

Therefore,

$$\Delta I = \delta I + \frac{1}{2!} \delta^2 I + \cdots \tag{3.16}$$

If I represents the total potential of a structure and we look for a stable equilibrium configuration, then we wish to find $u(x)$ that minimises I. Since $u(x)$ minimises I, $\Delta I \geq 0$. As ε is reduced, ΔI approaches zero and when $\overline{u}(x) = u(x)$, I attains a minimum and $\Delta I \equiv 0$.

By studying the relative orders of magnitude of the various terms, it is possible to show that δI is of the order $O(\varepsilon)$ while $\delta^2 I$ is $O(\varepsilon^2)$, etc. When ε is sufficiently small, $\delta^2 I$ and higher variations become negligible compared to δI, and thus the condition for I to be stationary becomes $\delta I = 0$. We propose to show that the variational form given by $\delta I = 0$ reduces to the governing differential equation of the problem. From Eqs. (3.10)–(3.14), we have

$$\delta I = \int_a^b \left(\frac{\partial F}{\partial u} \delta u + \frac{\partial F}{\partial u'} \delta u' \right) dx \tag{3.17}$$

From Eq. (3.6),

$$\delta(u') = (\delta u)' = \frac{d}{dx}(\delta u) \tag{3.18}$$

Thus,

$$\int_a^b \frac{\partial F}{\partial u'} \delta u' \, dx = \int_a^b \frac{\partial F}{\partial u'} d(\delta u) \tag{3.19}$$

Performing integration by parts, we obtain

$$\int_a^b \frac{\partial F}{\partial u'} d(\delta u) = \left[\frac{\partial F}{\partial u'} \delta u \right]_a^b - \int_a^b (\delta u) \frac{d}{dx} \left(\frac{\partial F}{\partial u'} \right) dx \tag{3.20}$$

Substituting Eq. (3.20) in Eq. (3.17), we get

$$\delta I = \int_a^b \left[\frac{\partial F}{\partial u} - \frac{d}{dx} \left(\frac{\partial F}{\partial u'} \right) \right] \delta u \, dx + \left(\frac{\partial F}{\partial u'} \delta u \right)_a^b = 0 \tag{3.21}$$

Since all the trial functions $u(x)$ satisfy the end conditions at $x = a$ and b, we have

$$\delta u(a) = 0 = \delta u(b) \tag{3.22}$$

For arbitrary δu, we therefore have, from Eq. (3.21), the relation

$$\boxed{\left[\frac{\partial F}{\partial u} - \frac{d}{dx} \left(\frac{\partial F}{\partial u'} \right) \right] = 0} \tag{3.23}$$

This is known as the Euler–Lagrange equation. For example, if

$$F = \left(\frac{1}{2} \right) AE \left(\frac{du}{dx} \right)^2 - (q)(u)$$

then we have

$$\frac{\partial F}{\partial u} = -q$$

$$\frac{\partial F}{\partial u'} = AE \frac{du}{dx} \tag{3.24}$$

Substituting in Eq. (3.23), we get

$$AE \frac{d^2 u}{dx^2} + q = 0 \tag{3.25}$$

which is the governing differential equation for a bar subjected to distributed loading $q(x)$.

Let us now rewrite Eq. (3.21) for the specific case of axial deformation of a rod. We have

$$\int_a^b \left[-q - AE\frac{d^2u}{dx^2} \right] \delta u \, dx + \left[AE\frac{du}{dx}\delta u \right]_0^L = 0 \tag{3.26}$$

with $\delta u(0) = 0$; $\delta u(L) = 0$ in view of prescribed boundary conditions at $x = 0, L$.

Recapitulating the weighted residual statement for this problem, we have

$$\int_a^b \left[q + AE\frac{d^2u}{dx^2} \right] W \, dx = 0 \tag{3.27}$$

While developing the weak form, we demanded that $W(x)$ be zero at those points where u is prescribed. Thus, $W(0) = 0 = W(L)$.

Comparing the form of the equations obtained using the weak form and the present discussion, we observe that the weighting function $W(x)$ used earlier in weighted residual methods has the connotation of a variation in $u(x)$, i.e. $\delta u(x)$ here. Since all the variations of $u(x)$ being considered satisfy the essential boundary conditions, $\delta u(x)$ at $x = a$ and b is zero. In the weak form, recall that we required the weighting function to be zero at these points. Further, for structural mechanics problems, if $u(x)$ denotes the equilibrium configuration, $\delta u(x)$ (and hence W) has the connotation of a virtual displacement consistent with the boundary conditions (i.e., $\delta u(a) = 0 = \delta u(b)$). $\delta I = 0$, *in essence, is a statement of the familiar virtual work principle.*

We will leave it as an exercise to the reader to show that if

$$I = \int_a^b F(u, u'\, u'', x) \, dx \tag{3.28}$$

the corresponding Euler–Lagrange equation is

$$\frac{\partial F}{\partial u} - \frac{d}{dx}\left(\frac{\partial F}{\partial u'} \right) + \frac{d^2}{dx^2}\left(\frac{\partial F}{\partial u''} \right) = 0 \tag{3.29}$$

As an example, let $F = \left(\dfrac{1}{2}\right)(EI)\left(\dfrac{d^2w}{dx^2}\right)^2 - (q)(w)$, which is the case for a beam subjected to distributed transverse loading $q(x)$. The resulting Euler–Lagrange equation is the familiar form of the governing differential equation of a beam.

Similarly, for a one-dimensional heat transfer problem satisfying the differential equation

$$k\frac{d^2T}{dx^2} + q_0 = 0 \tag{3.30}$$

subject to $T(0) = T_0$, $q\big|_{x=L} = h(T_L - T_\infty)$ is equivalent to minimising the functional

$$I = \int_0^L \frac{1}{2}k\left(\frac{dT}{dx}\right)^2 dx - \int_0^L q_0 T \, dx + \frac{1}{2}h(T - T_\infty)^2 \tag{3.31}$$

subject to $T(0) = T_0$.

Similarly, for two-dimensional heat conduction, satisfying the differential equation

$$k\frac{\partial^2 T}{\partial x^2} + k\frac{\partial^2 T}{\partial y^2} + q_0 = 0 \tag{3.32}$$

with the boundary conditions

(i) $T = T_0$ on S_T,

(ii) $q_n = \overline{q}$ on S_q,

(iii) $q_h = h(T - T_\infty)$ on S_c

is equivalent to minimising the functional

$$I = \frac{1}{2}\iint_A \left(k\left[\left(\frac{\partial T}{\partial x}\right)^2 + \left(\frac{\partial T}{\partial y}\right)^2\right] - 2q_0 T \right) dA + \int_{S_1} \overline{q}\, T\, ds + \int_{S_c} \frac{1}{2} h(T - T_\infty)^2\, ds$$

subject to

$$T = T_0 \quad \text{on } S_T \tag{3.33}$$

Thus, given a functional I, using the variational principle $\delta I = 0$, we can derive the governing differential equation. Therefore, both forms (differential and functional) are equivalent.

We will now discuss an approach to find approximate solutions to the problem of minimisation of a functional, e.g., $u(x)$ that minimises I in the example cited above. We will use, for our discussion, the popular principle of structural mechanics, viz., the Principle of Stationary Total Potential, and present the Rayleigh–Ritz method of finding approximate solutions based on this principle. We will initially use a single composite trial function valid over the entire solution domain, and then introduce piece-wise continuous trial functions leading to the formulation of the finite element equations starting from a given functional.

3.3 Principle of Stationary Total Potential (PSTP)

Consider a spring-mass system shown in Figure 3.3. Let the position of free length of the spring be at a height H from the datum. Now, when the mass is attached to the spring, it causes extension. We wish to illustrate the application of the Principle of Stationary Total Potential in determining the equilibrium position of the spring-mass system. At any extension of the spring given by x, the strain energy stored in the spring is $(1/2)(kx^2)$. The gravitational potential energy of the mass with respect to the datum is $mg(H - x)$. So we define the total potential of the system in this configuration to be

$$\Pi_p = \left(\frac{1}{2}\right)(kx^2) + mg(H - x) \tag{3.34}$$

Being the total potential of the system, this quantity represents the total amount of energy that the system can give to an external agency. The principle of stationary total potential demands that the equilibrium value of x be such that Π_p is stationary with respect to x in that configuration, i.e.,

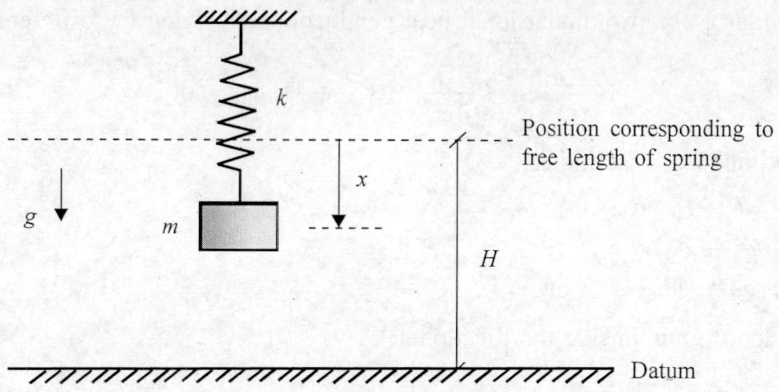

Fig. 3.3   Typical spring-mass system.

$$\frac{d\Pi_p}{dx} = 0 \tag{3.35}$$

i.e.,

$$(k)(x_{eq}) - (mg) = 0 \tag{3.36}$$

Thus, x_{eq} is given by (mg/k), as expected. This is a trivial example to introduce the basic idea. We observe that the choice of the datum is arbitrary and in no way affects the final result. We thus choose to drop the datum altogether and rewrite Π_p as

$$\Pi_p = \left(\frac{1}{2}\right)(kx^2) - mgx \tag{3.37}$$

This quantity represents the work to be done by an external observer in bringing the system to this configuration from $x = 0$. The weight in this example can as well be any external (conservative) force P, and we write the general expression for the total potential of a system as

$$\Pi_p = \text{Internal strain energy in the structure} + \text{Potential of external forces} \tag{3.38}$$

The potential of external forces is, by convention, defined in such a way that its derivative (with a negative sign) yields the force. If δ represents the generic displacement, then we have

$$\frac{\partial \Pi_p}{\partial \delta} = 0 \tag{3.39}$$

Thus, for a general multidegree of freedom structure, if δ_1, δ_2, δ_3, ..., δ_n are the displacements, then we have

$$\frac{\partial \Pi_p}{\partial \delta_i} = 0, \qquad i = 1, 2, ..., n \tag{3.40}$$

Thus, based on this principle, we can formulate a method of finding an approximate solution to the problem at hand as follows. Just as we did in the weighted residual method discussed in Chapter 2, we begin with an assumed trial function solution to the problem. For the deformation of the structure as per this assumed displacement field, we determine the total potential Π_p. Based on the PSTP, we set up the necessary equations to solve for the unknown coefficients. We will now discuss the details of the Rayleigh–Ritz method, which is popularly used in structural mechanics.

3.3.1 Rayleigh–Ritz Method

The Rayleigh–Ritz (R–R) method consists of three basic steps:

Step 1: *Assume a displacement field.* Let the displacement field be given by $\{\phi(x) + \Sigma\, c_i N_i\}$, $i = 1, 2, \ldots, n$, where N_i are the shape functions and c_i are the as yet undetermined coefficients. It is observed that the assumed displacement field should satisfy both the essential boundary conditions of the problem and internal compatibility, i.e. there should be no kinks, voids, etc. within the structure.

Step 2: *Evaluation of the total potential.* For the system under consideration, evaluate the total potential Π_p consistent with the assumed displacement field in Step 1 above.

Step 3: *Set up and solve the system of equations.* By virtue of the PSTP, the total potential will be stationary with respect to small variations in the displacement field. The variations in the displacement field in our case are attained by small variations in the coefficients c_i. Thus we have

$$\frac{\partial \Pi_p}{\partial c_i} = 0, \qquad i = 1, 2, \ldots, n \tag{3.41}$$

which will yield the necessary equations to be solved for the coefficients c_i.

This method was first formulated by Rayleigh (1877) and later refined and generalised by Ritz (1908). Rayleigh worked with just one term, viz., $c_1 N_1$ while Ritz extended the technique to an n-term approximation. However, both of them used shape functions N which were single composite functions valid over the entire domain of the problem. Further extension of their technique by using piece-wise defined shape functions N_i will lead us to the variational formulation of the finite element method.

We will now illustrate the basic scheme of R–R method with some simple example problems.

Example 3.1. *A bar under uniform load.* Consider a bar clamped at one end and left free at the other end and subjected to a uniform axial load q_0 as shown in Figure 3.4. The governing differential equation is given by

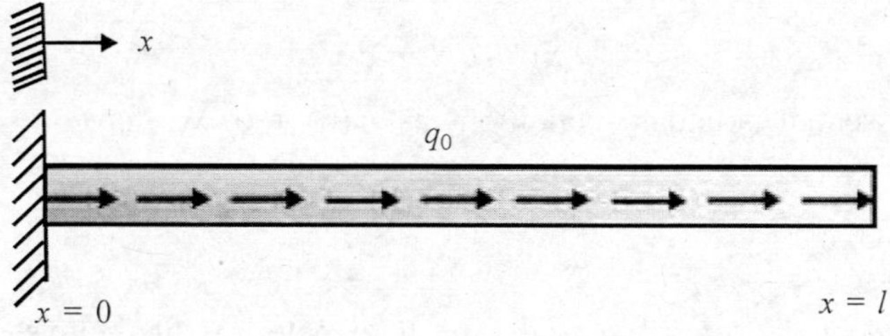

Fig. 3.4   Rod under axial load (Example 3.1).

$$AE \frac{d^2u}{dx^2} + q = 0 \tag{3.42}$$

with the boundary conditions $u(0) = 0$; $\left. \dfrac{du}{dx} \right|_{x=L} = 0$.

We discussed the solution using the Galerkin weighted residual method in Example 3.1. We will now illustrate the solution using the R–R method. For a general deformation $u(x)$,

$$\text{Strain energy stored in the bar, } U = \int_0^L \left[\frac{1}{2} AE \left(\frac{du}{dx} \right)^2 \right] dx \tag{3.43}$$

$$\text{Potential of the external forces, } V = -\int_0^L q_0 u \, dx - P_L u(L) \tag{3.44}$$

where, for the sake of generality, a tip load $P_L = AE \left. \dfrac{du}{dx} \right|_{x=L}$ has been included in the potential of external forces. In view of the second boundary condition, the expression for the potential of external forces reduces (for this example) to the following:

$$V = -\int_0^L q_0 u \, dx \tag{3.45}$$

Thus we need to find $u(x)$ that minimises the total potential of the system given by

$$\Pi_P = \int_0^L \left[\frac{1}{2} AE \left(\frac{du}{dx} \right)^2 - q_0 u \right] dx \tag{3.46}$$

subject to the essential boundary condition that $u(0) = 0$. It is to be observed that the force boundary condition at the free end ($du/dx \big|_{x=L} = 0$) has been incorporated in the expression for the functional itself and thus the trial function assumed need only satisfy the essential boundary condition. Also, we observe that the continuity demanded on $u(x)$ is lower (i.e. need to be differentiable only once) compared to the differential equation form.

Let us now solve for the displacement field using the R–R method.

Step 1:   *Assume a displacement field*. Let us assume that

$$u(x) \approx c_1 x + c_2 x^2 \tag{3.47}$$

This satisfies the essential boundary condition that $u(0) = 0$. We have

$$\frac{du}{dx} = c_1 + 2c_2 x \tag{3.48}$$

Step 2:   *Evaluation of the total potential*. The total potential of the system is given as

$$\Pi_p = \int_0^L \left[\frac{AE}{2}(c_1 + 2c_2 x)^2 - q_0(c_1 x + c_2 x^2) \right] dx$$

$$= \frac{AE}{2}\left[c_1^2 L + \frac{4c_2^2}{3L^3} + 2c_1 c_2 L^2 \right] - q_0 \frac{c_1 L^2}{2} - q_0 c_2 \frac{L^3}{3} \tag{3.49}$$

Step 3: *Set up and solve the system of equations.* From the principle of stationary total potential (PSTP), we have

$$\frac{\partial \Pi_p}{\partial c_i} = 0, \qquad i = 1, 2 \tag{3.50}$$

Therefore,

$$\frac{\partial \Pi_p}{\partial c_i} = 0 \Rightarrow \frac{AE}{2}(2c_1 L + 2c_2 L^2) - \frac{q_0 L^2}{2} = 0$$

$$\frac{\partial \Pi_p}{\partial c_2} = 0 \Rightarrow \frac{AE}{2}(8c_2 L^3/3 + 2c_1 L^2) - \frac{q_0 L^3}{3} = 0 \tag{3.51}$$

Solving, we obtain

$$c_1 = \frac{q_0 L}{AE}, \qquad c_2 = -\frac{q_0}{2AE}$$

Thus,

$$u(x) = \frac{q_0}{AE}x(L - x/2) = \frac{q_0}{2AE}(2Lx - x^2) \tag{3.52}$$

which is observed to be the same as that obtained in Example 2.1.

Example 3.2. *A simply supported beam under uniform load.* Consider a simply supported beam under uniformly distributed load q_0 as shown in Figure 3.5. For a deformation $v(x)$, we have

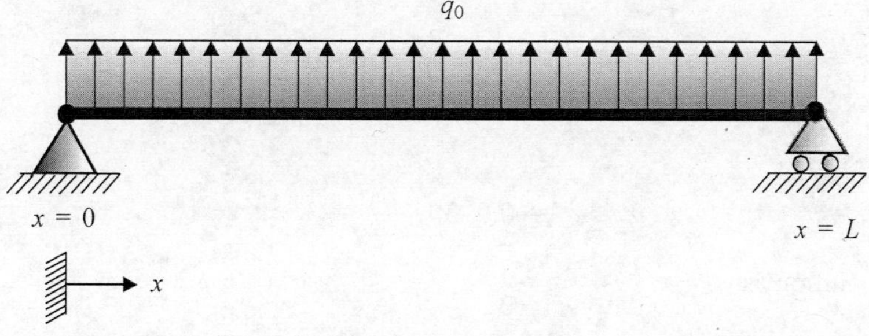

Fig. 3.5   Simply supported beam under load (Example 3.2).

The strain energy

$$U = \int_0^L \frac{1}{2} EI \left(\frac{d^2 v}{dx^2} \right)^2 dx \tag{3.53}$$

The potential of the external forces is

$$V = - \int_0^L q_0 v \, dx \tag{3.54}$$

Thus we have the total potential

$$\Pi_p = \int_0^L \left[\frac{EI}{2} \left(\frac{d^2 v}{dx^2} \right)^2 - q_0 v \right] dx \tag{3.55}$$

Step 1: *Assume a displacement field.* Let us assume $v(x) \approx c_1 \sin (\pi x / L)$. This satisfies the essential boundary conditions $v(0) = 0 = v(L)$. We have

$$\frac{d^2 v}{dx^2} = -c_1 \left(\frac{\pi}{L} \right)^2 \sin \frac{\pi x}{L} \tag{3.56}$$

Step 2: *Evaluation of the total potential.* The total potential of the system is given by

$$\Pi_p = \int_0^L \left[\frac{EI}{2} \left(-c_1 \left(\frac{\pi}{L} \right)^2 \sin \frac{\pi x}{L} \right)^2 dx - q_0 c_1 \sin \frac{\pi x}{L} \right] dx$$

$$= \frac{\pi^4 EI}{4 L^3} c_1^2 - \frac{2 q_0 L}{\pi} c_1 \tag{3.57}$$

Step 3: *Set up and solve the system of equations.* From the principle of stationary total potential, we have

$$\frac{\partial \Pi_p}{\partial c_1} = 0 \tag{3.58}$$

Therefore,

$$\frac{\pi^4 EI}{2 L^3} c_1 - \frac{2 q_0 L}{\pi} = 0$$

i.e.

$$c_1 = 0.01307 \, \frac{q_0 L^4}{EI} \tag{3.59}$$

Thus the final solution is

$$v(x) = 0.01307 \, \frac{q_0 L^4}{EI} \sin \frac{\pi x}{L} \tag{3.60}$$

We observe that this solution is identical to the solution obtained from the Galerkin method (ref. Example 2.4). In the Galerkin method, the original Weighted Residual statement was equivalent to only the differential equation and, therefore, required that the trial solution assumed should satisfy all the boundary conditions of the problem (essential and natural). The R–R method is based on the functional in which the external forces, if any, applied at the boundaries are taken into account through the potential of the external forces. Thus the trial solution for the R–R method need only satisfy the essential boundary conditions. In general, if the differential equation form and the functional form of a given problem are both available, then the Galerkin method and the R–R method yield identical solutions when the problem involves only essential boundary conditions and when they both use the same shape functions.

Example 3.3.   *Temperature distribution in a pin-fin.* Consider a 1 mm diameter, 50 mm long aluminium pin-fin as shown in Figure 3.6, used to enhance the heat transfer from a surface wall maintained at 300°C. Let $k = 200$ W/m/°C for aluminium; $h = 20$ W/m^2/°C, $T_\infty = 30$°C.

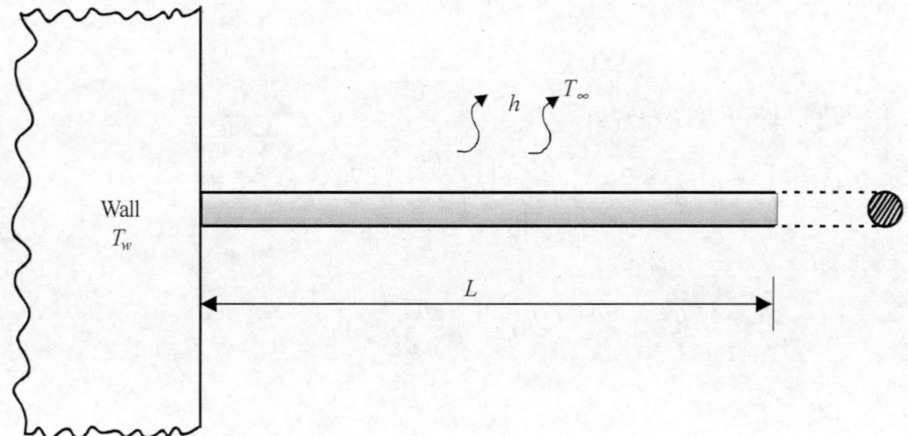

Fig. 3.6   A pin-fin (Example 3.3).

The temperature distribution in the fin was obtained using the Galerkin method in Example 2.8 based on the governing differential equation. The governing differential equation and the boundary conditions we used are repeated here for convenience:

$$k\frac{d^2T}{dx^2} = \frac{Ph}{A_c}(T - T_\infty)$$

$$T(0) = T_w = 300°C \tag{3.61}$$

$$q_L = kA_c\frac{dT}{dx}(L) = 0 \qquad \text{(insulated tip)}$$

where k is the coefficient of thermal conductivity, P is the perimeter, A_c is the cross-sectional area, h is the convective heat transfer coefficient, T_w is the wall temperature, and T_∞ the ambient temperature. The equivalent functional representation is given by

$$\Pi = \int_0^L \frac{1}{2} k \left(\frac{dT}{dx} \right)^2 dx + \int_0^L \frac{1}{2} \frac{Ph}{A_c} (T - T_\infty)^2 \, dx - q_L T_L \tag{3.62}$$

where, in general, the heat flux q_L is included. However, in view of the insulated tip boundary condition, the equivalent statement of the problem is to find $T(x)$ that minimises the functional

$$\Pi = \int_0^L \frac{1}{2} k \left(\frac{dT}{dx} \right)^2 dx + \int_0^L \frac{1}{2} \frac{Ph}{A_c} (T - T_\infty)^2 \, dx \tag{3.63}$$

subject to the boundary condition that $T(0) = T_w = 300°C$.

Step 1: *Assume a trial solution.* Let

$$T(x) \approx \hat{T}(x) = c_0 + c_1 x + c_2 x^2$$

From the boundary conditions, $c_0 = 300$. Therefore,

$$\hat{T}(x) = 300 + c_1 x + c_2 x^2 \tag{3.64}$$

Step 2: *Compute the functional*

$$\Pi = \int_0^L \frac{1}{2} k [c_1 + 2c_2 x]^2 \, dx + \int_0^L \left(\frac{Ph}{2A_c} \right) [270 + c_1 x + c_2 x^2]^2 \, dx \tag{3.65}$$

Step 3: *Minimise the functional*

$$\frac{\partial \Pi}{\partial c_1} = 0, \qquad \frac{\partial \Pi}{\partial c_2} = 0 \tag{3.66}$$

From Eq. (3.65),

$$k(2c_1 L + 2c_2 L^2) + \frac{Ph}{A_c} \left(\frac{2L^3}{3} c_1 + 270 L^2 + \frac{c_2}{2} L^4 \right) = 0$$

$$k \left(\frac{8}{3} c_2 L^3 + 2c_1 L^2 \right) + \frac{Ph}{A_c} \left(\frac{2}{5} L^5 c_2 + \frac{c_2}{2} L^4 + 180 L^3 \right) = 0 \tag{3.67}$$

On substituting the numerical values and solving for the two coefficients c_1 and c_2, we get

$$c_1 = -3923.36, \qquad c_2 = 40{,}498.44 \tag{3.68}$$

Thus our approximate solution for temperature, based on the minimisation of the functional, is given as follows:

$$T(x) = 300 - 3923.36 x + 40{,}498.44 x^2 \tag{3.69}$$

The Galerkin WR solution obtained in Example 2.8 is reproduced here from Eq. (2.69) for ready reference:

$$T(x) = 300 + 38{,}751.43(x^2 - 2Lx) \tag{3.70}$$

The two approximations are compared with the exact solution in Table 3.1. Even though both trial solutions are quadratic, the boundary conditions are different. The R–R trial solution was $c_1x + c_2x^2$ (with two *independent* parameters c_1 and c_2) while the original Galerkin WR method used $c_2(x^2 - 2Lx)$ after satisfying both the boundary conditions.

Table 3.1   Comparison of Galerkin WR and R–R Solutions (Example 3.3)

Axial location	Exact solution	Galerkin WR solution	R–R solution
0	300	300	300
0.005	280.75	281.59	281.40
0.01	264.02	265.12	264.82
0.015	249.62	250.59	250.26
0.02	237.43	238.00	237.73
0.025	227.31	227.34	227.23
0.03	219.16	218.62	218.75
0.035	212.91	211.84	212.29
0.04	208.49	207.00	207.86
0.045	205.85	204.09	205.46
0.05	204.97	203.12	205.08

3.4   Piece-wise Continuous Trial Functions—Finite Element Method

In the examples so far, we have used a single composite trial function solution for the entire domain. When we consider the essence of the R–R method, i.e., assuming a trial function solution and matching it as closely as possible to the exact solution, we realise that this is essentially a process of "curve fitting". It is well known that curve fitting is best done piece-wise—the more the number of pieces, the better the fit. We can thus benefit by using piece-wise defined trial functions rather than a set of functions valid over the entire domain of the problem. We will now discuss this technique leading to the formulation of the finite element method starting from a variational principle. We recall that the evaluation of the total potential Π_p involved integration over the entire domain of the problem. With the discretisation of the structure into sub-domains and use of piece-wise defined trial functions, we will evaluate these integrals over each sub-domain ("finite element") and sum up over all the elements. We will illustrate the procedure on the one-dimensional bar problem and demonstrate that we end up with exactly the same set of equations as in Chapter 2.

3.4.1   Bar Element Formulated from the Stationarity of a Functional

Consider a typical bar element as shown in Figure 3.7 with two nodes and the axial displacement u as the nodal d.o.f. We will use the same interpolation functions as used in Chapter 2, i.e., at any point within the element we have

$$u = \left(1 - \frac{x}{\ell}\right)u_1 + \left(\frac{x}{\ell}\right)u_2 \tag{3.71}$$

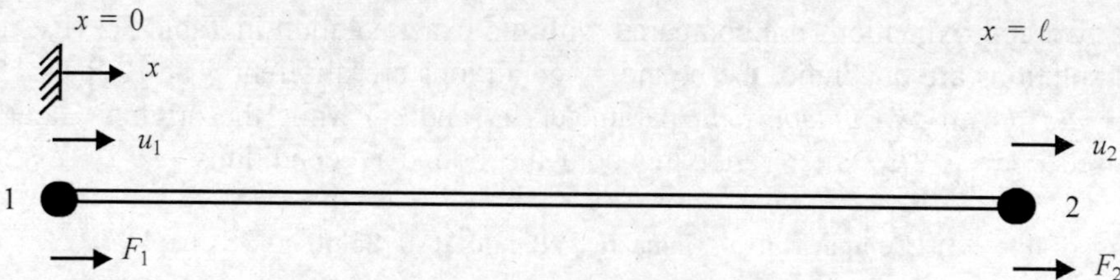

Fig. 3.7 Typical bar element.

Thus the strain energy stored within the element is given by

$$U^e = \int_0^\ell \frac{AE}{2}\left(\frac{du}{dx}\right)^2 dx = \frac{AE}{2}\frac{(u_2 - u_1)^2}{\ell} \tag{3.72}$$

If there is a distributed force q_0 acting at each point on the element and concentrated forces F at the nodes, the potential of the external forces is given by

$$V^e = -\int_0^\ell q_0 u\,dx - F_1 u_1 - F_2 u_2 = -q_0\frac{\ell}{2}(u_1 + u_2) - F_1 u_1 - F_2 u_2 \tag{3.73}$$

Thus the total potential for the kth element is given by

$$\Pi_p^e = U^e + V^e = \frac{AE}{2}\frac{(u_2 - u_1)^2}{\ell} - \frac{q_0\ell}{2}(u_1 + u_2) - F_1 u_1 - F_2 u_2 \tag{3.74}$$

For the entire structure, summing up the contributions of all elements, we have

$$\Pi_p = \sum_{k=1}^{n} \Pi_p^e \tag{3.75}$$

By virtue of the PSTP, the total potential will be stationary with respect to small variations in the displacement field. The variations in the displacement field in our case are attained by small variations in the nodal d.o.f., u_i. Thus,

$$\frac{\partial \Pi_p}{\partial u_i} = 0, \qquad i = 1, 2, ..., n \tag{3.76}$$

When applied to Eq. (3.75) above, this will lead to a set of simultaneous linear algebraic equations in the nodal d.o.f., in general, designated as $[K]\{\delta\} = \{f\}$. The contributions of the kth element to these system level equations can be obtained by applying Eq. (3.76) to the element total potential as given by Eq. (3.74). Thus we have the element level contributions as

$$\frac{\partial \Pi_p^e}{\partial u_1} = 0 \Rightarrow \frac{AE}{\ell}(u_1 - u_2) = \frac{q_0\ell}{2} + F_1$$

$$\tag{3.77}$$

$$\frac{\partial \Pi_p^e}{\partial u_2} = 0 \Rightarrow \frac{AE}{\ell}(u_2 - u_1) = \frac{q_0\ell}{2} + F_2$$

In matrix form, we write the element level equations as follows:

$$\frac{AE}{\ell}\begin{bmatrix} 1 & -1 \\ -1 & 1 \end{bmatrix}\begin{Bmatrix} u_1 \\ u_2 \end{Bmatrix} = \begin{Bmatrix} q_0\ell/2 \\ q_0\ell/2 \end{Bmatrix} + \begin{Bmatrix} F_1 \\ F_2 \end{Bmatrix} \tag{3.78}$$

We observe that these are identical to those equations obtained by the weak form of the Galerkin weighted residual statement, starting from the governing differential equation of the problem. (Please note that the sense of F_1 is opposite to that of P_0 in Eq. (2.134).) The process of assembling together individual element contributions is identical to the assembly procedure we used earlier.

3.4.2 One-dimensional Heat Transfer Element Based on the Stationarity of a Functional

Consider a typical one-dimensional heat conduction element as shown in Figure 3.8 with two nodes and the temperature T as the nodal d.o.f. The governing functional is given by

Fig. 3.8 Typical heat transfer element.

$$\Pi = \frac{1}{2}\int_0^\ell k\left(\frac{dT}{dx}\right)^2 dx - \int q_0 T\, dx - Q_1 T_1 - Q_2 T_2 \tag{3.79}$$

where q_0 is internal heat source and Q_1, Q_2 represent heat flux at the nodes. We will use the same interpolation functions as we used previously, i.e., at any point within the element we have

$$T(x) = \left(1 - \frac{x}{\ell}\right)T_1 + \left(\frac{x}{\ell}\right)T_2 \tag{3.80}$$

Thus the element level functional is given by

$$\Pi^e = \frac{1}{2}\left(\frac{k}{\ell}(T_2 - T_1)^2\right) - \frac{q_0\ell}{2}(T_1 + T_2) - Q_1 T_1 - Q_2 T_2 \tag{3.81}$$

For the entire system, summing up the contributions of all elements, we have

$$\Pi = \sum_{k=1}^{n} \Pi^e \tag{3.82}$$

The functional will be made stationary with respect to small variations in the temperature field.

The variations in the temperature field in our case are attained by small variations in the nodal d.o.f. T_i., Thus,

$$\frac{\partial \Pi}{\partial T_i} = 0, \qquad i = 1, 2, ..., n \tag{3.83}$$

When applied to Eq. (3.82) above, this will lead to a set of simultaneous linear algebraic equations in the nodal d.o.f., in general designated as $[K]\{\delta\} = \{f\}$. The contributions of the kth element to these system level equations can be obtained by applying Eq. (3.83) to the element level functional as given by Eq. (3.81). Thus we have the element level contributions as

$$\frac{k}{\ell}\begin{bmatrix} 1 & -1 \\ -1 & 1 \end{bmatrix}\begin{Bmatrix} T_1 \\ T_2 \end{Bmatrix} = \begin{Bmatrix} q_0\ell/2 \\ q_0\ell/2 \end{Bmatrix} + \begin{Bmatrix} Q_1 \\ Q_2 \end{Bmatrix} \tag{3.84}$$

We observe that these contributions are identical to those obtained by the weak form of the Galerkin weighted residual statement, starting from the governing differential equation of the problem (please note that the sense of Q_1 here is opposite to that of Q_0 in Eq. (2.162)).

3.5   Meaning of Finite Element Equations

Let us look at the element level equations for a bar element. We have, from Eq. (3.78), with $q_0 = 0$,

$$\frac{AE}{\ell}\begin{bmatrix} 1 & -1 \\ -1 & 1 \end{bmatrix}\begin{Bmatrix} u_1 \\ u_2 \end{Bmatrix} = \begin{Bmatrix} F_1 \\ F_2 \end{Bmatrix} \tag{3.85}$$

i.e.

$$[K]\{u\} = \{F\}$$

The individual elements of the element stiffness matrix have a direct physical meaning. Suppose we consider a deformation state for the element wherein we have $u_1 = 1$ and $u_2 = 0$. From elementary mechanics of materials we know that the force required to cause such a deformation will be AE/L to be applied at node 1 and, since node 2 is to be held at rest, an equal and opposite reaction force develops at node 2. From the above equation if we solve for the forces F_1, F_2 for this prescribed displacement, we get the same result. We observe that the elements of the first column of the stiffness matrix are identical to these forces. Similarly, if we consider a deformation state wherein $u_2 = 1$ and $u_1 = 0$, the forces required will be given in the second column of the stiffness matrix. **Thus elements of each column of a stiffness matrix actually represent the forces required to cause a certain deformation pattern— the ith column standing for a deformation pattern, where the ith d.o.f. is given unit displacement (translational or rotational) and all other d.o.f. are held zero.**

We can actually attempt to derive the finite element equations based on this direct physical correlation, called the **Direct method of formulation** of finite element equations. We will illustrate the usefulness of such an approach for a beam finite element (to be discussed in detail in Section 4.5). Our beam element (Figure 3.9) is a line element having two nodes just like the

Fig. 3.9   Typical beam element.

bar element, but the d.o.f. permitted at each node are the transverse displacement v, and the slope (rotational d.o.f.) dv/dx. We thus expect to derive the beam element equations as

$$\begin{bmatrix} k_{11} & k_{12} & k_{13} & k_{14} \\ k_{21} & k_{22} & k_{23} & k_{24} \\ k_{31} & k_{32} & k_{33} & k_{34} \\ k_{41} & k_{42} & k_{43} & k_{44} \end{bmatrix} \begin{Bmatrix} v_1 \\ \theta_1 \\ v_2 \\ \theta_2 \end{Bmatrix} = \begin{Bmatrix} F_1 \\ M_1 \\ F_2 \\ M_2 \end{Bmatrix} \tag{3.86}$$

Based on the physical interpretation of the columns of the stiffness matrix above, we observe that the columns of the stiffness matrix should correspond to the forces (moments) required to cause the deformation states shown in Figure 3.10. For example, the first deformation state requires that the beam act as a clamped-free beam ($x = L$ being the clamped end) with a force F and a moment M applied at the end $x = 0$ such that it undergoes unit transverse deformation but zero rotation. From elementary mechanics of materials involving the Euler–Bernoulli beam theory, we find the forces (moments) for causing these deformation states as shown in Figure 3.10. Thus the beam element stiffness matrix derived using the "direct" method is found to be

$$[k] = \begin{bmatrix} \dfrac{12EI}{\ell^3} & & & \text{Symmetric} \\[2ex] \dfrac{6EI}{\ell^2} & \dfrac{4EI}{\ell} & & \\[2ex] \dfrac{-12EI}{\ell^3} & \dfrac{-6EI}{\ell^2} & \dfrac{12EI}{\ell^3} & \\[2ex] \dfrac{6EI}{\ell^2} & \dfrac{2EI}{\ell} & \dfrac{-6EI}{\ell^2} & \dfrac{4EI}{\ell} \end{bmatrix} \tag{3.87}$$

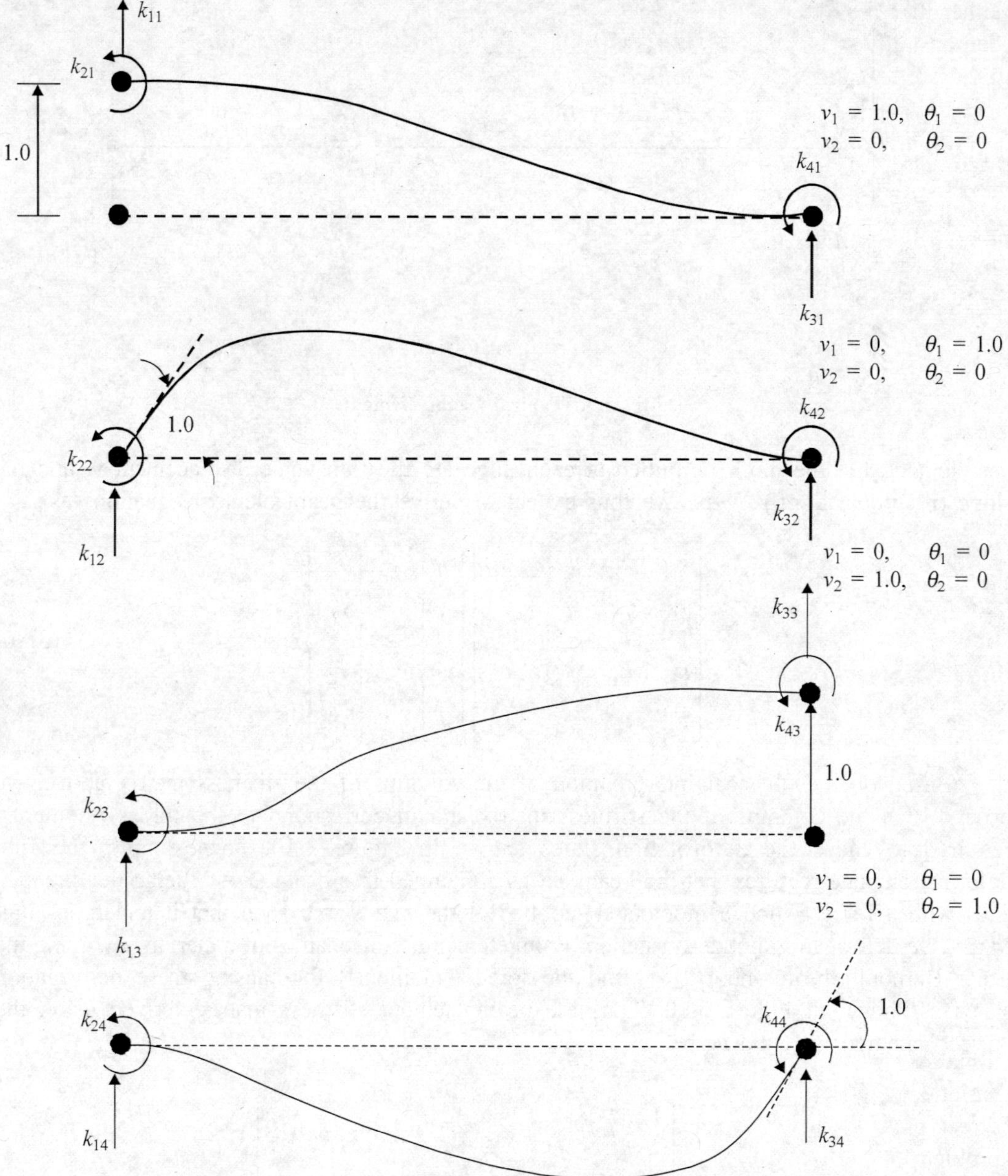

Fig. 3.10 Characteristic deformation shapes for a beam element.

We will derive the beam element equations from the PSTP and show that we indeed get the same equations (see Section 4.5). Even though the direct physical meaning of the individual columns of the stiffness matrix provides useful insight, its use in actually deriving the element matrices is limited to only simple cases. When we discuss two- and three-dimensional finite elements and higher order elements with internal nodes, etc., it is not possible to derive the stiffness matrices based on this interpretation alone. We need a systematic procedure based on either the Galerkin WR or the Variational method, to formulate these finite elements. We have demonstrated that both differential and functional forms yield identical finite element equations.

We shall now modify our general procedure for finite element analysis given in Chapter 2 and recast it as follows:

General Procedure for Finite Element Analysis

Step 1

➢ Using the governing differential equation, write down the weighted residual form. Perform integration by parts sufficient number of times and develop the weak form of the differential equation. Rewrite the weak form as a summation over n elements.

OR

➢ Write the functional representation as a summation over n elements.

Step 2

➢ Define the finite element, i.e. the geometry of the element, its nodes, and nodal d.o.f.
➢ Derive the shape or interpolation functions.

Step 3

➢ Compute the element level equations by substituting these shape functions in the weak form or the functional representation.

Step 4

➢ Discretise the given domain into a finite element mesh. For a given topology of finite element mesh, build up the system of equations by assembling together element level equations.
➢ Substitute the prescribed boundary conditions and solve for the unknowns.

SUMMARY

We have discussed the formulation of the finite element equations starting from either the weak form of the Galerkin weighted residual statement or the condition of stationarity for a functional and demonstrated that they yield identical results. For long, it has been common practice to use energy methods in structural mechanics and, thus, for most problems of structural mechanics, it may be convenient to use the PSTP and formulate the necessary finite element equations. The variational principle dealing with the strain energy and the work potential of external forces has direct physical meaning for the structural analyst. However, given the functional for any problem (such as the heat transfer example above), we will be able to develop the necessary finite element equations. The functional for some problems, such as some problems of heat transfer and fluid flow, may not have any direct physical meaning, and so it may be more convenient to state the problem in terms of the governing differential equations. It must be appreciated that the finite element equations can be derived systematically with equal ease, starting from either form of representation of governing equations—differential or functional. In the subsequent chapters of the text, we will use both these forms interchangeably for developing the necessary finite elements.

PROBLEMS

3.1 Redo Example 3.1 with a tip load P.

[*Hint:* Use $V = -Pu(L)$].

3.2 Redo Example 3.1 with a linearly varying load $q = \alpha x$. Do we still get an exact solution?

3.3 Redo Example 3.2, using $v(x) \approx c_1(x)(L - x)$.

3.4 Repeat Example 3.2, using $v(x) \approx c_0 + c_1 x + c_2 x^2 + c_3 x^3 + c_4 x^4$.

(*Note:* Observe that this field does not automatically satisfy the essential boundary conditions of the problem. Thus we need to set $v(0) = 0 = v(L)$, yielding two relations amongst c_i. We then get the remaining three equations from the PSTP.)

3.5 Solve for the displacement field of a cantilever beam under uniformly distributed load q_0. Use $v(x) \approx c_1[1 - \cos(\pi x/2L)]$.

Problems to Investigate

3.6 The functional, for a one-dimensional problem in structural mechanics, is given by

$$I = \frac{EI}{2} \int_0^L \left(\frac{d^2 y}{dx^2} \right)^2 dx - Py(L)$$

$$y(0) = \frac{dy}{dx}(0) = 0$$

By performing the variation $\delta I = 0$, derive the governing differential equation of the problem and the boundary conditions at $x = L$. What is the physical situation represented by this functional?

3.7 For the functional given below, obtain the corresponding differential (Euler–Lagrange) equation and boundary conditions:

$$I = \frac{1}{2} \int_0^L \left[\alpha \left(\frac{dy}{dx} \right)^2 - \beta y^2 + ryx^2 \right] dx - y(L)$$

subject to $y(0) = 0$. Assume α, β to be known constants.

CHAPTER 4

One-dimensional Finite Element Analysis

We have already discussed three different methods of deriving finite element equations, viz., direct method, weak form, and stationarity of a functional. The direct method was useful for very simple structural elements only, and we illustrated the equivalence of other two methods with the example of a one-dimensional bar element and heat transfer element. In this chapter we will study the finite element analysis of one-dimensional problems in greater detail. We will use the stationarity of a functional approach (i.e., the Principle of Stationary Total Potential) to develop the finite element equations for structural mechanics problems. We will first recast the earlier procedure in a generic form and then work out more example problems using the bar element. Subsequently, we will discuss the beam and frame elements. For heat transfer problems, we will use the weak form equation.

4.1   General Form of the Total Potential for 1-d

A one-dimensional structure (such as a bar or a beam) can, in general, be subjected to distributed loading in the form of body forces (such as gravity), forces distributed over a surface (such as a pressure) or concentrated forces lumped at certain points. The first two types of forces for one-dimensional problems can be reduced to force per unit length, q. It may also be subjected to initial strains and stresses (such as those due to pre-loading, and thermal loading). Under the action of all these loads, the structure, when properly supported (so as to prevent rigid body motion), undergoes deformation and stores internal strain energy. We will now develop an expression for the total potential of such a structure assuming linear elastic isotropic material obeying Hooke's law. We also assume that all loads and other parameters are time-invariant.

$$\text{Stress } \sigma_x = E(\varepsilon_x - \varepsilon_{x0}) + \sigma_{x0} \tag{4.1}$$

where ε_{x0} is the initial strain and σ_{x0} is the initial stress. This stress (σ_x) does work in causing an increment in strain ($d\varepsilon_x$), which is further stored in the elastic body as an increment in strain energy. Consider a cube of unit edge length in the body, i.e. unit volume of material. The increment in strain energy in this unit volume, dU is given by the incremental work which equals the product of force (F) and incremental displacement ($d\alpha$). Here, "α" stands for the generic displacement e.g. axial displacement u for a bar. Thus

$$dU = [F][d\alpha] = [(\sigma_x)(1)(1)][(d\varepsilon_x)(1)] = \sigma_x \, d\varepsilon_x \tag{4.2}$$

Substituting for stress from Eq. (4.1) and integrating with respect to strain yields the strain energy in this unit volume of material. This, when integrated over the whole structure, yields the total strain energy. Thus,

$$U = \int \left[\left(\frac{1}{2}\right)E\varepsilon_x^2\right] A \, dx - \int \varepsilon_x E\varepsilon_{x0} A \, dx + \int \varepsilon_x \sigma_{x0} \, A \, dx \tag{4.3}$$

The potential V of an external force in any generalised coordinate α is defined, by convention, in such a way that the generalised force in that coordinate is given as $F_\alpha = -(\partial V/\partial\alpha)$. The forces on a one-dimensional structure may be distributed (q per unit length) or concentrated (P). Therefore, the potential of external forces is given by

$$V = -\left(\int \alpha q \, dx + \sum \alpha_i P_i\right) \tag{4.4}$$

where P_i is the concentrated force acting at a point i , α_i is its displacement and the summation is taken over all the points where concentrated forces are present. The total potential of the structure is obtained as

$$\Pi_p = U + V \tag{4.5}$$

We will use this generic form in our further discussion while attempting to derive the finite element equations in their generic form.

4.2   Generic Form of Finite Element Equations

Let the displacement α at any point P within the element be interpolated from the nodal values of displacement $\{\delta\}^e$ using the shape functions N as follows:

$$\alpha = [N]\{\delta\}^e \tag{4.6}$$

The interpretation of Eq. (4.6) is as follows: For a one-dimensional problem, we have one possible displacement at each point (axial displacement u or transverse displacement v), and $\{\delta\}^e$ is a column vector of nodal d.o.f. of the element and is of size equal to the product of number of nodes per element (NNOEL) and the d.o.f. per node (NFREE).

For a bar element (Figure 4.1),

$$\text{NNOEL} = 2$$
$$\text{NFREE} = 1$$

$$\{\delta\}^e = \begin{Bmatrix} u_1 \\ u_2 \end{Bmatrix} \tag{4.7}$$

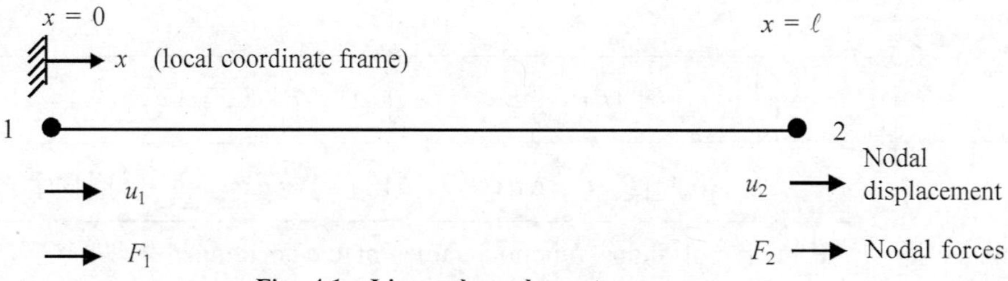

Fig. 4.1   Linear bar element.

For a beam element

$$\text{NNOEL} = 2$$
$$\text{NFREE} = 2$$

$$\{\delta\}^e = \begin{Bmatrix} v_1 \\ dv_1/dx \\ v_2 \\ dv_2/dx \end{Bmatrix} \tag{4.8}$$

The shape functions $[N]$ represent the interpolation formulae used to obtain the displacement field within the element, and the size of the matrix $[N]$ will be consistent with the size of $\{\delta\}^e$. For example, for the bar element, $[N]$ is given by

$$[N] = \left[\left(1 - \frac{x}{\ell}\right) \quad \left(\frac{x}{\ell}\right) \right] \tag{4.9}$$

The nonzero strain at a point P within the element (i.e., ε_x) is given in terms of the derivatives of the displacement field and is again represented in terms of nodal d.o.f. as

$$\varepsilon_x = [\partial](\alpha) = [\partial][N]\{\delta\}^e = [B]\{\delta\}^e \tag{4.10}$$

where $[\partial]$ is a differential operator and $[B] = [\partial][N]$, representing the derivatives of shape functions. For a bar element, for example, $[\partial]$ is simply d/dx and $[B] = \begin{bmatrix} \dfrac{-1}{\ell} & \dfrac{1}{\ell} \end{bmatrix}$.

We can now rewrite the generic expression for the total potential as follows from Eq. (4.3). For an element,

$$U^e = \int \frac{1}{2}\{\delta\}^{e^T}[B]^T[B]\{\delta\}^e EA\,dx - \int \{\delta\}^{e^T}[B]^T E\varepsilon_{x0}A\,dx + \int \{\delta\}^{e^T}[B]^T \sigma_{x0}A\,dx \tag{4.11}$$

Again from Eq. (4.4),

$$V^e = -\int \{\delta\}^{e^T}[N]^T q\,dx - \sum \{\delta\}^{e^T}[N]_i^T P_i \tag{4.12}$$

Thus,

$$\Pi_p^e = \int \frac{1}{2}\{\delta\}^{e^T} [B]^T [B]\{\delta\}^e EA\, dx - \int \{\delta\}^{e^T} [B]^T E\varepsilon_{x0}A\, dx$$
$$+ \int \{\delta\}^{e^T} [B]^T \sigma_{x0}A\, dx - \int \{\delta\}^{e^T} [N]^T q\, dx - \sum \{\delta\}^{e^T} [N]_i^T P_i \tag{4.13}$$

where $[N]_i$ is the matrix of shape function values at the location i.

Since $\{\delta\}^e$ is a vector of nodal d.o.f., and is not a function of the spatial coordinate x, it can be taken outside the integration. Thus we get

$$\Pi_p^e = \frac{1}{2}\{\delta\}^{e^T} \int \left([B]^T [B]EA\, dx\right)\{\delta\}^e - \{\delta\}^{e^T} \int [B]^T E\varepsilon_{x0}A\, dx + \{\delta\}^{e^T} \int [B]^T \sigma_{x0}A\, dx$$
$$- \{\delta\}^{e^T} \int [N]^T q\, dx - \{\delta\}^{e^T} \sum [N]_i^T P_i \tag{4.14}$$

Defining the element stiffness matrix $[k]^e$ and the load vector $\{f\}^e$ as

$$[k]^e = \int [B]^T [B]EA\, dx \tag{4.15}$$

$$\{f\}^e = \int [B]^T E\varepsilon_{x0}A\, dx - \int [B]^T \sigma_{x0}A\, dx + \int [N]^T q\, dx + \sum [N]_i^T P_i \tag{4.16}$$

we can rewrite Eq. (4.14) in the form

$$\Pi_p^e = \frac{1}{2}\{\delta\}^{e^T} [k]^e\{\delta\}^e - \{\delta\}^{e^T} \{f\}^e \tag{4.17}$$

Potential π_p^e being a scalar quantity, the total potential for a mesh of finite elements can be obtained by simple addition. The total potential of the system is given as

$$\Pi_p = \sum_1^{NOELEM} \Pi_p^e = \frac{1}{2}\{\delta\}^T [K]\{\delta\} - \{\delta\}^T \{F\} \tag{4.18}$$

wherein the global stiffness matrix of the structure $[K]$ and the global load vector $\{F\}$ are obtained as

$$[K] = \sum_{n=1}^{NOELEM} [k]^e \tag{4.19}$$

$$\{F\} = \sum_1^{NOELEM} \{f\}^e \tag{4.20}$$

"NOELEM" represents the number of elements in the mesh and $\{\delta\}$ contains all the nodal d.o.f. variables for the entire finite element mesh. The summations indicating assembly imply that the individual element matrices have been appropriately placed in the global matrices following the standard procedure of assembly.

Using the PSTP, we set the total potential stationary with respect to small variations in the nodal d.o.f., i.e.*,

$$\frac{\partial \Pi_p}{\partial \{\delta\}^T} = 0 \tag{4.21}$$

Thus the system level equations are given by

$$[K]\{\delta\} = \{F\} \tag{4.22}$$

We will now illustrate the details of the procedure through the familiar bar element and then derive the element matrices for a beam element. We will follow this generic notation in our derivation of two-dimensional elements also. Thus the reader is advised to become fully conversant with this notation, which will be extensively used in the rest of the book.

4.3   The Linear Bar Finite Element

We recall that our linear bar element (Figure 4.1) is a one-dimensional line element with two nodes. Each point P within the element is permitted to move only along the axis of the element and the displacement of the entire cross-section is assumed to be same. Therefore, we have

$$\text{Displacement field } \{\alpha\} = u \tag{4.23}$$

$$\text{Nodal d.o.f. vector } \{\delta\}^e = \begin{Bmatrix} u_1 \\ u_2 \end{Bmatrix} \tag{4.24}$$

$$\text{Shape functions } [N] = [N_1 \ N_2] = \left[\left(1 - \frac{x}{\ell}\right) \quad \left(\frac{x}{\ell}\right) \right] \tag{4.25}$$

$$\text{Strain } \{\varepsilon\} = \varepsilon_x = \frac{du}{dx} \tag{4.26}$$

Thus the differential operator $[\partial]$ is simply d/dx. Hence,

$$[B] = \left[\frac{dN_1}{dx} \ \frac{dN_2}{dx} \right] = \left[\frac{-1}{\ell} \ \frac{1}{\ell} \right] \tag{4.27}$$

Thus the element stiffness matrix is obtained as

$$[k]^e = \int_0^\ell [B]^T [B] EA \, dx \tag{4.28}$$

$$= \int_0^l \begin{bmatrix} -1/\ell \\ 1/\ell \end{bmatrix} [-1/\ell \quad 1/\ell] EA \, dx \tag{4.29}$$

$$= \frac{AE}{\ell} \begin{bmatrix} 1 & -1 \\ -1 & 1 \end{bmatrix} \tag{4.30}$$

---

*Though this equation is written in compact matrix notation, it simply states that the total potential is stationary with respect to small variations in each and every nodal d.o.f.

which is the same as the result obtained earlier (see Eq. 3.78). In the examples that follow, we will illustrate the use of the bar element in several situations.

Example 4.1.   *A bar subjected to self-weight.* Consider a vertically hanging rod (Figure 4.2) of length L, uniform cross-sectional area A, density ρ, and Young's modulus E. We wish to find the state of deformation and stress in the rod subjected to gravity load, using the bar element just described. Through this example, we wish to bring out several insights into the nature of finite element approximate solutions.

For a bar element, the stiffness matrix is given by Eq. (4.30). Thus,

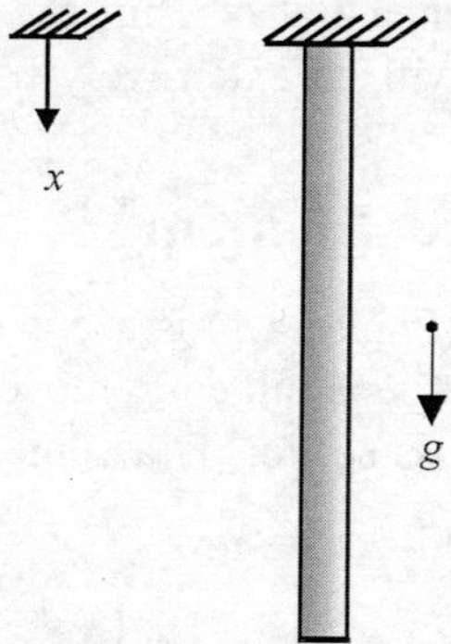

Fig. 4.2   A vertically hanging rod (Example 4.1).

$$[k]^e = \frac{AE}{l} \begin{bmatrix} 1 & -1 \\ -1 & 1 \end{bmatrix} \tag{4.31}$$

For the present example, the body force (i.e., self-weight) expressed as "q" per unit length is given by $(\rho A g)$. In the absence of any other force, the element nodal force vector is given by Eq. (4.16):

$$\{f\}^e = \int [N]^T q \, dx = \int_0^l \begin{bmatrix} 1 - x/l \\ x/l \end{bmatrix} (\rho g)(A) \, dx \tag{4.32}$$

$$= \begin{Bmatrix} \rho A l g/2 \\ \rho A l g/2 \end{Bmatrix} \tag{4.33}$$

Since $(\rho A l g)$ represents the weight of the element, the equivalent nodal force vector just obtained is thus simply half the weight of the element distributed at either node. We will now use one, two, four, eight and 16 finite elements and, in each case, compare our solution with the exact solution. The exact solution is readily verified to be given by

$$\text{Displacement } u(x) = \frac{\rho g}{E}\left(Lx - \frac{x^2}{2}\right) \tag{4.34}$$

$$\text{Stress } \sigma(x) = \rho g(L - x) \tag{4.35}$$

We observe that the exact solution corresponds to linear variation of stress and quadratic variation of displacement, whereas within each bar element we permit only linear displacement field (Eq. 4.25) and, consequently, constant stress. Thus we expect our solution to be poor when the number of elements is very small and should become more and more accurate as we increase the number of elements.

One-element solution. We use a single bar element to model the entire length of the rod as shown in Figure 4.3(a). We have

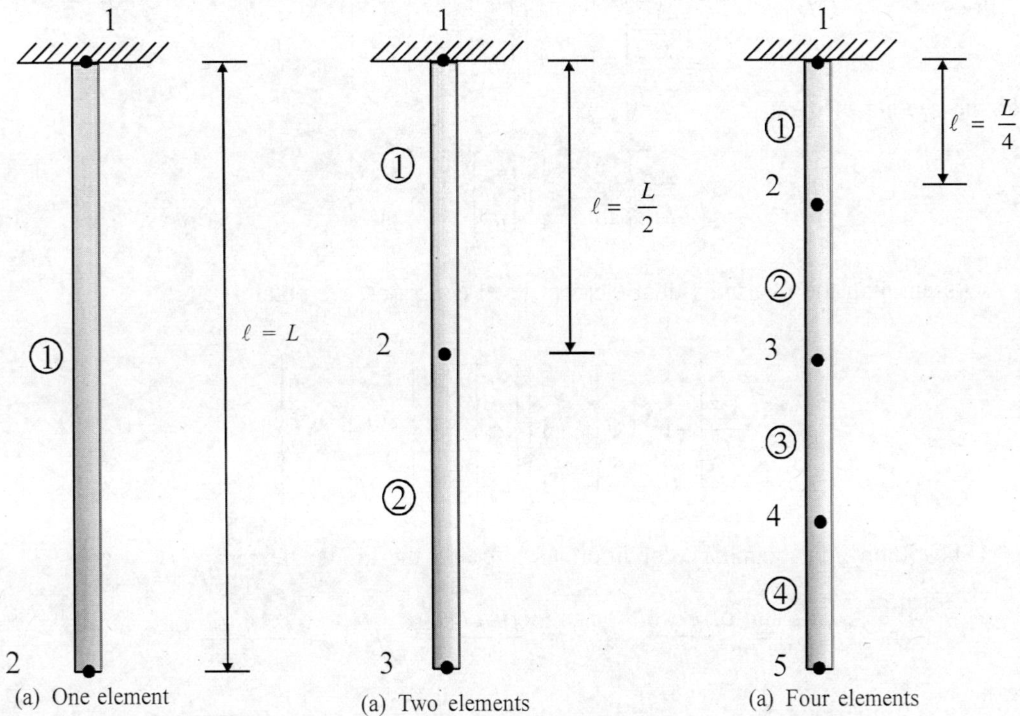

(a) One element   (a) Two elements   (a) Four elements

Fig. 4.3   Finite element mesh for bar under self-weight (Example 4.1).

$$\frac{AE}{l}\begin{bmatrix} 1 & -1 \\ -1 & 1 \end{bmatrix}\begin{Bmatrix} u_1 \\ u_2 \end{Bmatrix} = \begin{Bmatrix} F_1 \\ F_2 \end{Bmatrix} \tag{4.36}$$

Using $\ell = L$ and substituting for the boundary condition $u_1 = 0$ and for the known force $F_2 = \rho AgL/2$, we get

$$u_2 = \left(\frac{\rho AgL}{2}\right)\frac{L}{AE} = \frac{\rho g L^2}{2E} \tag{4.37}$$

Thus,

$$u(x) = \left(\frac{\rho g L}{2E}\right)x \tag{4.38}$$

$$\text{Strain } \varepsilon_x = \frac{\rho g L}{2E} \tag{4.39}$$

$$\text{Stress } \sigma_x = E\varepsilon_x = \frac{\rho g L}{2} \tag{4.40}$$

Two-element solution. Let us discretise the bar into two equal length elements ($\ell = L/2$) as shown in Figure 4.3(b). For the first element, we have

$$\frac{AE}{L/2}\begin{bmatrix} 1 & -1 \\ -1 & 1 \end{bmatrix}\begin{Bmatrix} u_1 \\ u_2 \end{Bmatrix} = \begin{Bmatrix} f_1^{(1)} \\ f_2^{(1)} \end{Bmatrix} \tag{4.41}$$

For the second element,

$$\frac{AE}{L/2}\begin{bmatrix} 1 & -1 \\ -1 & 1 \end{bmatrix}\begin{Bmatrix} u_2 \\ u_3 \end{Bmatrix} = \begin{Bmatrix} f_1^{(2)} \\ f_2^{(2)} \end{Bmatrix} \tag{4.42}$$

Assembling the two individual element level equations, we obtain

$$\frac{AE}{L/2}\begin{bmatrix} 1 & -1 & 0 \\ -1 & 1+1 & -1 \\ 0 & -1 & 1 \end{bmatrix}\begin{Bmatrix} u_1 \\ u_2 \\ u_3 \end{Bmatrix} = \begin{Bmatrix} f_1^{(1)} \\ f_2^{(1)} + f_1^{(2)} \\ f_2^{(2)} \end{Bmatrix} \tag{4.43}$$

Substituting the boundary condition, $u_1 = 0$, and the known forces

$$f_2^{(1)} = \frac{\rho A g L/2}{2}, \quad f_1^{(2)} = \frac{\rho A g L/2}{2}, \quad f_2^{(2)} = \frac{\rho A g L/2}{2}$$

we get

$$\frac{AE}{L/2}\begin{bmatrix} 2 & -1 \\ -1 & 1 \end{bmatrix}\begin{Bmatrix} u_2 \\ u_3 \end{Bmatrix} = \begin{Bmatrix} \rho A g L/2 \\ \rho A g L/4 \end{Bmatrix} \tag{4.44}$$

Solving, we obtain

$$\begin{Bmatrix} u_2 \\ u_3 \end{Bmatrix} = \frac{\rho g L^2}{4E}\begin{Bmatrix} 3/2 \\ 2 \end{Bmatrix} \tag{4.45}$$

The element level strains and stresses are computed as

$$\varepsilon_x^{(1)} = [B]\left\{\begin{matrix} u_1 \\ u_2 \end{matrix}\right\}^{(1)} = \left[\dfrac{-1}{L/2} \quad \dfrac{1}{L/2}\right]\left\{\begin{matrix} 0 \\ \dfrac{3\rho g L^2}{8E} \end{matrix}\right\} = \dfrac{3}{4}\dfrac{\rho g L}{E} \tag{4.46}$$

$$\varepsilon_x^{(2)} = [B]\left\{\begin{matrix} u_1 \\ u_2 \end{matrix}\right\}^{(2)} = \left[\dfrac{-1}{L/2} \quad \dfrac{1}{L/2}\right]\left\{\begin{matrix} 3\rho g L^2/8E \\ \rho g L^2/2E \end{matrix}\right\} = \dfrac{1}{4}\dfrac{\rho g L}{E} \tag{4.47}$$

$$\sigma_x^{(1)} = E\varepsilon_x^{(1)} = \dfrac{3\rho g L}{4} \tag{4.48}$$

$$\sigma_x^{(2)} = E\varepsilon_x^{(2)} = \dfrac{\rho g L}{4} \tag{4.49}$$

Four-element solution. We further subdivide each of the two elements used earlier by half Figure 4.3(c), i.e., the total rod is discretised into four elements, each of length ($L/4$). For each element, we have

$$\dfrac{AE}{L/4}\begin{bmatrix} 1 & -1 \\ -1 & 1 \end{bmatrix}\left\{\begin{matrix} u_1 \\ u_2 \end{matrix}\right\} = \left\{\begin{matrix} F_1 \\ F_2 \end{matrix}\right\} \tag{4.50}$$

Assembling all the individual element matrices and substituting for the boundary condition and the known forces, we obtain

$$\dfrac{AE}{L/4}\begin{bmatrix} 1+1 & -1 & 0 & 0 \\ -1 & 1+1 & -1 & 0 \\ 0 & -1 & 1+1 & -1 \\ 0 & 0 & -1 & 1 \end{bmatrix}\left\{\begin{matrix} u_2 \\ u_3 \\ u_4 \\ u_5 \end{matrix}\right\} = \left\{\begin{matrix} 1/2+1/2 \\ 1/2+1/2 \\ 1/2+1/2 \\ 1/2 \end{matrix}\right\}\dfrac{\rho A L g}{4} \tag{4.51}$$

Solving, we get

$$u_2 = \left(\dfrac{7}{16}\right)\dfrac{\rho g L^2}{2E}, \qquad u_3 = \left(\dfrac{12}{16}\right)\dfrac{\rho g L^2}{2E},$$

$$u_4 = \left(\dfrac{15}{16}\right)\dfrac{\rho g L^2}{2E}, \qquad u_5 = \dfrac{\rho g L^2}{2E} \tag{4.52}$$

The corresponding stresses in individual elements are obtained as

$$\sigma^{(1)} = \left(\frac{7}{32}\right)\rho g L, \qquad \sigma^{(2)} = \left(\frac{5}{32}\right)\rho g L,$$

$$\sigma^{(3)} = \left(\frac{3}{32}\right)\rho g L, \qquad \sigma^{(4)} = \left(\frac{1}{32}\right)\rho g L \qquad (4.53)$$

On similar lines, we can obtain the solution with eight and 16 bar elements, each time subdividing the existing elements into two. For a one metre long steel rod ($E = 2 \times 10^{11}$ N/m^2, $\rho = 7{,}800$ kg/m^3) of cross-sectional area 5 cm^2, Table 4.1 (Figures 4.4 and 4.5) compares the displacements (stresses) obtained using the various finite element models against the exact solution.

Table 4.1   Rod under Self-weight—comparison of deflections (deflections are in micrometers)

x	1 element	2 elements	4 elements	8 elements	16 elements	Exact
0	0	0	0	0	0	0
0.0625					0.0231	0.0231
0.125				0.0448	0.0448	0.0448
0.1875					0.0649	0.0649
0.25			0.0836	0.0836	0.0836	0.0836
0.3125					0.101	0.101
0.375				0.116	0.116	0.116
0.4375					0.131	0.131
0.5		0.143	0.143	0.143	0.143	0.143
0.5625					0.155	0.155
0.625				0.164	0.164	0.164
0.6875					0.172	0.172
0.75			0.179	0.179	0.179	0.179
0.8125					0.184	0.184
0.875				0.188	0.188	0.188
0.9375					0.190	0.190
1	0.191	0.191	0.191	0.191	0.191	0.191

We observe some very significant points—for example, no matter how few the elements we take are, our solution for the displacement at the nodes is exact! Even with just one element (i.e., a single straight line variation of displacement within the entire rod), we get the exact value for tip displacement, whereas the exact solution is clearly a quadratic variation of displacement!* However, we must note here that the nodal solution is not exact for most finite element solutions.

A point of immediate interest is "With nodal displacements being exact, how could the stresses be inaccurate?". The calculation of stress depends not just on the nodal values but on the variation of displacement within the element. Since the element used permits only linear

---

*The interested reader is referred to P. Tong, Exact solution of certain problems by the finite element method, *AIAA Journal*, **7**, 1, 178–180, 1969.

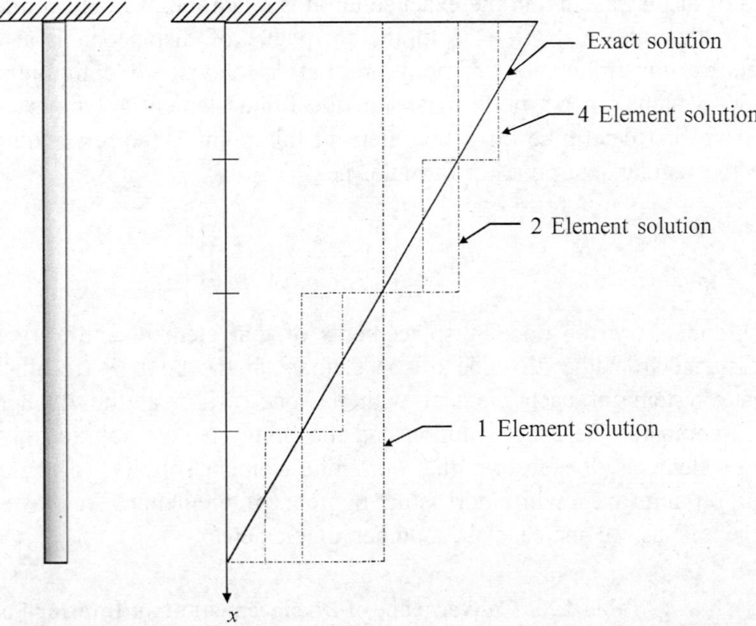

Fig. 4.4   Rod under self-weight—Comparison of stresses (1, 2 and 4 element solutions).

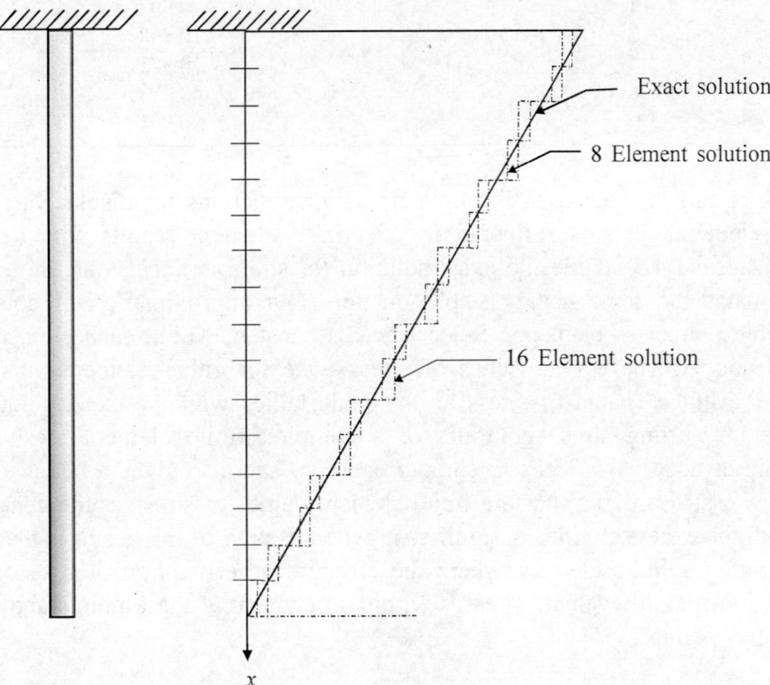

Fig. 4.5   Rod under self-weight—Comparison of stresses (with 8 and 16 elements).

variation of displacement and the exact solution is quadratic, we expect differences in the stress result. To illustrate this, let us compute the value of displacement at an interior point as predicted by our finite element solution. Let us choose to compute the displacement at $x = 0.33L$, which is not a node in any of the finite element solutions we obtained. For this purpose, we first determine on which element this point P (corresponding to $x = 0.33L$) lies, and use the regular interpolation formula, i.e.,

$$u_p = \left[\left(1 - \frac{x_p}{\ell} \right)\left(\frac{x_p}{\ell} \right) \right] \begin{Bmatrix} u_i \\ u_j \end{Bmatrix} \tag{4.54}$$

where u_i and u_j are the nodal displacements of that element and x_p in this equation refers to the distance from the ith node of the element to the point P (recall that we used a local coordinate system for each element with its local origin at the ith node, see Figure 4.1). Table 4.2 compares the exact solution and the finite element solution obtained with different number of elements. We observe that our finite element solution is in great error initially but our solution improves with increasing number of elements. We have achieved excellent "convergence" as we increase the number of elements.

Table 4.2   Convergence of Displacement at an Interior Point

No. of elements	Normalised displacement at $x = 0.33$ (FE displacement/Exact solution)
1	0.598
2	0.898
4	0.975
8	0.993
16	0.998

Since this element uses linear interpolation functions for displacement, it is a "constant stress" element. This is reflected in our finite element results for stresses as shown in Figures 4.4 and 4.5. While the exact solution for stress is linear, our finite element prediction approximates the same using "step" function. Our approximation of course improves with increasing number of elements. Since stress is constant within each element, we observe that at each node we have two values of stress—one for either element. A significant result is that the "average" value of stress at any node tallies with the exact solution! It is common practice in plotting stress contours of a complex finite element mesh to perform stress averaging at nodes; this leads to smooth contours and also gives a better estimate of stress at a node. We also observe that the finite element values of stress at the centre of each element tally with the exact solution. A significant generalisation of this result is that there exist certain points within a finite element where the error in the estimation of stresses is the least. It is common practice to evaluate stresses within an element at these points and then extrapolate to the desired points.

A summary of our observations based on the above discussion is now given.

➢ Depending on the variation of the field variable (displacement in this case) and the shape functions used in the finite element, we may require many elements to model the field accurately, especially at the interior points.

➢ The field variable itself may be reasonably well captured, but the error in the derivatives (e.g. stress here) could be considerable.

➢ Stress averaging will be necessary to get a more realistic value of stress at a node, jointly shared by many elements.

The bar element is capable of modelling only a linear variation of displacement, and hence we require many such elements to model a physical system accurately. We can improve this element by permitting quadratic variation of displacement within the element. We will now discuss a typical higher order element, viz., quadratic bar element. We cannot achieve this (i.e., a *complete* quadratic polynomial) with just two nodes and so we have to add another node to the element.

It must be observed that we could have used two nodes only but permitted an *incomplete* quadratic polynomial shape function, e.g. $u(x) = c_0 + c_2 x^2$. For such an element, $\varepsilon_x = du/dx = 2c_2 x$. Thus, strain and stress will always be zero at one end of each element (at $x = 0$)!

4.4 The Quadratic Bar Element

4.4.1 Determination of Shape Functions

The three noded quadratic bar element is shown in Figure 4.6. We number the two end nodes as 1 and 2 and the middle node is given number 3. The displacement at any point within the

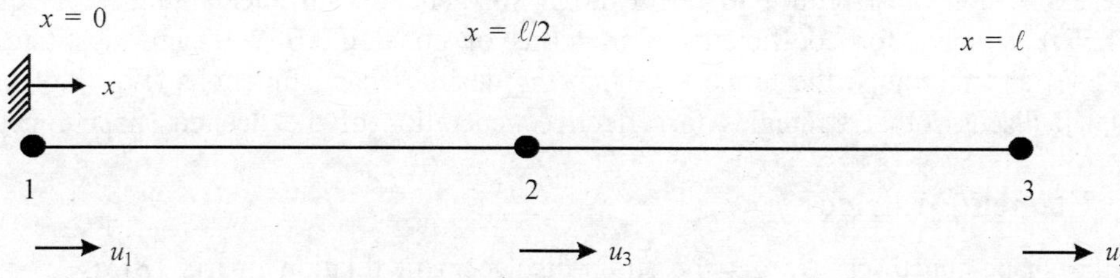

Fig. 4.6 Quadratic bar element.

element is now interpolated from the three nodal displacements using the shape functions as follows:

$$u(x) = N_1(x)u_1 + N_2(x)u_2 + N_3(x)u_3 \qquad (4.55)$$

The shape functions N_i vary quadratically within the element and we now discuss a systematic procedure to derive such shape functions. Let $u(x)$ be given by the complete quadratic polynomial

$$u(x) = c_0 + c_1 x + c_2 x^2 \qquad (4.56)$$

We know that $u(0) = u_1$, $u(\ell) = u_2$ and $u(\ell/2) = u_3$. Hence, from Eq. (4.56),

$$u_1 = c_0, \qquad u_2 = c_0 + c_1\ell + c_2\ell^2$$

$$u_3 = c_0 + c_1\left(\frac{\ell}{2}\right) + c_2\left(\frac{\ell}{2}\right)^2 \tag{4.57}$$

Solving for c_i, we obtain

$$c_0 = u_1$$
$$c_1 = (4u_3 - u_2 - 3u_1)/\ell \tag{4.58}$$
$$c_2 = (2u_1 + 2u_2 - 4u_3)/\ell^2$$

Thus we have

$$u(x) = u_1 + \left(\frac{4u_3 - u_2 - 3u_1}{\ell}\right)x + \left(\frac{2u_1 + 2u_2 - 4u_3}{\ell^2}\right)x^2 \tag{4.59}$$

Rearranging the terms, we get

$$u(x) = u_1\left(1 - \frac{3x}{\ell} + \frac{2x^2}{\ell^2}\right) + u_2\left(-\frac{x}{\ell} + \frac{2x^2}{\ell^2}\right) + u_3\left(\frac{4x}{\ell} - \frac{4x^2}{\ell^2}\right) \tag{4.60}$$

Comparing Eqs. (4.55) and (4.60), we obtain

$$\boxed{N_1 = 1 - \frac{3x}{\ell} + \frac{2x^2}{\ell^2}, \qquad N_2 = \frac{-x}{\ell} + \frac{2x^2}{\ell^2}, \qquad N_3 = \frac{4x}{\ell} - \frac{4x^2}{\ell^2}} \tag{4.61}$$

This is a general technique for determining the necessary shape functions N_i. We first assume a general polynomial field (Eq. (4.56)). Substituting the nodal coordinates and requiring that the expression must reduce to nodal d.o.f., we generate sufficient number of equations (Eq. (4.57)) to solve for coefficients c_i in terms of nodal d.o.f. We then substitute for c_i (Eq. (4.59)) and rearrange the terms to obtain the desired shape functions (Eq. (4.60)). Later on we will discuss other strategies for effective generation of the desired shape function.

4.4.2 Element Matrices

Given the shape functions, we get the strain-displacement relation matrix $[B]$ as

$$\varepsilon_x = \frac{du}{dx} = \frac{d}{dx}[N]\{\delta\}^e = [B]\{\delta\}^e \tag{4.62}$$

where

$$[B] = \left[\frac{dN_1}{dx} \quad \frac{dN_2}{dx} \quad \frac{dN_3}{dx}\right] = \left[\frac{2}{l^2}\left(2x - \frac{3\ell}{2}\right) \quad \frac{2}{\ell^2}\left(2x - \frac{\ell}{2}\right) \quad \frac{-4}{\ell^2}(2x - \ell)\right] \tag{4.63}$$

Hence we have the element stiffness matrix given by from Eq. (4.15) and Eq. (4.63):

$$[k]^e = \int [B]^T [B] EA \, dx \tag{4.64}$$

$$= \int_0^l \begin{bmatrix} \dfrac{2}{\ell^2}\left(2x - \dfrac{3\ell}{2}\right) \\[2ex] \dfrac{2}{\ell^2}\left(2x - \dfrac{\ell}{2}\right) \\[2ex] \dfrac{-4}{\ell^2}(2x - \ell) \end{bmatrix} \begin{bmatrix} \dfrac{2}{\ell^2}\left(2x - \dfrac{3\ell}{2}\right) & \dfrac{2}{\ell^2}\left(2x - \dfrac{\ell}{2}\right) & \dfrac{-4}{\ell^2}(2x - \ell) \end{bmatrix} EA \, dx \tag{4.65}$$

On evaluating all the integrals, we obtain the element stiffness matrix as

$$[k]^e = \frac{AE}{3\ell} \begin{bmatrix} 7 & 1 & -8 \\ 1 & 7 & -8 \\ -8 & -8 & 16 \end{bmatrix} \tag{4.66}$$

We now redo Example 4.1 with this quadratic element and compare the results obtained using typical lower order (linear bar) and higher order (quadratic bar) finite elements.

Example 4.2. *A vertically hanging rod (Example 4.1 Reworked).* For the present example, the body force due to gravity expressed as force q per unit length is given by $(\rho A g)$. Thus, the element nodal force vector (in the absence of any other force) is given by

$$\{f\}^e = \int_0^\ell [N]^T q \, dx = \int_0^\ell \begin{bmatrix} 1 - \dfrac{3x}{\ell} + \dfrac{2x^2}{\ell^2} \\[2ex] -\dfrac{x}{\ell} + \dfrac{2x^2}{\ell^2} \\[2ex] \dfrac{4x}{\ell} - \dfrac{4x^2}{\ell^2} \end{bmatrix} \rho A g \, dx \tag{4.67}$$

$$= \rho A g l \begin{Bmatrix} 1/6 \\ 1/6 \\ 4/6 \end{Bmatrix} \tag{4.68}$$

Using just one element spanning the entire length of the rod, as shown in Figure 4.7, we obtain the equations

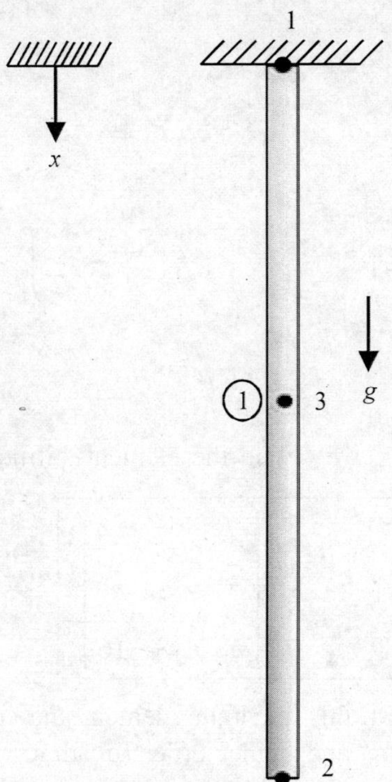

Fig. 4.7   **Rod under self-weight—Quadratic element solution (Example 4.2).**

$$\frac{AE}{3\ell}\begin{bmatrix} 7 & -8 \\ -8 & 16 \end{bmatrix}\begin{Bmatrix} u_2 \\ u_3 \end{Bmatrix} = \begin{Bmatrix} 1/6 \\ 4/6 \end{Bmatrix}\rho Ag\ell \tag{4.69}$$

where the boundary condition $u_1 = 0$, and the known forces have already been substituted and $\ell = L$.

Solving, we get

$$u_2 = \frac{\rho g L^2}{2E}, \qquad u_3 = \frac{3\rho g L^2}{8E} \tag{4.70}$$

Substituting these equations in Eq. (4.60), we obtain

$$u(x) = \left(\frac{-x}{\ell} + \frac{2x^2}{\ell^2}\right)\frac{\rho g \ell^2}{2E} + \left(\frac{4x}{\ell} - \frac{4x^2}{\ell^2}\right)\left(\frac{3\rho g \ell^2}{8E}\right)$$

$$= \frac{\rho g \ell^2}{2E}\left(\frac{2x}{\ell} - \frac{x^2}{\ell^2}\right) = \frac{\rho g}{E}\left(Lx - \frac{x^2}{2}\right) \tag{4.71}$$

Comparing with Eq. (4.34), we observe that our one-element solution ($\ell = L$) tallies with the exact solution completely (i.e. not just at the nodes but interior points also). It is left as an exercise for the reader to show that the stresses also match the exact solution.

We thus observe that we could get the exact solution with just one quadratic bar element where as we needed several linear bar elements to get a reasonably accurate solution.

Conclusion

A few higher order elements are far superior to several lower order elements.

Example 4.3.   *Effect of node numbering on assembled stiffness matrix.* We consider the rod shown in Figure 4.8 to highlight an important aspect of typical finite element analysis. We use six bar elements each of length ℓ but in the two cases considered, the node numbering is different. For simplicity, let us assume $AE/\ell = 1$. Thus the element stiffness matrix is simply given as

$$[k]^e = \begin{bmatrix} 1 & -1 \\ -1 & 1 \end{bmatrix} \tag{4.72}$$

For the two finite element meshes shown in Figure 4.8, we now perform the assembly and obtain the global stiffness matrix. In either case, since there are seven nodes, we expect

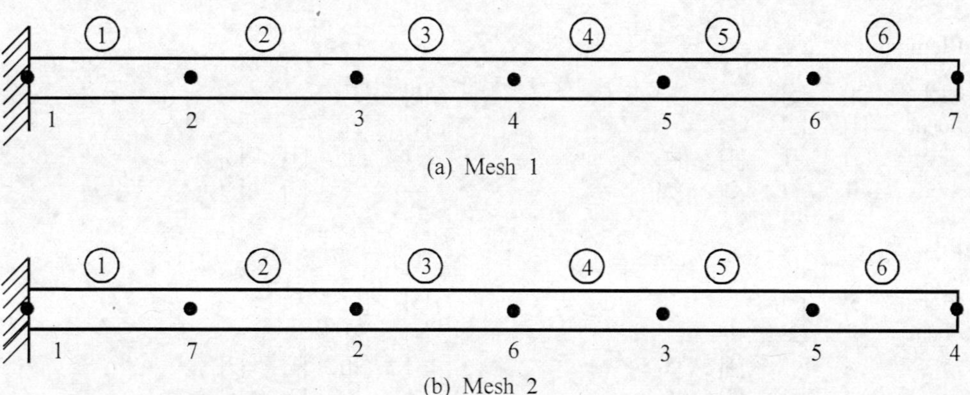

(a) Mesh 1

(b) Mesh 2

Fig. 4.8   Bar element mesh—Different node numbering (Example 4.3).

to get a (7×7) global stiffness matrix for the entire structure. For the sake of convenience, the individual element matrices; the local node numbers, and the global node numbers indicating the appropriate locations for the coefficients of the matrices in the assembled matrix are all indicated.

	Mesh 1		Mesh 2

Element 1

						Local	Global
Global	[1]	[3]			[1]	[7]	
Local	[1]	[2]			[1]	[2]	
	1	−1	[1]	[1]	1	−1	[1]	[1]
	−1	1	[2]	[2]	−1	1	[2]	[7]

Element 2

Global	[2]	[3]			[7]	[2]
Local	[1]	[2]			[1]	[2]
	1	−1	[1]	[2]	1	−1	[1]	[7]
	−1	1	[2]	[3]	−1	1	[2]	[2]

Element 3

Global	[3]	[4]			[2]	[6]
Local	[1]	[2]			[1]	[2]
	1	−1	[1]	[3]	1	−1	[1]	[2]
	−1	1	[2]	[4]	−1	1	[2]	[6]

Element 4

Global	[4]	[5]			[6]	[3]
Local	[1]	[2]			[1]	[2]
	1	−1	[1]	[4]	1	−1	[1]	[6]
	−1	1	[2]	[5]	−1	1	[2]	[3]

Element 5

Global	[5]	[6]			[3]	[5]
Local	[1]	[2]			[1]	[2]
	1	−1	[1]	[5]	1	−1	[1]	[3]
	−1	1	[2]	[6]	−1	1	[2]	[5]

Element 6

Global	[6]	[7]			[5]	[4]
Local	[1]	[2]			[1]	[2]
	1	−1	[1]	[6]	1	−1	[1]	[5]
	−1	1	[2]	[7]	−1	1	[2]	[4]

$$\tag{4.73}$$

For the two cases of node numbering, the global assembled matrices obtained are shown in Figures 4.9 and 4.10.

We observe the following with respect to the assembled stiffness matrix of the structure:

$$
\begin{bmatrix}
1 & -1 & 0 & 0 & 0 & 0 & 0 \\
-1 & 2 & -1 & 0 & 0 & 0 & 0 \\
0 & -1 & 2 & -1 & 0 & 0 & 0 \\
0 & 0 & -1 & 2 & -1 & 0 & 0 \\
0 & 0 & 0 & -1 & 2 & -1 & 0 \\
0 & 0 & 0 & 0 & -1 & 2 & -1 \\
0 & 0 & 0 & 0 & 0 & -1 & 1
\end{bmatrix}
$$

Fig. 4.9   Assembled stiffness matrix $[K]$ for Mesh 1

$$
\begin{bmatrix}
1 & 0 & 0 & 0 & 0 & 0 & -1 \\
0 & 2 & 0 & 0 & 0 & -1 & -1 \\
0 & 0 & 2 & 0 & -1 & -1 & 0 \\
0 & 0 & 0 & 1 & -1 & 0 & 0 \\
0 & 0 & -1 & -1 & 2 & 0 & 0 \\
0 & -1 & -1 & 0 & 0 & 2 & 0 \\
-1 & -1 & 0 & 0 & 0 & 0 & 2
\end{bmatrix}
$$

Fig. 4.10   Assembled stiffness matrix $[K]$ for Mesh 2.

Many elements in the assembled stiffness matrix $[K]$ are zero. Recall our interpretation for the elements of a column of an element stiffness matrix $[k]^e$—the ith column represents the forces required to cause a deformation state such that $u_i = 1$, and all other nodal d.o.f. are zero. Thus a zero element implies that no force is required to be applied at that node. With respect to the node numbering in case (1), for example, since elements are connected only at their nodes, it is easy to see that no force need be applied at node 2 (for example) to cause a unit deformation at node 6 with all other d.o.f. arrested.

The assembled matrix $[K]$ in both the cases is symmetric. Thus we need only to store half the matrix, i.e. either the upper or the lower triangle, saving valuable computer memory. All the nonzero coefficients in case 1 (Figure 4.9) are clustered around the leading diagonal, while they are all well spread out in case 2 (Figure 4.10). If we employ typical Gauss elimination to solve for the deflections from $[K]\{\delta\} = \{F\}$, most of these zero coefficients in $[K]$ could remain zero throughout the solution process. Thus there is no reason to even store them, and we look for efficient schemes of storage for $[K]$ to optimally use computer memory. For example, it will be sufficient to simply store the following as the stiffness matrix for Mesh 1

$$\begin{bmatrix} 1 & -1 \\ 2 & -1 \\ 2 & -1 \\ 2 & -1 \\ 2 & -1 \\ 2 & -1 \\ 1 & 0 \end{bmatrix} \tag{4.74}$$

It is usually a simple computer program that relates these stiffness coefficients to their appropriate locations in the full, global stiffness matrix.

The node numbering employed in case (1) leads to minimal size arrays for storage of $[K]$. Node numbering, therefore, has a significant effect on the distribution of zero and nonzero coefficients in the global matrix $[K]$, and *efficient node numbering schemes* are an integral part of any pre-processor used for modelling complex real-life finite element problems.

Example 4.4   *A bar subjected to thermal stresses*. Consider a uniform rod of length L, area of cross-section A, Young's modulus E, and the coefficient of thermal expansion α, heated uniformly through a temperature rise of $T°C$. Determine the thermal stresses in the bar for the end conditions shown in Figure 4.11(a).

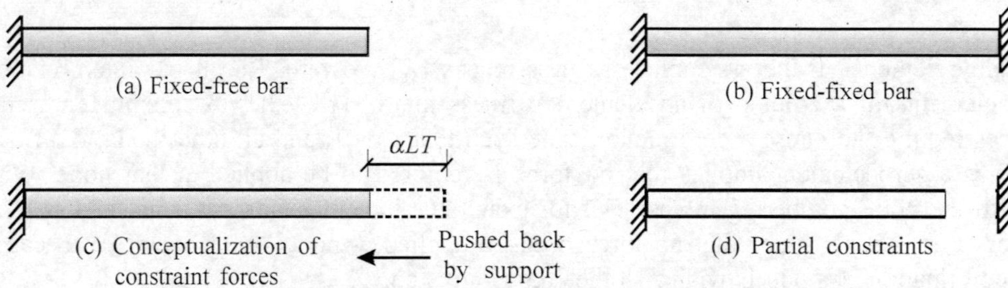

Fig. 4.11   Thermal stress analysis (Example 4.4).

This example problem is indicative of how problems of initial strain $\{\varepsilon_0\}$ are solved. The initial strain in this example is due to heating of the bar. It could as well be due to any other reasons such as bolt pretensioning. While we work out this example on a bar element, the general methodology remains identical even when we consider situations involving more complex elements.

Before we analyse the finite element solution to the problem, we will briefly review the concept of thermal stresses. When a rod is heated through a temperature rise T, the rod would tend to expand by an amount (αLT). This corresponds to an initial strain $\varepsilon_0 = (\alpha T)$. If the support restraints on the rod are such that this expansion is possible, then the rod freely undergoes the thermal expansion (αLT), and thus the mechanical strain in the rod is given by $\varepsilon = \alpha LT/L = (\alpha T)$. This is the case in Figure 4.11(a). As there are no forces restraining this thermal expansion, there will be no internal stresses developed. This is reflected in the material constitutive equation given by

$$\sigma = E(\varepsilon - \varepsilon_0) = E(\alpha T - \alpha T) = 0 \tag{4.75}$$

However, as in Figure 4.11(b), if the rod is entirely prevented from undergoing the free thermal expansion as dictated by the initial strain ε_0, then the mechanical strain in the rod is $\varepsilon = 0$. Thus the rod is subjected to a compressive stress as

$$\sigma = E(\varepsilon - \varepsilon_0) = E(0 - \alpha T) = -E\alpha T \tag{4.76}$$

This can be visualised as shown in Figure 4.11(c), i.e., the rod can be imagined to have expanded through the support by the amount (αLT), and the support develops reaction forces which push the rod back (i.e. compress) all the way to its original position. If, in general, the support is a little distance away from the end of the rod as shown in Figure 4.11(d) so that the rod is allowed partial free expansion, then internal stresses arise only corresponding to the part of initial strain that was not actually realised. The structure depicted in Figure 4.11(d) is for the purpose of illustration only. In real structures, when a part of the structure undergoes thermal expansion, it will be against the elasticity of the other parts. Therefore, only a part of the total thermal expansion will be permitted.

Thus our general method of modelling initial strain will be as follows: we first determine the mechanical loads corresponding to the initial strain, i.e. the loads that would be required in order to cause a mechanical strain equal to the initial strain. Under these loads, we then find the actual deformation of the structure consistent with its support restraint conditions. From these actual deformations, we find the actual mechanical strain. The difference between the actual mechanical strain and the originally imposed initial strain leads to the internal stresses.

From our general expression for nodal force vector given in Eq. (4.16), the nodal loads due to initial strain are given by

$$\{f\}^e = \int_v [B]^T E\varepsilon_{x0} A\, dx \tag{4.77}$$

For the linear bar element,

$$[B] = \begin{bmatrix} \dfrac{-1}{l} & \dfrac{1}{l} \end{bmatrix} \tag{4.78}$$

$$\varepsilon_{x0} = \alpha T \tag{4.79}$$

Thus the nodal force vector corresponding to initial strain is given by

$$\{f\}^e = \int_0^l \begin{bmatrix} -1/l \\ 1/l \end{bmatrix} \alpha ETA\, dx = \begin{Bmatrix} -AE\alpha T \\ AE\alpha T \end{Bmatrix} \tag{4.80}$$

We will use, for the purpose of illustration, two linear bar elements, each of length $(L/2)$ for modelling the entire rod. For element 1, we have, in the absence of any other load,

$$\frac{AE}{L/2}\begin{bmatrix} 1 & -1 \\ -1 & 1 \end{bmatrix}\begin{Bmatrix} u_1 \\ u_2 \end{Bmatrix}^{(1)} = \begin{Bmatrix} f_1^{(1)} \\ f_2^{(1)} \end{Bmatrix} = \begin{Bmatrix} -AE\alpha T \\ AE\alpha T \end{Bmatrix} \tag{4.81}$$

For element 2, we have

$$\frac{AE}{L/2}\begin{bmatrix} 1 & -1 \\ -1 & 1 \end{bmatrix}\begin{Bmatrix} u_1 \\ u_2 \end{Bmatrix}^{(2)} = \begin{Bmatrix} f_1^{(2)} \\ f_2^{(2)} \end{Bmatrix} = \begin{Bmatrix} -AE\alpha T \\ AE\alpha T \end{Bmatrix} \tag{4.82}$$

It is readily seen that these forces, acting on a free bar element, will cause strains corresponding to the initial strain $\varepsilon_0 = \alpha T$. Assembling the element matrices, we get

$$\frac{AE}{L/2}\begin{bmatrix} 1 & -1 & 0 \\ -1 & 1+1 & -1 \\ 0 & -1 & 1 \end{bmatrix}\begin{Bmatrix} u_1 \\ u_2 \\ u_3 \end{Bmatrix} = \begin{Bmatrix} f_1^{(1)} \\ f_2^{(1)} + f_1^{(2)} \\ f_2^{(2)} \end{Bmatrix} \tag{4.83}$$

If the end $x = L$ is free, then we have, with $u_1 = 0$,

$$\frac{AE}{L/2}\begin{bmatrix} 2 & -1 \\ -1 & 1 \end{bmatrix}\begin{Bmatrix} u_2 \\ u_3 \end{Bmatrix} = \begin{Bmatrix} AE\alpha T - AE\alpha T \\ AE\alpha T \end{Bmatrix} = \begin{bmatrix} 0 \\ AE\alpha T \end{bmatrix} \tag{4.84}$$

Solving, we get

$$u_2 = L\alpha T/2, \qquad u_3 = L\alpha T \tag{4.85}$$

For element 1, the actual mechanical strain is

$$\{\varepsilon\} = [B]\{\delta\}^e = \begin{bmatrix} \dfrac{-1}{l} & \dfrac{1}{l} \end{bmatrix}\begin{Bmatrix} 0 \\ L\alpha T/2 \end{Bmatrix} = \alpha T \tag{4.86}$$

The internal thermal stress in element 1 is given by

$$\{\sigma\} = E(\varepsilon - \varepsilon_0) = 0 \tag{4.87}$$

Similarly, for element 2, we can show that the internal stresses will be zero. It is left as an exercise for the reader to show that internal stresses equal to $(-\alpha T)$ will be developed if the end of the rod $x = L$ is also clamped.

Summary: Finite Element Modelling of Initial Strains

Step 1: Compute the element nodal force vector corresponding to the initial strain $\{\varepsilon_0\}$.

Step 2: Solve for the deformations of the structure when subjected to the "loads due to initial strain" along with any other loads on the structure, consistent with the boundary conditions.

Step 3: From the deformations obtained in Step 2 above, evaluate the element level strains $\{\varepsilon\}$ using $\{\varepsilon\} = [B]\{\delta\}^e$.

Step 4: Evaluate the internal stresses developed using the constitutive equation of the material, viz., $\{\sigma\} = E(\{\varepsilon\} - \{\varepsilon_o\})$.

Example 4.5. *A three-bar truss network.* Let us consider the statically indeterminate problem of three trusses shown in Figure 4.12. Unlike all the previous examples discussed so far, in this case the members of the structure are not all in one line. We use this example to

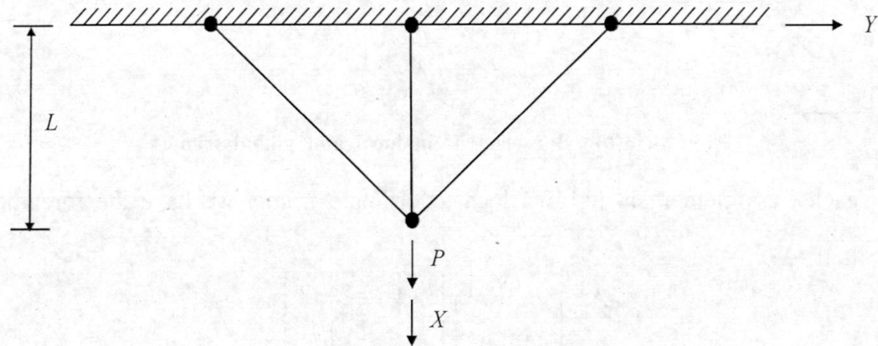

Fig. 4.12   A three-bar truss (Example 4.5).

bring out the issue of "transforming" element level equations in local coordinates into a common global reference frame before assembly. Since the structure is subjected to only a concentrated tip load, we expect the deflection to vary linearly within each member, and hence we expect that one linear bar element would suffice for each truss member. Figure 4.13(a)

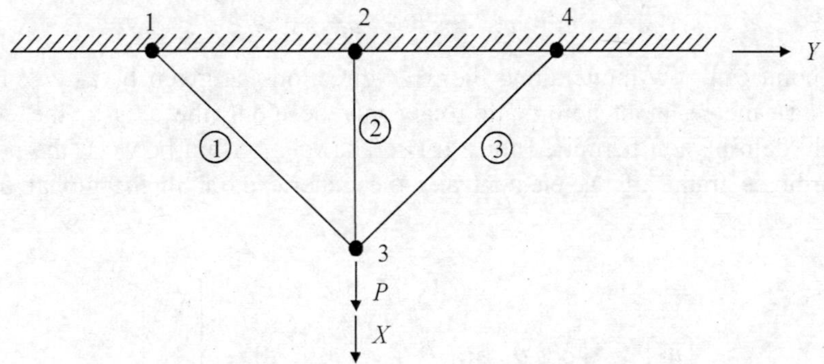

Fig. 4.13(a)   Three-element model of indeterminate truss (Example 4.5).

shows a three-element model of the structure indicating the element numbers and the node numbers. We also show a generic truss element in Figure 4.13(b), indicating its own local axis and the nodal d.o.f. along its axis, which are not necessarily along the global axes.

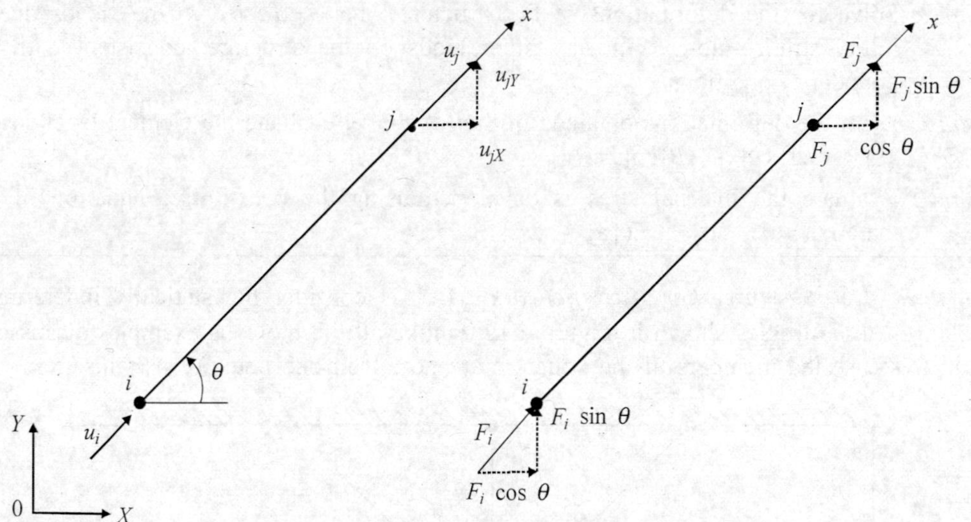

Fig. 4.13(b)   Bar element in local and global frames.

For each truss element in its own local coordinate frame, we have the force-deflection relations, viz.

$$\frac{AE}{l}\begin{bmatrix} 1 & -1 \\ -1 & 1 \end{bmatrix}\begin{Bmatrix} u_i \\ u_j \end{Bmatrix} = \begin{Bmatrix} F_i \\ F_j \end{Bmatrix} \tag{4.88}$$

In our particular example, the length of the central element is L while that of the inclined rods is $L/\cos \alpha$. Thus the individual element equations will be essentially similar to Eq. (4.88) above except for the appropriate value of element length. Considering the components of deflection as shown in Figure 4.13(b), we can write

$$u_i = u_{iX} \cos \theta + u_{iY} \sin \theta$$

$$u_j = u_{jX} \cos \theta + u_{jY} \sin \theta \tag{4.89}$$

Thus a point can now move along the X, Y directions as given by u_X, u_Y, but these are not independent—the resultant deformation can only be along the axis of the element since that is the only deformation permitted for a truss. The relationship between the nodal d.o.f. in the local coordinate frame of the element and the generic d.o.f. in the global frame can be written as

$$\begin{Bmatrix} u_i \\ u_j \end{Bmatrix} = \begin{bmatrix} \cos \theta & \sin \theta & 0 & 0 \\ 0 & 0 & \cos \theta & \sin \theta \end{bmatrix}\begin{Bmatrix} u_{iX} \\ u_{iY} \\ u_{jX} \\ u_{jY} \end{Bmatrix} \tag{4.90}$$

Thus we have

$$\boxed{\{\delta\}_\ell^e = [T]\{\delta\}_g^e} \tag{4.91}$$

where the subscript "ℓ" stands for local and the subscript "g" stands for global coordinate frame.

If we write the element level equations as

$$[k]_\ell^e\{\delta\}_\ell^e = \{F\}_\ell^e \tag{4.92}$$

then

$$[k]_\ell^e[T]\{\delta\}_g^e = \{F\}_\ell^e \tag{4.93}$$

Premultiplying by $[T]^T$, we get

$$\left([T]^T[k]_\ell^e[T]\right)\{\delta\}_g^e = [T]^T\{F\}_\ell^e \tag{4.94}$$

The RHS can be expanded as

$$[T]^T\{F\}_\ell^e = \begin{bmatrix} \cos\theta & 0 \\ \sin\theta & 0 \\ 0 & \cos\theta \\ 0 & \sin\theta \end{bmatrix} \begin{Bmatrix} F_i \\ F_j \end{Bmatrix} = \begin{Bmatrix} F_i\cos\theta \\ F_i\sin\theta \\ F_j\cos\theta \\ F_j\sin\theta \end{Bmatrix} \tag{4.95}$$

From Figure 4.13(b), we see that these are generic forces at the nodes in the global directions. Thus we have

$$[T]^T\{F\}_\ell^e = \{F\}_g^e \tag{4.96}$$

If we define the element stiffness matrix in global reference frame as

$$\boxed{[k]_g^e = [T]^T[k]_\ell^e[T]} \tag{4.97}$$

then Eq. (4.94) can be rewritten as

$$[k]_g^e[\delta]_g^e = \{F\}_g^e \tag{4.98}$$

Figure 4.14 contrasts the element level equations in local and global coordinates for a generic element. Equation (4.97), though derived in the context of this specific example, is a generic equation, and can be used to transform any finite element stiffness matrix from local to global coordinate frame.

We can derive Eq. (4.97) in another way, based on the equivalence of work done. We observe that the nodal forces acting on the nodal deflection must perform the same amount of work whether we consider them in their own local coordinate frame or a global coordinate system. Hence the corresponding strain energy in the structure must also be the same and be independent of the coordinate reference frame. Thus we have

$$U_{\text{local}} = U_{\text{global}} \tag{4.99}$$

$$\frac{1}{2}\left(\{\delta\}_\ell^e\right)^T[k]_\ell^e\{\delta\}_\ell^e = \frac{1}{2}\left(\{\delta\}_g^e\right)^T[k]_g^e\{\delta\}_g^e \tag{4.100}$$

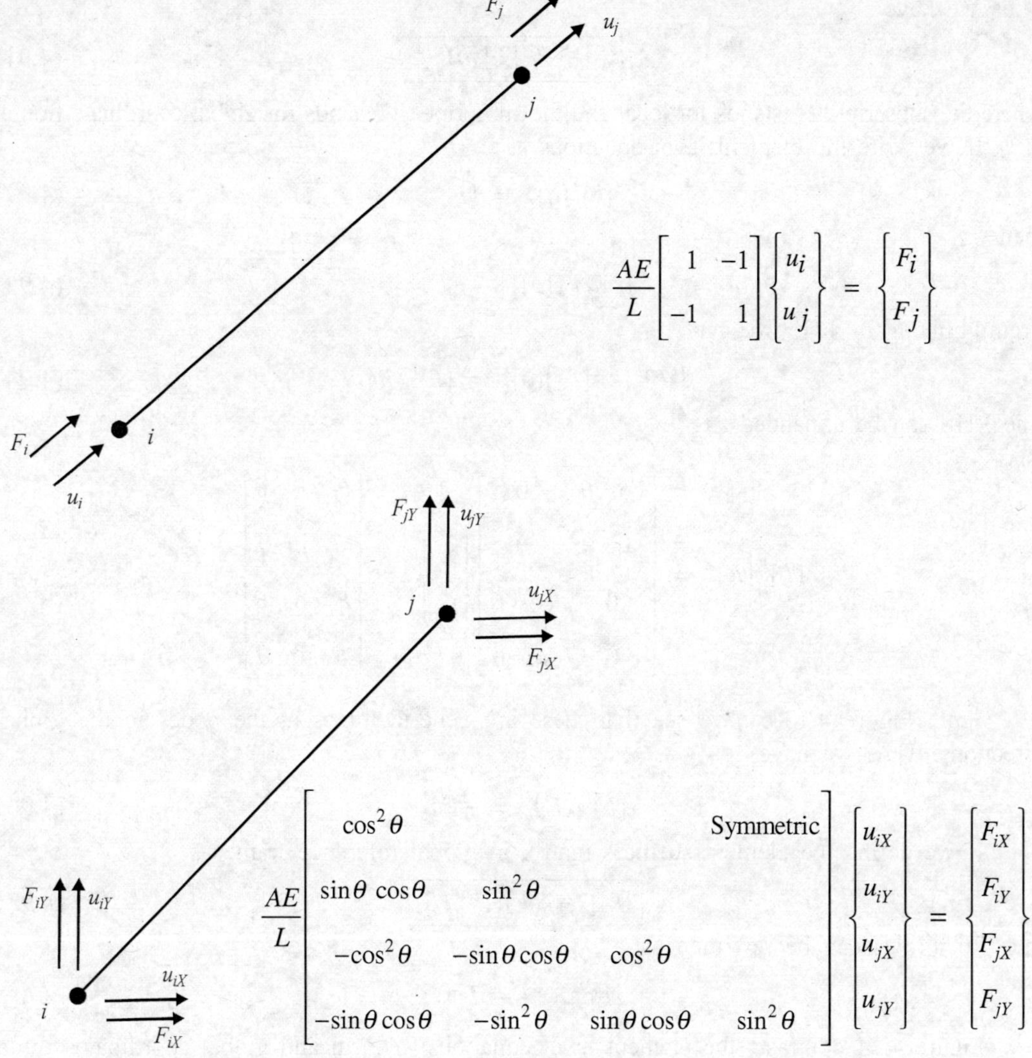

$$\frac{AE}{L}\begin{bmatrix} 1 & -1 \\ -1 & 1 \end{bmatrix}\begin{Bmatrix} u_i \\ u_j \end{Bmatrix} = \begin{Bmatrix} F_i \\ F_j \end{Bmatrix}$$

$$\frac{AE}{L}\begin{bmatrix} \cos^2\theta & & & \text{Symmetric} \\ \sin\theta\cos\theta & \sin^2\theta & & \\ -\cos^2\theta & -\sin\theta\cos\theta & \cos^2\theta & \\ -\sin\theta\cos\theta & -\sin^2\theta & \sin\theta\cos\theta & \sin^2\theta \end{bmatrix}\begin{Bmatrix} u_{iX} \\ u_{iY} \\ u_{jX} \\ u_{jY} \end{Bmatrix} = \begin{Bmatrix} F_{iX} \\ F_{iY} \\ F_{jX} \\ F_{jY} \end{Bmatrix}$$

Fig. 4.14   Bar element equations in local and global frames.

Using the transformation

$$\{\delta\}^e_\ell = [T]\{\delta\}^e_g \tag{4.101}$$

we have

$$\frac{1}{2}\left(\{\delta\}^e_g\right)^T \left([T]^T[k]^e_\ell[T]\right)\{\delta\}^e_g = \frac{1}{2}\left(\{\delta\}^e_g\right)^T [k]^e_g \{\delta\}^e_g \tag{4.102}$$

Therefore,

$$\{k\}^e_g = [T]^T[k]^e_\ell[T] \tag{4.103}$$

which is the same as the result obtained earlier.

We will now use this result to transform the element stiffness matrices in our example.

Element 1. The element stiffness matrix in its own coordinate frame (xy) is given by

$$[k]_l^e = \frac{AE}{L/\cos\alpha}\begin{bmatrix} 1 & -1 \\ -1 & 1 \end{bmatrix} \tag{4.104}$$

In the global reference frame (XY), we have, with $\theta = \alpha$,

| *Global* | $(1X)$ | $(1Y)$ | $(3X)$ | $(3Y)$ | |
| *Local* | (iX) | (iY) | (jX) | (jY) | Local Global |

$$[k]_g^{(1)} = \frac{AE\cos\alpha}{L}\begin{bmatrix} \cos^2\alpha & \sin\alpha\cos\alpha & -\cos^2\alpha & -\sin\alpha\cos\alpha \\ & \sin^2\alpha & -\sin\alpha\cos\alpha & -\sin^2\alpha \\ & & \cos^2\alpha & \sin\alpha\cos\alpha \\ \text{Symmetric} & & & \sin^2\alpha \end{bmatrix}\begin{matrix}(iX) & (1X) \\ (iY) & (1Y) \\ (jX) & (3X) \\ (jY) & (3Y)\end{matrix} \tag{4.105}$$

Element 2. The local frame (xy) and global frame (XY) are identical and hence there is no transformation involved. We have

$$
\begin{matrix}(2) & (3) \\ (i) & (j)\end{matrix}
$$

$$[k]_g^{(2)} = \frac{AE}{L}\begin{bmatrix} 1 & -1 \\ -1 & 1 \end{bmatrix}\begin{matrix}(i) & (2) \\ (j) & (3)\end{matrix} \tag{4.106}$$

Equation (4.106), when rewritten as a (4×4) matrix, gets modified as

| *Global* | $(2X)$ | $(2Y)$ | $(3X)$ | $(3Y)$ | |
| *Local* | (iX) | (iY) | (jX) | (jY) | Local   Global |

$$[k]_g^{(2)} = \frac{AE}{L}\begin{bmatrix} 1 & 0 & -1 & 0 \\ 0 & 0 & 0 & 0 \\ -1 & 0 & 1 & 0 \\ 0 & 0 & 0 & 0 \end{bmatrix}\begin{matrix}(iX) & (2X) \\ (iY) & (2Y) \\ (jX) & (3X) \\ (jY) & (3Y)\end{matrix} \tag{4.107}$$

Element 3. The transformation matrix would be $[T(-\alpha)]$, i.e.,

$$[T] = \begin{bmatrix} \cos\alpha & -\sin\alpha & 0 & 0 \\ 0 & 0 & \cos\alpha & -\sin\alpha \end{bmatrix} \tag{4.108}$$

Therefore,

$$
\begin{array}{cccc}
(4X) & (4Y) & (3X) & (3Y) \\
(iX) & (iY) & (jX) & (jY)
\end{array}
$$

$$
[k]_g^{(3)} = \frac{AE}{L/\cos\alpha}
\begin{bmatrix}
\cos^2\alpha & -\sin\alpha\cos\alpha & -\cos^2\alpha & \sin\alpha\cos\alpha \\
& \sin^2\alpha & \sin\alpha\cos\alpha & -\sin^2\alpha \\
& & \cos^2\alpha & -\sin\alpha\cos\alpha \\
\text{Symmetric} & & & \sin^2\alpha
\end{bmatrix}
\begin{array}{l}
(iX)\ (4X) \\
(iY)\ (4Y) \\
(jX)\ (3X) \\
(jY)\ (3Y)
\end{array}
\tag{4.109}
$$

Having obtained the element matrices in a common global frame XY, we are ready to assemble the equations at the structure level. For the convenience of assembly, we have marked the element level and structure level locations of the matrix coefficients. Since we have four nodes and u_X and u_Y at each node, we have eight variables, and our global stiffness matrix is of size (8×8). The assembled equations are given in Figure 4.15.

$$
\frac{AE}{L}
\begin{bmatrix}
\cos^3\alpha & \sin\alpha\cos^2\alpha & 0 & 0 & -\cos^3\alpha & -\sin\alpha\cos^2\alpha & 0 & 0 \\
& \sin^2\alpha\cos\alpha & 0 & 0 & -\sin\alpha\cos^2\alpha & -\sin^2\alpha\cos\alpha & 0 & 0 \\
& & 1 & 0 & -1 & 0 & 0 & 0 \\
& & & 0 & 0 & 0 & 0 & 0 \\
& & & & 1+2\cos^3\alpha & 0 & -\cos^3\alpha & \sin\alpha\cos^2\alpha \\
& & & & & 2\sin^2\alpha\cos\alpha & \sin\alpha\cos^2\alpha & -\sin^2\alpha\cos\alpha \\
& & & & & & \cos^3\alpha & -\sin\alpha\cos^2\alpha \\
\text{Symmetric} & & & & & & & \sin^2\alpha\cos\alpha
\end{bmatrix}
\begin{Bmatrix}
u_{1X} \\ u_{1Y} \\ u_{2X} \\ u_{2Y} \\ u_{3X} \\ u_{3Y} \\ u_{4X} \\ u_{4Y}
\end{Bmatrix}
=
\begin{Bmatrix}
0 \\ 0 \\ 0 \\ 0 \\ P \\ 0 \\ 0 \\ 0
\end{Bmatrix}
$$

Fig. 4.15   Assembled equations (Example 4.5).

By virtue of the boundary conditions due to supports at nodes 1, 2 and 3, we have

$$
u_{1X} = 0 = u_{1Y}, \qquad u_{2X} = 0 = u_{2Y}, \qquad u_{4X} = 0 = u_{4Y}
\tag{4.110}
$$

Thus from the 5th and 6th equations of Figure 4.15,

$$
\frac{AE}{L}
\begin{bmatrix}
1 + 2\cos^3\alpha & 0 \\
0 & 2\sin^2\alpha\cos\alpha
\end{bmatrix}
\begin{Bmatrix}
u_{3X} \\ u_{3Y}
\end{Bmatrix}
=
\begin{Bmatrix}
P \\ 0
\end{Bmatrix}
\tag{4.111}
$$

Solving, we get

$$
u_{3X} = \frac{PL}{AE}\,\frac{1}{1+2\cos^3\alpha}, \qquad u_{3Y} = 0
\tag{4.112}
$$

which tallies with the well-known result.

Having obtained the deflection of the structure, we can use the first six equations of Figure 4.15 to find the support reactions at nodes 1–3, which ensure that the displacements at these nodes are zero. We now summarise the general procedure for assembling element level equations.

Summary: General Method of Assembly

Step 1: Identify the local (or element) level and global (on structure) level coordinate frames of reference.

Step 2: Obtain the element matrices in the local reference frame.

Step 3: Obtain the coordinate transformation matrix $[T]$ between the local and global frames of reference.

Step 4: Transform the element matrices into the common global reference frame.

Step 5: Identify the global locations of the individual coefficients of the element matrices based on local and global node numbers.

Step 6: Assemble the element matrices by placing the coefficients of the element matrices in their appropriate places as identified in Step 5.

4.5 Beam Element

4.5.1 Selection of Nodal d.o.f.

A truss or rod undergoes only axial deformation and we assumed that the entire cross-section undergoes the same displacement u. Thus the bar element was just a line element with two nodes and each node having u as the nodal d.o.f. A beam, on the other hand, undergoes transverse deflection denoted by v. Figure 4.16 shows a typical beam section undergoing transverse deflection. We assume that the cross-section is doubly symmetric and the bending

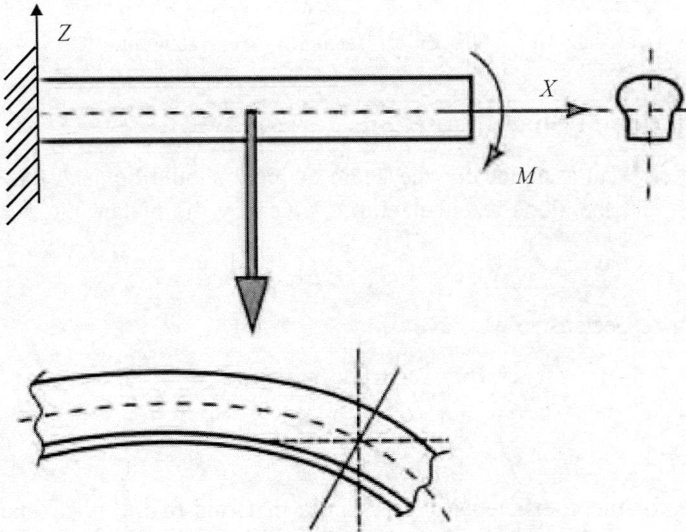

Fig. 4.16 Typical beam section and its deformation.

takes place in a plane of symmetry. According to the classical theory of beam bending (Euler–Bernoulli theory), the entire cross-section has the same transverse deflection v as the neutral axis; sections originally perpendicular to the neutral axis remain so after bending. Hence the deflections are small and we assume that the rotation of each section is the same as the slope of the deflection curve at that point, i.e., (dv/dx). Therefore, the axial deformation of any point P away from the neutral axis (due to bending alone) is given by $u_P = -(z)(dv/dx)$. *Thus, if $v(x)$ is determined, then the entire state of deformation of the body is completely determined.*

Our beam element is therefore also a simple line element, representing the *neutral axis* of the beam. However, in order to ensure the continuity of deformation at any point (i.e. *those on the neutral axis as well as away from it*), we have to ensure that v and (dv/dx) are continuous. We achieve this by taking two nodal d.o.f., viz., v and θ $(= dv/dx)$. Thus when two beam elements are joined together, they share a common node and so have the same transverse deflection and slope. If we did not have the slope d.o.f., then we would only have been able to ensure that transverse deflection is continuous across elements, but not guarantee that the slope also would be continuous. Thus the points away from the neutral axis could actually separate (thereby creating a void) or overlap (creating a kink). We recall that for this problem (fourth order differential equation in v), the essential boundary conditions consist of both v and dv/dx. Thus, taking these nodal d.o.f. permits us to readily assign prescribed essential boundary conditions (e.g. fixed end $v = 0 = dv/dx$). Moreover, a prescribed value of moment load can readily be taken into account with the rotational d.o.f θ. The Euler–Bernoulli beam element is shown in Figure 4.17.

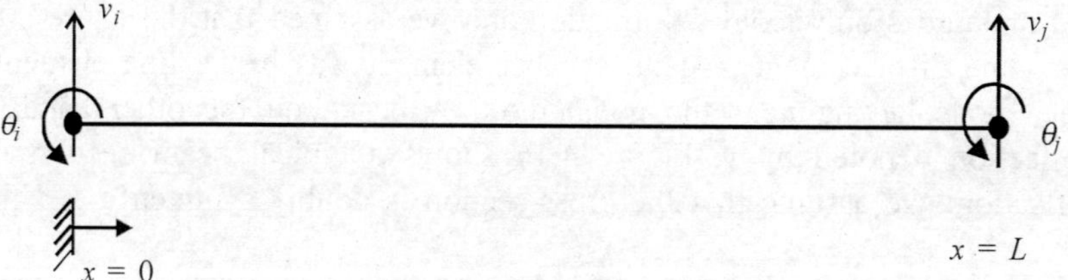

Fig. 4.17   The Euler–Bernoulli beam element.

4.5.2   Determination of Shape Functions

The displacement field $v(x)$ assumed for the beam element should be such that it takes on the values of deflection and the slope at either end as given by the nodal values v_i, θ_i, v_j, θ_j. Let $v(x)$ be given by

$$v(x) = c_0 + c_1 x + c_2 x^2 + c_3 x^3 \tag{4.113}$$

Differentiating with respect to x, we have

$$dv/dx = c_1 + 2c_2 x + 3c_3 x^2 \tag{4.114}$$

At $x = 0$ and L, we have

$$v_i = c_0, \qquad \theta_i = c_1, \qquad v_j = c_0 + c_1 L + c_2 L^2 + c_3 L^3, \qquad \theta_j = c_1 + 2c_2 L + 3c_3 L^2 \tag{4.115}$$

We can solve for the coefficients c_0, c_1, c_2, c_3 in terms of the nodal d.o.f. v_i, θ_i, v_j, θ_j and substitute in Eq. (4.113) to get

$$v(x) = N_1 v_i + N_2 \theta_i + N_3 v_j + N_4 \theta_j \tag{4.116}$$

where the shape functions N are given by

$$\boxed{\begin{array}{ll} N_1 = 1 - 3x^2/L^2 + 2x^3/L^3, & N_2 = x - 2x^2/L + x^3/L^2 \\ N_3 = 3x^2/L^2 - 2x^3/L^3, & N_4 = -x^2/L + x^3/L^2 \end{array}} \tag{4.117}$$

(*Note:* The reader is advised to carry out the steps to gain familiarity with this method of deriving the shape functions.)

Rewriting Eq. (4.11), we get

$$v(x) = [N]\{\delta\}^e = [N_1 \; N_2 \; N_3 \; N_4] \begin{Bmatrix} v_i \\ \theta_i \\ v_j \\ \theta_j \end{Bmatrix} \tag{4.118}$$

4.5.3 Element Matrices

The strain at any point in the beam is given by

$$\{\varepsilon\} = \varepsilon_x = \frac{du}{dx} = \frac{d}{dx}\left(-z\frac{dv}{dx}\right) = -z\frac{d^2v}{dx^2} \tag{4.119}$$

From Eq. (4.118), we have

$$\begin{aligned} \varepsilon_x &= -z\left[\frac{d^2N_1}{dx^2} \; \frac{d^2N_2}{dx^2} \; \frac{d^2N_3}{dx^2} \; \frac{d^2N_4}{dx^2}\right]\{\delta\}^e \\ &= -z\left[\left(\frac{-6}{L^2}+\frac{12x}{L^3}\right)\left(\frac{-4}{L}+\frac{6x}{L^2}\right)\left(\frac{6}{L^2}-\frac{12x}{L^3}\right)\left(\frac{-2}{L}+\frac{6x}{L^2}\right)\right]\{\delta\}^e \end{aligned} \tag{4.120}$$

Thus the strain–displacement relation matrix $[B]$ is given by

$$[B] = -z\left[\left(\frac{12x}{L^3}-\frac{6}{L^2}\right)\left(\frac{6x}{L^2}-\frac{4}{L}\right)\left(\frac{6}{L^2}-\frac{12x}{L^3}\right)\left(\frac{6x}{L^2}-\frac{2}{L}\right)\right] \tag{4.121}$$

We can obtain the element stiffness matrix as

$$[k]^e = \int_v [B]^T[B]E\,dv$$

$$= \int_A \int_0^l (-z) \begin{bmatrix} \dfrac{12x}{L^3} - \dfrac{6}{L^2} \\[2mm] \dfrac{6x}{L^2} - \dfrac{4}{L} \\[2mm] \dfrac{6}{L^2} - \dfrac{12x}{L^3} \\[2mm] \dfrac{6x}{L^2} - \dfrac{2}{L} \end{bmatrix} (E)(-z) \left[\left(\dfrac{12x}{L^3} - \dfrac{6}{L^2} \right)\left(\dfrac{6x}{L^2} - \dfrac{4}{L} \right)\left(\dfrac{6}{L^2} - \dfrac{12x}{L^3} \right)\left(\dfrac{6x}{L^2} - \dfrac{2}{L} \right) \right] (dA)\, dx$$

$$(4.122)$$

Noting that $\int_A z^2 \, dA = I$, the second moment of area of the cross-section and carrying out the integration with respect to x, we get

$$[k]^e = \frac{EI}{L^3} \begin{bmatrix} 12 & 6L & -12 & 6L \\ 6L & 4L^2 & -6L & 2L^2 \\ -12 & -6L & 12 & -6L \\ 6L & 2L^2 & -6L & 4L^2 \end{bmatrix} \tag{4.123}$$

The element nodal force vectors are now obtained for specific cases of loading.

Gravity loading

Gravity loading is a typical body force and is given by (ρg) per unit volume or $(\rho A g)$ per unit length, where ρ is the mass density of the material. The equivalent nodal force vector for the distributed body force can be obtained as

$$\{f\}^e = \int_v [N]^T (\rho g) \, dv = \int_0^L [N]^T (\rho g) A \, dx = (\rho A g) \begin{Bmatrix} \ell/2 \\[1mm] L^2/12 \\[1mm] \ell/2 \\[1mm] -L^2/12 \end{Bmatrix} \tag{4.124}$$

Any other distributed load on a beam specified in the form of "q" per unit length can be addressed in a similar manner. For a specified $q(x)$, one can perform the integration $\int [N]^T q(x) \, dx$ and obtain $\{f\}^e$. Thus, for a uniformly distributed force q_0, we can write the equivalent nodal force vector as

$$\{f\}^e = \int_0^L [N]^T q_0 \, dx = \begin{Bmatrix} q_0 L/2 \\ q_0 L^2/12 \\ q_0 L/2 \\ -q_0 L^2/12 \end{Bmatrix} \tag{4.125}$$

We observe that our equivalent nodal forces include nodal moments! We term this as equivalent to the distributed force in the sense that these nodal forces acting through the nodal displacements do the same amount of work as the distributed force.* Hence such a force vector is termed **consistent nodal force vector**. We could have simply replaced the entire distributed force on the element ($q_0 L$) by lumping half each at the two nodes. Such a nodal force vector is termed **lumped nodal force vector** and is given by

$$\{f\}^e = \begin{Bmatrix} q_0 L/2 \\ 0 \\ q_0 L/2 \\ 0 \end{Bmatrix} \tag{4.126}$$

Since the lumped force vector does not guarantee work equivalence, in general, we expect the results to be inferior. Moreover, for more complex two- and three-dimensional elements with internal nodes, an appropriate scheme of lumping is not entirely intuitive. Thus, normally, we use only the consistent nodal force vector. The positive sign convention for the forces and moments is shown in Figure 4.18, where the consistent and lumped nodal loads are illustrated for two typical external load cases.

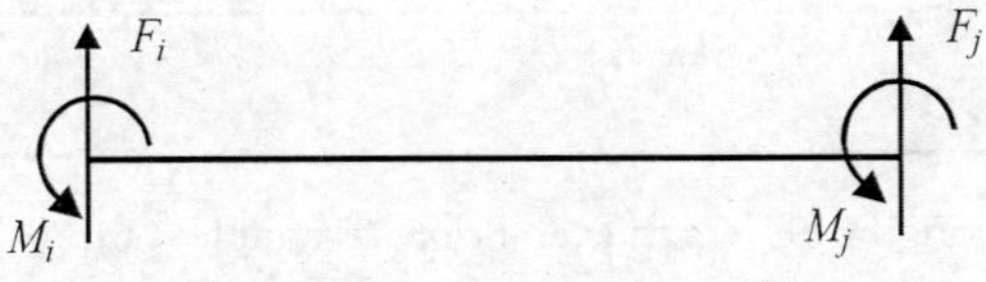

(a) Positive forces and moments

Fig. 4.18 Beam element forces. (cont.)

---

*Recall that work done = (force) (displacement). Thus work done by the distributed force

$$\int \underbrace{(q \, dx)}_{\text{elemental force}} v = \int_0^\ell (q \, dx)\{\delta\}^{e^T} [N]^T = \{\delta\}^{e^T} \int_0^\ell [N]^T q \, dx = \{\delta\}^{e^T} \{f\}_{\text{consistent}}$$

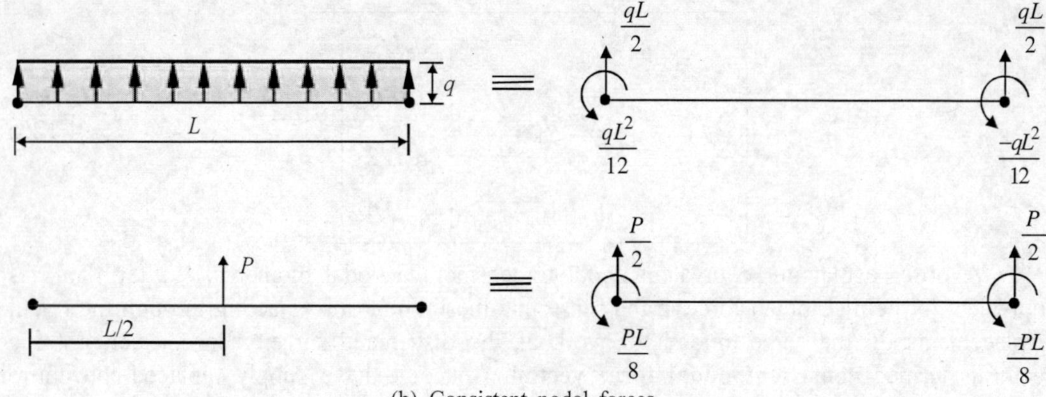

(b) Consistent nodal forces

Fig. 4.18 Beam element forces.

Example 4.6. *Cantilever beam under transverse load.* Consider the cantilever beam loaded as shown in Figure 4.19. Using one-beam element, we get

$$\frac{EI}{L^3}\begin{bmatrix} 12 & 6L & -12 & 6L \\ 6L & 4L^2 & -6L & 2L^2 \\ -12 & -6L & 12 & -6L \\ 6L & 2L^2 & -6L & 4L^2 \end{bmatrix}\begin{Bmatrix} v_1 \\ \theta_1 \\ v_2 \\ \theta_2 \end{Bmatrix} = \begin{Bmatrix} F_1 \\ M_1 \\ F_2 \\ M_2 \end{Bmatrix} \qquad (4.127)$$

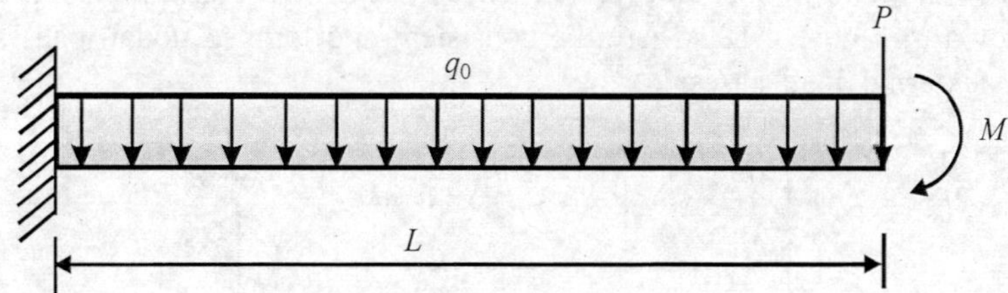

Fig. 4.19 Cantilever beam (Example 4.6).

At the clamped end, the boundary condition stipulates that $v = \theta = 0$. At $x = L$, taking into account concentrated force/moment as well as nodal forces equivalent to the distributed forces, we have

$$F_2 = \frac{-q_0 L}{2} - P \qquad (4.128)$$

$$M_2 = \frac{+q_0 L^2}{12} - M \qquad (4.129)$$

From the last two rows of Eq. (4.127),

$$\frac{EI}{L^3}\begin{bmatrix} 12 & -6L \\ -6L & 4L^2 \end{bmatrix}\begin{Bmatrix} v_2 \\ \theta_2 \end{Bmatrix} = \begin{Bmatrix} \dfrac{-q_0 L}{2} - P \\[2ex] \dfrac{+q_0 L^2}{12} - M \end{Bmatrix} \tag{4.130}$$

Solving, we obtain the following tip deflection and slope:

$$\begin{Bmatrix} v_2 \\ \theta_2 \end{Bmatrix} = \frac{L}{12EI}\begin{Bmatrix} \dfrac{-3q_0 L^3}{2} - 4PL^2 - 6ML \\[2ex] -2q_0 L^2 - 6PL - 12M \end{Bmatrix} \tag{4.131}$$

Thus the displacement field $v(x)$ over the element is given by

$$v(x) = [N]\{\delta\}^e = (N_1)(0) + (N_2)(0) + (N_3)(v_2) + (N_4)(\theta_2)$$

$$= \left(\frac{3x^2}{L^2} - \frac{2x^3}{L^3}\right)\left(\frac{-3q_0 L^3}{2} - 4PL^2 - 6ML\right)\left(\frac{L}{12EI}\right)$$

$$+ \left(\frac{x^3}{L^2} - \frac{x^2}{L}\right)\left(-2q_0 L^2 - 6PL - 12M\right)\left(\frac{L}{12EI}\right)$$

$$= \frac{P}{6EI}(x^3 - 3x^2 L) - \frac{Mx^2}{2EI} + \frac{q_0}{24EI}(2x^3 L - 5x^2 L^2) \tag{4.132}$$

From standard text books on the Mechanics of Materials, we know that

$$v(x) = \frac{P}{6EI}(x^3 - 3x^2 L) - \frac{Mx^2}{2EI} + \frac{q_0}{24EI}(-x^4 + 4x^3 L - 6x^2 L^2) \tag{4.133}$$

The beam element admits cubic variation of transverse deflection and thus we can get the exact solution for the concentrated load P and the moment M. However, the uniformly distributed force q_0 causes a quadratic curve and our one-element solution is in error. We expect that our solution will improve with more number of elements.

Example 4.7. *A continuous beam.* Figure 4.20 shows a typical continuous beam. We wish to obtain the deflection of the beam using the beam element just described. For simplicity we assume $EI = 1$.

We will take a five-element mesh as shown in the figure. Subdividing the span AB into two elements with a node at the load point has the advantage that we can very easily specify the nodal forces. On the other hand, if we had not placed a node at the load point, then it would have been necessary to find equivalent nodal forces. Since we have just now seen in the

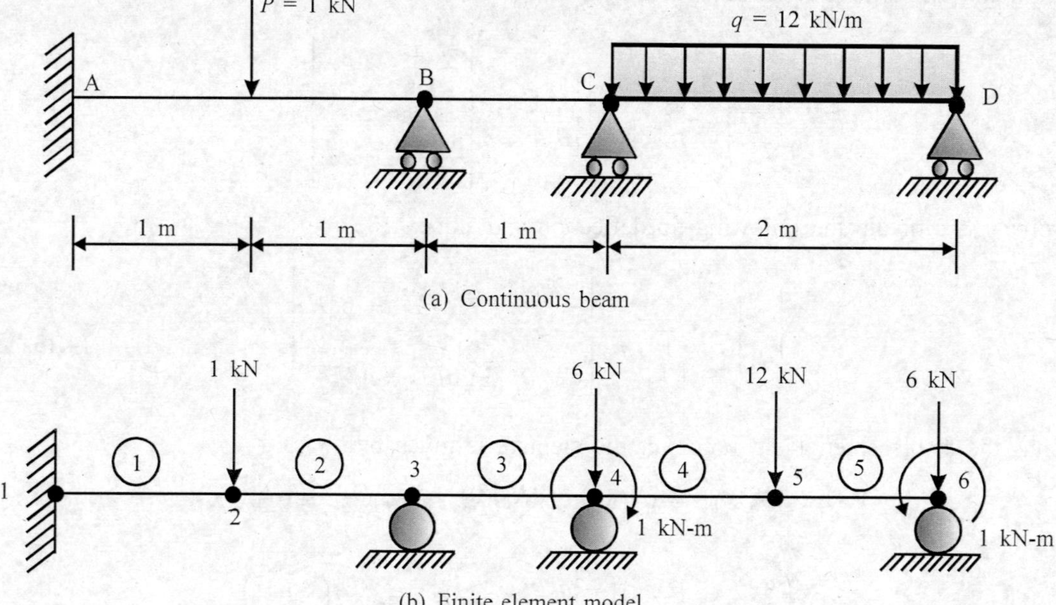

Fig. 4.20 Typical continuous beam (Example 4.7).

previous example that one beam element is inadequate for a uniformly distributed force, we try with two elements in the span CD. Our meshing has also ensured that all elements are of uniform size, for easy hand calculation. Following the standard procedure, we obtain the global stiffness matrix and force vector as

$$[K]_{12\times12}\{\delta\}_{12\times1} = \{F\}_{12\times1} \tag{4.134}$$

$$[K] =
\begin{bmatrix}
12 & 6 & -12 & 6 & 0 & 0 & 0 & 0 & 0 & 0 & 0 & 0 \\
 & 4 & -6 & 2 & 0 & 0 & 0 & 0 & 0 & 0 & 0 & 0 \\
 & & 12+12 & -6+6 & -12 & 6 & 0 & 0 & 0 & 0 & 0 & 0 \\
 & & & 4+4 & -6 & 2 & 0 & 0 & 0 & 0 & 0 & 0 \\
 & & & & 12+12 & -6+6 & -12 & 6 & 0 & 0 & 0 & 0 \\
 & & & & & 4+4 & -6 & 2 & 0 & 0 & 0 & 0 \\
 & & & & & & 12+12 & -6+6 & -12 & 6 & 0 & 0 \\
 & & & & & & & 4+4 & -6 & 2 & 0 & 0 \\
 & & & & & & & & 12+12 & -6+6 & -12 & 6 \\
 & & & & & & & & & 4+4 & -6 & 2 \\
 & & & & & & & & & & 12 & -6 \\
\text{Symmetric} & & & & & & & & & & & 4
\end{bmatrix} \tag{4.135}$$

$$\{F\} = \begin{Bmatrix} R_1 & M_1 & -1000 & 0 & R_3 & 0 & (R_4 - 600) & -1000 & -12000 & 0 & (R_6 - 6000) & 1000 \end{Bmatrix}^T \tag{4.136}$$

Since there are six nodes and two d.o.f. per node, the global stiffness matrix $[K]$ is of size (12×12) and $\{F\}$ is a column vector of size (12). The boundary conditions stipulate that the slope and deflection be zero at node 1, and the vertical deflection be zero at nodes 3, 4 and 6. Thus the reduced set of equations involving unknown nodal d.o.f. is obtained in matrix form as

$$\begin{bmatrix} 24 & 0 & 6 & 0 & 0 & 0 & 0 \\ & 8 & 2 & 0 & 0 & 0 & 0 \\ & & 8 & 2 & 0 & 0 & 0 \\ & & & 8 & -6 & 2 & 0 \\ & & & & 24 & 0 & 6 \\ & & & & & 8 & 2 \\ \text{Symmetric} & & & & & & 4 \end{bmatrix} \begin{Bmatrix} v_2 \\ \theta_2 \\ \theta_3 \\ \theta_4 \\ v_5 \\ \theta_5 \\ \theta_6 \end{Bmatrix} = \begin{Bmatrix} -1000 \\ 0 \\ 0 \\ -1000 \\ -12,000 \\ 0 \\ 1000 \end{Bmatrix} \qquad (4.137)$$

Solving, we get

$$\begin{aligned} v_2 &= -0.1570, & v_5 &= -1.4720 \\ \theta_2 &= -0.1153, & \theta_5 &= -0.3427 \\ \theta_3 &= 0.4612, & \theta_6 &= 2.6293 \\ \theta_4 &= -1.2586 \end{aligned} \qquad (4.138)$$

Using these values in the system equations (Eqs. (4.134)–(4.136)), we can obtain the support reaction forces. Following standard mechanics of materials approaches, we can obtain the exact solution as

$$\begin{aligned} v_2\big|_{\text{exact}} &= -0.1570 \\ v_5\big|_{\text{exact}} &= -1.4720 \end{aligned} \qquad (4.139)$$

From the element stiffness matrices and the nodal displacements, we can obtain the element level forces as

$$\{f\}^e = [K]^e \{\delta\}^e \qquad (4.140)$$

These are the forces on the element at its nodes when each element is viewed as a free body. The element level forces are shown in Figure 4.21. When the individual elements are put together to build up the entire beam, the forces are algebraically summed up as shown in the figure. At each node, the internal reaction forces annul each other and the resultant forces are the externally applied forces. If a node is supported, the resultant force is the support reaction.

4.6 Frame Element

The bar element can permit only axial deformation and the beam element can permit only transverse deflection. By combining the bar and the beam elements, we obtain the "frame

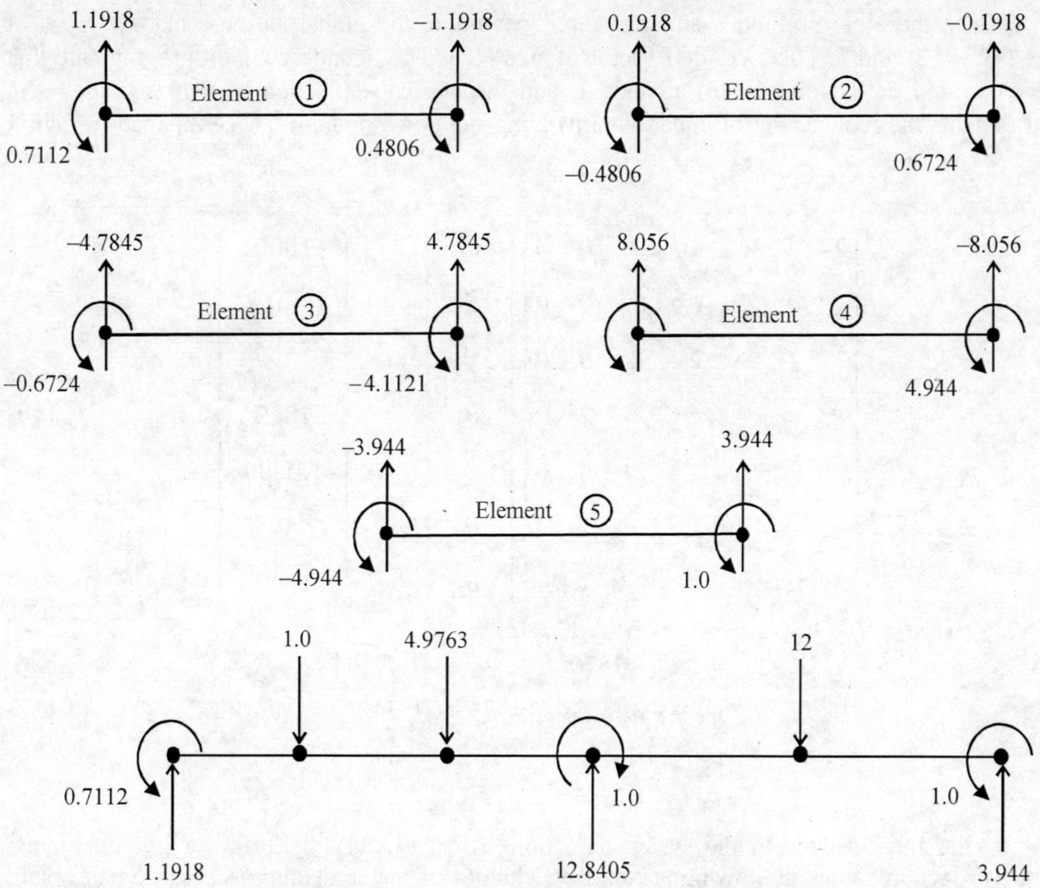

Fig. 4.21 Element and system level forces (Example 4.7).

element" which enables us to model typical problems of framed structures which in general involve both types of deformation. A typical planar frame element is shown in Figure 4.22, and the corresponding element equations are

$$[k]^e \{\delta\}^e = \{f\}^e \tag{4.141}$$

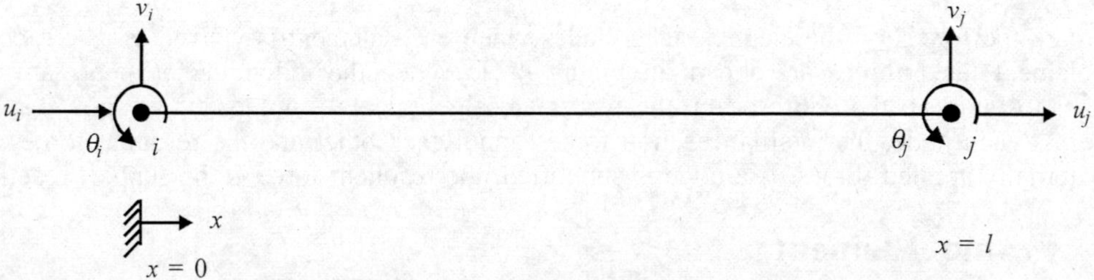

Fig. 4.22 Plane frame element.

where

$$\{\delta\}^e = \left\{u_i \;\; v_i \;\; \theta_i \;\; u_j \;\; v_j \;\; \theta_j\right\}^T \tag{4.142}$$

and the element stiffness matrix is

$$[k]^e = \begin{bmatrix} AE/L & 0 & 0 & -AE/L & 0 & 0 \\ 0 & 12EI/L^3 & 6EI/L^2 & 0 & -12EI/L^3 & 6EI/L^2 \\ 0 & 6EI/L^2 & 4EI/L & 0 & -6EI/L^2 & 2EI/L \\ -AE/L & 0 & 0 & AE/L & 0 & 0 \\ 0 & -12EI/L^3 & -6EI/L^2 & 0 & 12EI/L^3 & -6EI/L^2 \\ 0 & 6EI/L^2 & 2EI/L & 0 & -6EI/L^2 & 4EI/L \end{bmatrix} \tag{4.143}$$

Example 4.8.   *A frame structure.* Consider the simple framed structure made of steel shown in Figure 4.23(a). We wish to obtain the deflections using the frame element just described.

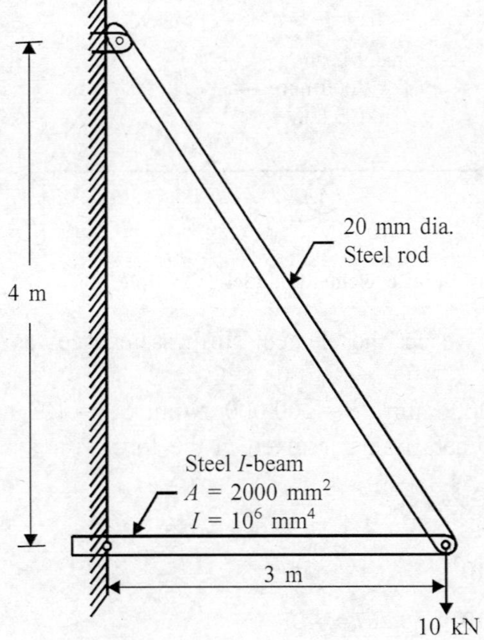

Fig. 4.23(a)   Structure details (Example 4.8).

Through this example, we will illustrate the procedure for transformation of frame element matrices from local to global reference frame. We will also study how to model pin joints in a structure. We will use two elements as shown in Figure 4.23(b). Using the expression for

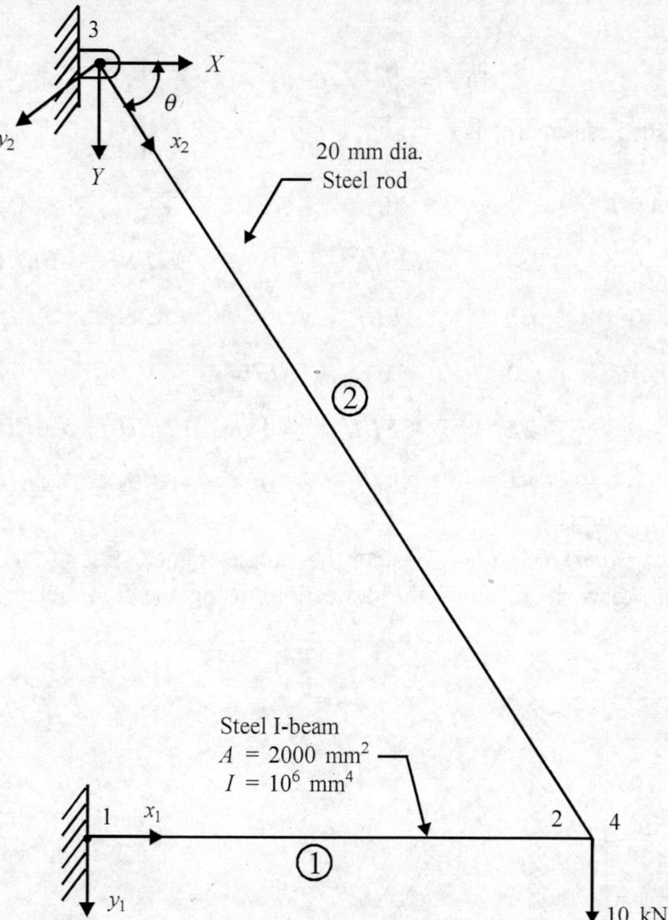

Fig. 4.23(b)   Finite element model (Example 4.8).

the element matrix given above, we get the element stiffness matrices as described now.

Element 1. $L = 3$ m, $A = 2,000$ mm^2, $E = 200,000$ N/mm^2, $I = 10^6$ mm^4. The element stiffness matrix in its local coordinate frame is given in the form

$$
[k]^{(1)} = \begin{bmatrix}
0.133 \times 10^6 & & & & & \\
0 & 0.889 \times 10^2 & & & & \text{Symmetric} \\
0 & 0.133 \times 10^6 & 0.267 \times 10^9 & & & \\
-0.133 \times 10^6 & 0 & 0 & 0.133 \times 10^6 & & \\
0 & -0.889 \times 10^2 & -0.133 \times 10^6 & 0 & 0.889 \times 10^2 & \\
0 & 0.133 \times 10^6 & 0.133 \times 10^6 & 0 & -0.133 \times 10^6 & 0.267 \times 10^9
\end{bmatrix}
$$

$$(4.144)$$

The local coordinate frame of element 1 is identical to the global frame and hence no transformation is required.

Element 2. $L = 5$ m, $A = 314.16$ mm^2, $E = 200,000$ N/mm^2, $I = 7,854$ mm^4. The element stiffness matrix in its local coordinate frame is obtained in the form

$$[k]_\ell^{(2)} = \begin{bmatrix} 0.126 \times 10^5 & & & & \text{Symmetric} & \\ 0 & 0.151 & & & & \\ 0 & 0.377 \times 10^3 & 0.126 \times 10^7 & & & \\ -0.126 \times 10^5 & 0 & 0 & 0.126 \times 10^5 & & \\ 0 & -0.151 & -0.377 \times 10^3 & 0 & 0.151 & \\ 0 & 0.377 \times 10^3 & 0.628 \times 10^6 & 0 & -0.377 \times 10^3 & 0.126 \times 10^7 \end{bmatrix}$$

$$(4.145)$$

The local coordinate frame is inclined with respect to global reference frame and hence we need to transform the element stiffness matrix into global frame. For any orientation θ, the transformation matrix relating local and global d.o.f. can be readily shown to be given by

$$[T] = \begin{bmatrix} \cos\theta & \sin\theta & 0 & 0 & 0 & 0 \\ -\sin\theta & \cos\theta & 0 & 0 & 0 & 0 \\ 0 & 0 & 1 & 0 & 0 & 0 \\ 0 & 0 & 0 & \cos\theta & \sin\theta & 0 \\ 0 & 0 & 0 & -\sin\theta & \cos\theta & 0 \\ 0 & 0 & 0 & 0 & 0 & 1 \end{bmatrix} \qquad (4.146)$$

Thus the element stiffness matrix in the global frame is written in the form

$$[k]_g^{(2)} = [T]^T [k]_l^{(2)} [T]$$

$$= \begin{bmatrix} 0.452 \times 10^4 & & & & & \\ 0.603 \times 10^4 & 0.804 \times 10^4 & & \text{Symmetric} & & \\ -0.302 \times 10^3 & 0.226 \times 10^3 & 0.126 \times 10^7 & & & \\ -0.452 \times 10^4 & -0.603 \times 10^4 & 0.302 \times 10^3 & 0.452 \times 10^4 & & \\ -0.603 \times 10^4 & -0.804 \times 10^4 & -0.226 \times 10^3 & 0.603 \times 10^4 & 0.804 \times 10^4 & \\ -0.302 \times 10^3 & 0.226 \times 10^3 & 0.628 \times 10^6 & 0.302 \times 10^3 & -0.226 \times 10^3 & 0.126 \times 10^7 \end{bmatrix}$$

$$(4.147)$$

Assembly

We observe that, at the pin joint between elements 1 and 2, we use two coincident nodes. These two nodes undergo the same translation but can have relative rotation because of the pin joint. This is reflected in assembly as the translations having the same equation numbers (rows in the matrix) and the rotational d.o.f. having different equation numbers (rows in the matrix). The boundary conditions require that at node 1, all the d.o.f. be arrested and at node 3, the translations be arrested. Thus the reduced system of equations is given as follows:

$$
\begin{bmatrix}
\underline{u_2,\,u_4} & \underline{v_2,\,v_4} & \underline{\theta_2} & \underline{\theta_3} & \underline{\theta_4} \\
0.138 \times 10^6 \\
0.603 \times 10^4 & 0.813 \times 10^4 & & \text{Symmetric} \\
0 & -0.133 \times 10^6 & 0.267 \times 10^9 \\
0.302 \times 10^3 & -0.226 \times 10^3 & 0 & 0.126 \times 10^7 \\
0.302 \times 10^3 & -0.226 \times 10^3 & 0 & 0.628 \times 10^6 & 0.126 \times 10^7
\end{bmatrix}
\begin{Bmatrix} u_2 \\ v_2 \\ \theta_2 \\ \theta_3 \\ \theta_4 \end{Bmatrix}
=
\begin{Bmatrix} 0 \\ 10{,}000 \\ 0 \\ 0 \\ 0 \end{Bmatrix}
\qquad (4.148)
$$

Solving, we get

$$u_2 = -0.12 \text{ mm}, \qquad v_2 = 2.564 \text{ mm}$$
$$\theta_2 = 0.0013 \text{ rad}, \qquad \theta_3 = \theta_4 = 3.26 \times 10^{-4} \text{ rad} \qquad (4.149)$$

These results can be easily verified based on classical mechanics. We observe that the rotations at node 3 and 4 are the same for element 2, in view of the pin joints and it remains a two-force member as expected.

Example 4.9. *Simply supported beam under combined axial and transverse loads.* Consider the simply supported beam shown in Figure 4.24 subjected to an axial force P_a and a transverse force P_t. We wish to use the frame element just developed to model this problem. We will use

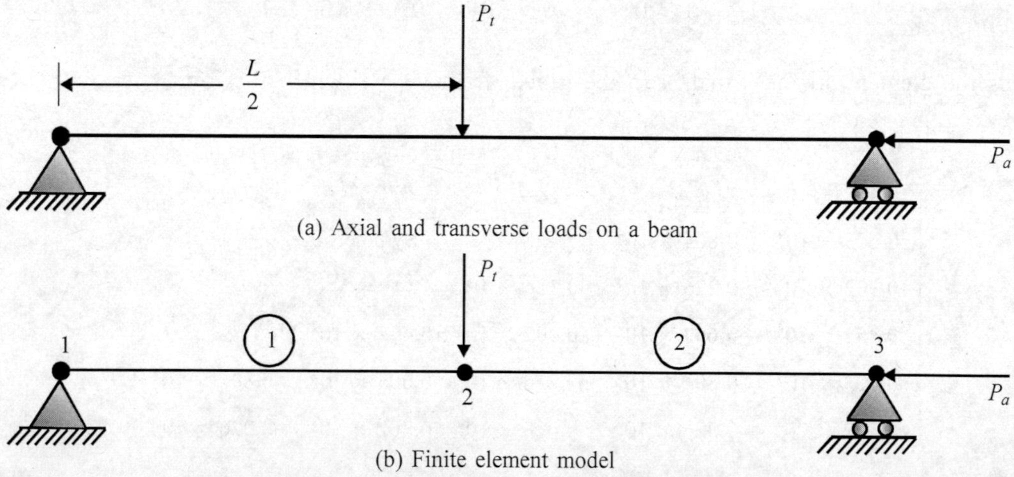

Fig. 4.24   Beam under combined loads.

just two elements. For simplicity, let $EI = 1$, $L = 2$, $AE = 1$. Incorporating the boundary conditions at nodes 1 and 3, we get the following equations for the unknown nodal d.o.f.:

$$
\begin{bmatrix}
4 & 0 & -6 & 2 & 0 & 0 \\
 & 1 & 0 & 0 & -1 & 0 \\
 & & 24 & 0 & 0 & 6 \\
 & & & 8 & 0 & 2 \\
 & & & & 1 & 0 \\
\text{Symmetric} & & & & & 4
\end{bmatrix}
\begin{Bmatrix}
\theta_1 \\ u_2 \\ v_2 \\ \theta_2 \\ u_3 \\ \theta_3
\end{Bmatrix}
=
\begin{Bmatrix}
0 \\ 0 \\ -P_t \\ 0 \\ -P_a \\ 0
\end{Bmatrix}
\tag{4.150}
$$

We observe that the solution of these equations is the same as the solution of the following equations separately:

$$
\begin{bmatrix}
4 & -6 & 2 & 0 \\
 & 24 & 0 & 6 \\
 & & 8 & 2 \\
\text{Symmetric} & & & 4
\end{bmatrix}
\begin{Bmatrix}
\theta_1 \\ v_2 \\ \theta_2 \\ \theta_3
\end{Bmatrix}
=
\begin{Bmatrix}
0 \\ -P_t \\ 0 \\ 0
\end{Bmatrix}
\tag{4.151}
$$

$$
\begin{bmatrix}
1 & -1 \\
\text{Symmetric} & 1
\end{bmatrix}
\begin{Bmatrix}
u_2 \\ u_3
\end{Bmatrix}
=
\begin{Bmatrix}
0 \\ -P_a
\end{Bmatrix}
\tag{4.152}
$$

We observe that our solution is the same as the one we would have obtained if the loads were individually applied. Thus there is no effect of the axial force on the transverse deflection in our finite element solution whereas the exact solution for this problem is given as

$$
V_{\max} = V\left(\frac{L}{2}\right) = \left(\frac{P_t L^3}{48EI}\right)\left(\frac{\tan\beta - \beta}{\frac{1}{3}\beta^3}\right)
\tag{4.153}
$$

where

$$
\beta^2 = \frac{P_a L^2}{4EI}
\tag{4.154}
$$

This is to be expected since in Eq. (4.144) there is no coupling between axial and transverse deflections (note the zeroes in the stiffness matrix $[k]^e$). More fundamentally, we have simply placed together the individual bar and beam elements and not really formulated a frame element starting from first principles. Therefore, we should have started with an appropriate expression for strain energy with the nonlinear coupling explicitly taken into account. Thus, even though our frame element has the nodal d.o.f. u, v and θ, it cannot model phenomena such as buckling or stiffening due to centrifugal forces on a rotating beam (such as a fan blade).

> ## Conclusion
>
> The finite element solution cannot capture any more physics of the problem than has been taken into account in the functional or the weak form.

We have thus far studied various aspects of one-dimensional finite element analysis. Though we used problems from the field of structural mechanics for the purpose of illustration, many of the key ideas are equally applicable to other fields too. We will now study some heat transfer problems. In Section 2.7, we studied the one dimensional heat conduction problem. We will extend it to include convection also. We will use the weak form approach, which is perhaps more pertinent for such problems.

4.7   One-dimensional Heat Transfer

The governing differential equation for the steady state one-dimensional conduction heat transfer with convective heat loss from lateral surfaces is given by

$$k\frac{d^2T}{dx^2} + q = \left(\frac{P}{A_c}\right)h(T - T_\infty) \tag{4.155}$$

where

$k =$ coefficient of thermal conductivity of the material,
$T =$ temperature,
$q =$ internal heat source per unit volume,
$P =$ perimeter,
$A_c =$ the cross-sectional area,
$h =$ convective heat transfer coefficient, and
$T_\infty =$ ambient temperature.

The weighted residual statement can be written as

$$\int_0^L W\left(k\frac{d^2T}{dx^2} + q - \left(\frac{P}{A_c}\right)h(T - T_\infty)\right) dx = 0 \tag{4.156}$$

By performing integration by parts, the weak form of the differential equation can be obtained as

$$\left[Wk\frac{dT}{dx}\right]_0^L - \int_0^L k\frac{dW}{dx}\frac{dT}{dx}\,dx + \int_0^L Wq\,dx - \int_0^L W\left(\frac{P}{A_c}\right)h(T - T_\infty)\,dx = 0 \tag{4.157}$$

i.e.

$$\int_0^L k\frac{dW}{dx}\frac{dT}{dx}\,dx + \int_0^L W\left(\frac{P}{A_c}\right)hT\,dx = \int_0^L Wq\,dx + \int_0^L W\frac{P}{A_c}h(T_\infty)\,dx + \left[Wk\frac{dT}{dx}\right]_0^L \tag{4.158}$$

The weak form, for a typical mesh of n finite elements, can be written as

$$\sum_{k=1}^{n} \left[\int_0^{\ell} k \frac{dW}{dx} \frac{dT}{dx}\, dx + \int_0^{\ell} W \frac{P}{A_c} hT\, dx \right] = \sum_{k=1}^{n} \left[\int_0^{\ell} W \left(q + \frac{P}{A_c} hT_\infty \right) dx + \left[Wk \frac{dT}{dx} \right]_0^{\ell} \right] \quad (4.159)$$

The attention of the reader is drawn to the similarity of the weak form in the above equation and that in Eq. (2.125) for the bar problem. We now propose to develop the one-dimensional heat transfer element following the procedure we used while developing the equations for the bar element. Figure 4.25 shows a typical heat transfer element, with its nodes

Fig. 4.25 Typical heat transfer element.

and nodal d.o.f. Following the similarity of the bar and heat transfer element, we simply use the same shape functions for interpolating the temperature within the element from the nodal temperature.

$$T(x) = \left(1 - \frac{x}{\ell} \right) T_1 + \left(\frac{x}{\ell} \right) T_2 \quad (4.160)$$

$$\frac{dT}{dx} = \frac{T_2 - T_1}{\ell} \quad (4.161)$$

$$W_1 = 1 - \frac{x}{\ell}, \qquad \frac{dW_1}{dx} = -\frac{1}{\ell}$$

$$W_2 = \frac{x}{\ell}, \qquad \frac{dW_2}{dx} = \frac{1}{\ell} \quad (4.162)$$

We now compute the element level contributions to Eq. (4.159) above.

LHS first term (Eq. (4.159)). Similar to the LHS of bar element (Eq. (2.125)), k replaces AE and T replaces u. Thus we have

$$\frac{k}{\ell} \begin{bmatrix} 1 & -1 \\ -1 & 1 \end{bmatrix} \begin{Bmatrix} T_1 \\ T_2 \end{Bmatrix}$$

LHS second term (Eq. (4.159)). With W_1,

$$\int_0^{\ell} \left(1 - \frac{x}{\ell} \right) \left(\frac{P}{A_c} \right) h \left[\left(1 - \frac{x}{\ell} \right) T_1 + \left(\frac{x}{\ell} \right) T_2 \right] dx$$

$$= \left(\frac{Ph}{A_c} \right) \left[\int_0^{\ell} \left(1 - \frac{x}{\ell} \right)^2 dx\, T_1 + \int_0^{\ell} \left(\frac{x}{\ell} \right) \left(1 - \frac{x}{\ell} \right) T_2 \right] dx$$

$$= \frac{Phl}{6A_c} \left[2T_1 + T_2 \right]$$

With W_2,

$$\int_0^\ell \left(\frac{x}{\ell}\right)\left(\frac{P}{A_c}\right)h\left[\left(1-\frac{x}{\ell}\right)T_1 + \frac{x}{\ell}T_2\right]dx = \frac{Phl}{6A_c}\left[T_1 + 2T_2\right]$$

We can now write the second term on the LHS in matrix notation as

$$\frac{Phl}{6A_c}\begin{bmatrix} 2 & 1 \\ 1 & 2 \end{bmatrix}\begin{Bmatrix} T_1 \\ T_2 \end{Bmatrix}$$

RHS first term.   Similar to the RHS first term of the bar element equation replacing q_0

by $\left(q_0 + \dfrac{PhT_\infty}{A_c}\right)$, we can write

$$\left(q_0 + \frac{P}{A_c}hT_\infty\right)\begin{Bmatrix} \ell/2 \\ \ell/2 \end{Bmatrix}$$

RHS second term. Similar to RHS second term of the bar element (Eq. (2.125), we have

$$\begin{Bmatrix} -Q_0 \\ Q_l \end{Bmatrix}$$

where Q_0, Q_ℓ represent the heat flux at the ends of the element (nodes). Thus the element level equations can be written as

$$\left(\frac{k}{\ell}\begin{bmatrix} 1 & -1 \\ -1 & 1 \end{bmatrix} + \frac{Phl}{6A_c}\begin{bmatrix} 2 & 1 \\ 1 & 2 \end{bmatrix}\right)\begin{Bmatrix} T_1 \\ T_2 \end{Bmatrix} = \left(q_0 + \frac{Ph}{A_c}T_\infty\right)\begin{Bmatrix} \ell/2 \\ \ell/2 \end{Bmatrix} + \begin{Bmatrix} -Q_0 \\ Q_l \end{Bmatrix} \tag{4.163}$$

The LHS may be considered to be the element conductance matrix, and the first term on the RHS is the equivalent nodal heat flux corresponding to the distributed heat source/sink. The second vector on the RHS refers to the net heat flux at the nodes and permits us to incorporate prescribed natural boundary conditions. When the individual elements are assembled together, these Q's algebraically sum up to the resultant externally applied heat flux at the section. Comparing Eq. (2.163) and (4.163), we observe that the latter contains certain additional terms corresponding to convention.

We now illustrate the use of this element through the following example.

Example 4.10.   *Temperature distribution in a pin-fin.* Consider a 1 mm diameter, 50 mm long aluminium pin-fin as shown in Figure 4.26 used to enhance the heat transfer from a surface wall maintained at 300°C. Use $k = 200$ W/m/°C for aluminium, $h = 20$ W/m^2 C, $T_\infty = 30$°C. We obtained the temperature distribution in the fin using the Galerkin weighted residual method in Example 2.8. We will now estimate the temperature in the pin using one, two, and four equal size elements, assuming that the tip of the fin is insulated. There is no internal heat source.

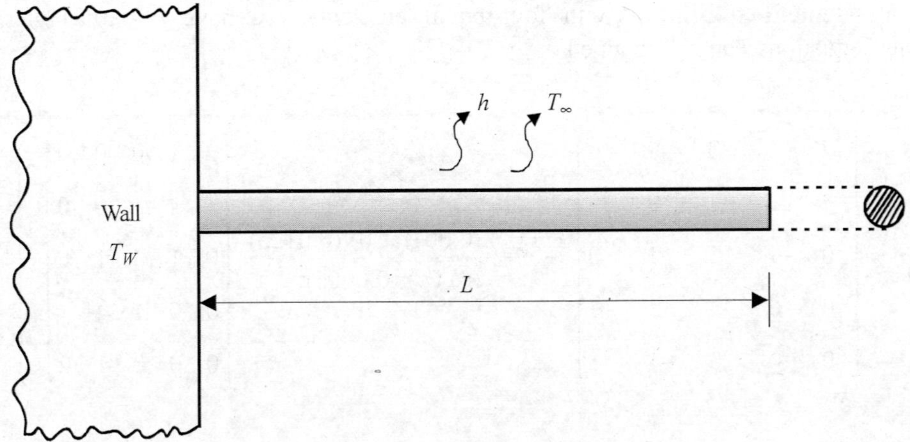

Fig. 4.26 A pin-fin.

One-element solution. Substituting the numerical values in Eq. (4.163), we get the element level equations as

$$\left(\frac{200}{0.05} \begin{bmatrix} 1 & -1 \\ -1 & 1 \end{bmatrix} + \frac{(\pi)(0.001)(20)(0.05)}{(6)(\pi)(0.0005)^2} \begin{bmatrix} 2 & 1 \\ 1 & 2 \end{bmatrix} \right) \begin{Bmatrix} T_1 \\ T_2 \end{Bmatrix}$$

$$= \frac{(\pi)(0.001)(20)}{(\pi)(0.0005)^2}(30) \begin{Bmatrix} 0.025 \\ 0.025 \end{Bmatrix} + \begin{Bmatrix} Q_{\text{wall}} \\ Q_{\text{tip}} \end{Bmatrix} \tag{4.164}$$

From the boundary conditions, we have $T_1 = 300°C$, $Q_{\text{tip}} = 0$. Thus we get $T_2 = 198.75°C$.

Two-element solution.   With two equal elements, we have $\ell = 0.025$ m. Assembling the two-element level equations in the same way as we did for the two bar elements, we get the system level equations as

$$\left(\frac{200}{0.025} \begin{bmatrix} 1 & -1 & 0 \\ -1 & 2 & -1 \\ 0 & -1 & 1 \end{bmatrix} + \frac{(\pi)(0.001)(20)(0.025)}{(6)(\pi)(0.0005)^2} \begin{bmatrix} 2 & 1 & 0 \\ 1 & 4 & 1 \\ 0 & 1 & 2 \end{bmatrix} \right) \begin{Bmatrix} T_1 \\ T_2 \\ T_3 \end{Bmatrix}$$

$$= \frac{(\pi)(0.001)(20)}{(\pi)(0.0005)^2}(30) \begin{Bmatrix} 0.0125 \\ 0.025 \\ 0.0125 \end{Bmatrix} + \begin{Bmatrix} Q_{\text{wall}} \\ 0 \\ Q_{\text{tip}} \end{Bmatrix} \tag{4.165}$$

From the boundary conditions we have $T_1 = 300°C$, $Q_{\text{tip}} = 0$. Substituting and solving the equations, we get

$$T_2 = 226.185°C, \qquad T_3 = 203.548°C \tag{4.166}$$

Four-element solution. With four equal elements, we have $\ell = 0.0125$ m. The assembled equations can be obtained as

$$
\left(\frac{200}{0.0125}
\begin{bmatrix}
1 & -1 & 0 & 0 & 0 \\
-1 & 2 & -1 & 0 & 0 \\
0 & -1 & 2 & -1 & 0 \\
0 & 0 & -1 & 2 & -1 \\
0 & 0 & 0 & -1 & 1
\end{bmatrix}
+ \frac{(\pi)\,(0.001)\,(20)\,(0.0125)}{(6)\,(\pi)\,(0.0005)^2}
\begin{bmatrix}
2 & 1 & 0 & 0 & 0 \\
1 & 4 & 1 & 0 & 0 \\
0 & 1 & 4 & 1 & 0 \\
0 & 0 & 1 & 4 & 1 \\
0 & 0 & 0 & 1 & 2
\end{bmatrix}
\right)
\begin{Bmatrix}
T_1 \\ T_2 \\ T_3 \\ T_4 \\ T_5
\end{Bmatrix}
$$

$$
= \frac{(\pi)\,(0.001)\,(20)}{(\pi)\,(0.0005)^2}\,(30)
\begin{Bmatrix}
0.00625 \\ 0.0125 \\ 0.0125 \\ 0.0125 \\ 0.00625
\end{Bmatrix}
+
\begin{Bmatrix}
Q_{\text{wall}} \\ 0 \\ 0 \\ 0 \\ Q_{\text{tip}}
\end{Bmatrix}
\tag{4.167}
$$

Substituting the boundary conditions and solving the equations, we get

$$
\begin{aligned}
T_2 &= 256.37°\text{C}, \qquad T_3 = 227.03°\text{C} \\
T_4 &= 210.14°\text{C}, \qquad T_5 = 204.63°\text{C}
\end{aligned}
\tag{4.168}
$$

The exact solution for this problem is given by

$$
T(x) = T_\infty + (T_w - T_\infty)\left[\frac{\cosh\,m(L-x)}{\cosh\,mL} \right]
\tag{4.169}
$$

where $m = \sqrt{\dfrac{hP}{kA_c}}$.

Figure 4.27 compares the exact solution with the various finite element solutions. It is observed that the exact solution is an exponential function while the finite element permits only linear variation in temperature within the element. The finite element approximation improves with increasing number of elements. A 10-element solution is shown to match very closely with the exact solution in Figure 4.28.

As observed with respect to the problem of a rod under self-weight (Example 4.1 and 4.2), we expect that fewer number of higher order elements, e.g. quadratic heat transfer element, may give much better solution than the linear element used here.

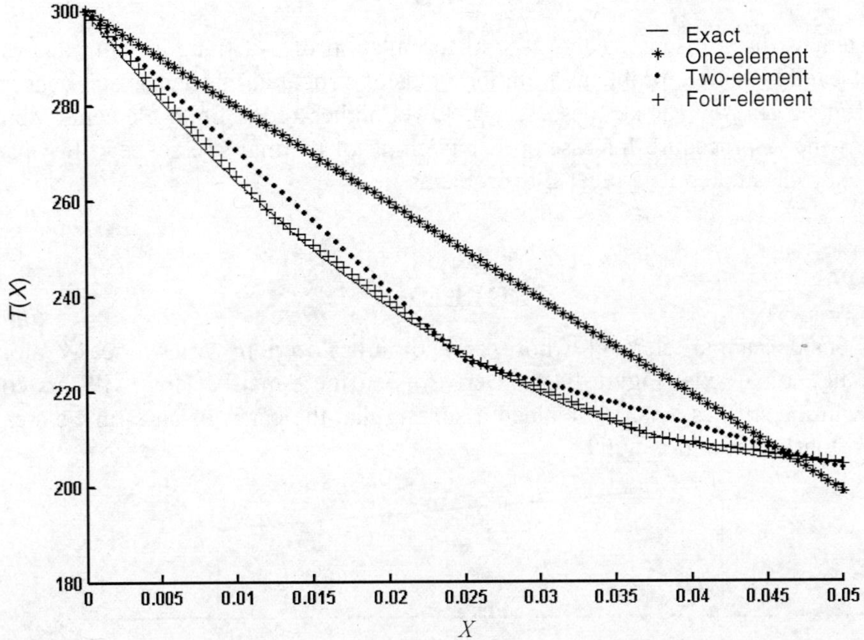

Fig. 4.27 Comparison of exact solution with finite element solutions (Example 4.10).

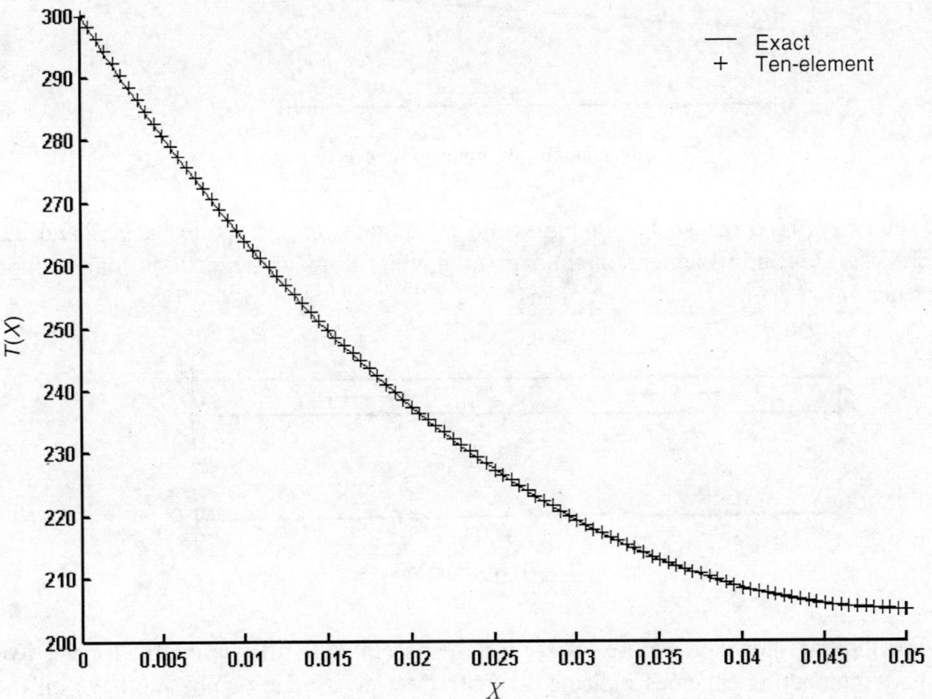

Fig. 4.28 Comparison of exact solution with 10-element solution (Example 4.10).

SUMMARY

In this chapter, we have discussed the general formulation of 1-d finite element analysis. We have solved several example problems from the fields of structural mechanics and heat transfer. We have brought out several key issues, e.g. lower/higher order finite elements, consistent nodal loads, generic procedure for assembly, modelling of thermal stresses, etc. In Chapter 5, we will extend our studies to typical 2-d problems.

PROBLEMS

4.1 Consider a bar element whose area of cross-section varies linearly along the longitudinal axis (Figure P4.1). Derive its stiffness matrix. How will this compare with the stiffness matrix obtained assuming that the bar is of uniform c/s area equal to that at its mid-length?

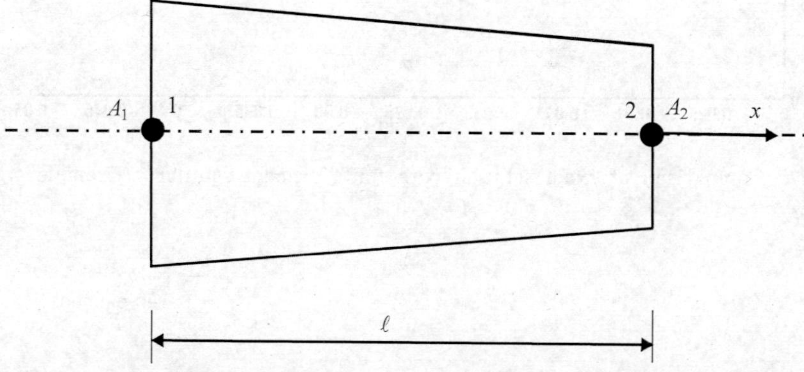

Fig. P4.1 Bar element with varying cross-section.

4.2 The fixed-fixed bar shown in Figure P4.2 has axial forces applied at $L/3$ and $2L/3$. Use the finite element method to compute the axial deflections and support reactions.

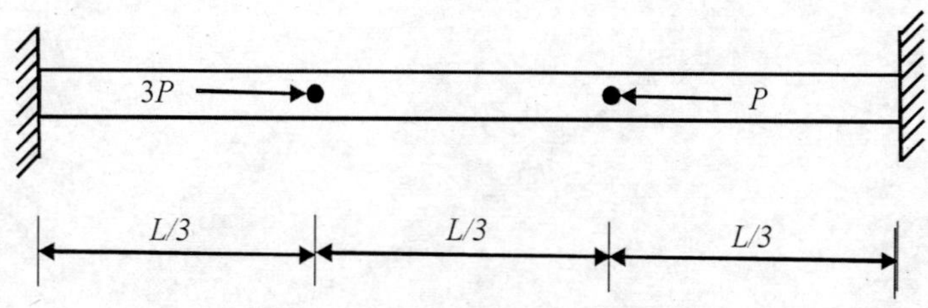

Fig. P4.2 Fixed-fixed bar.

4.3 For the bar shown in Figure P4.3, give the assembled stiffness matrix for the two node numbering schemes indicated. What effect does node numbering have on the assembled matrix? Area of c/s = A; Young's modulus E.

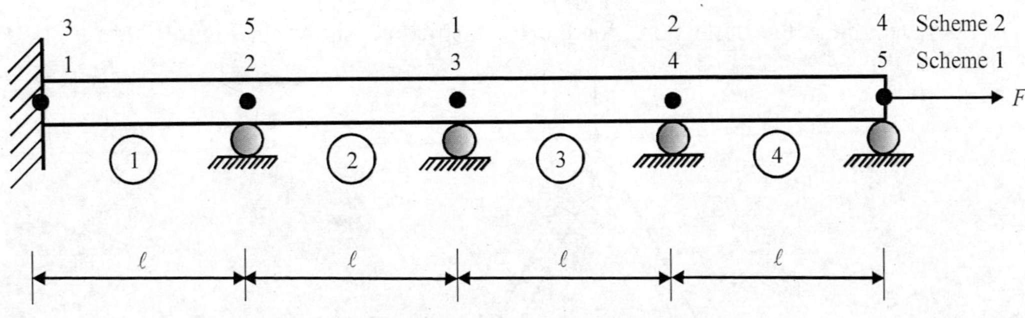

Fig. P4.3   Effect of node numbering.

4.4 Consider the spring mounted bar shown in Figure P4.4. Solve for the displacements of points P and Q using bar elements (assume AE = constant).

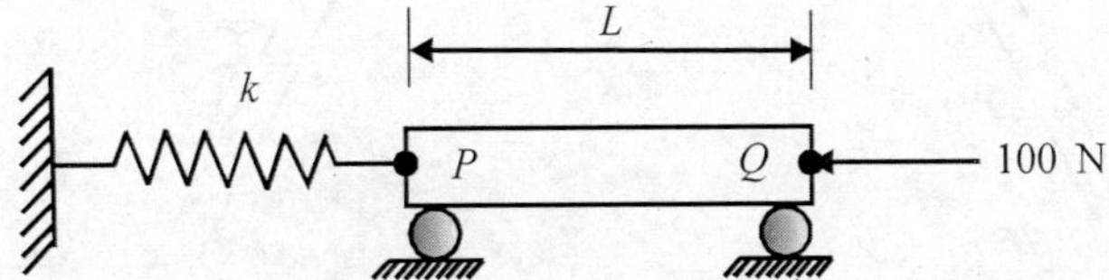

Fig. P4.4   Spring mounted bar.

4.5 The planar structure shown in Figure P4.5 is hinged at the supports A and B. The elements of the truss have identical cross-sectional properties. Find out the displacement of the node C under the action of the given loading.

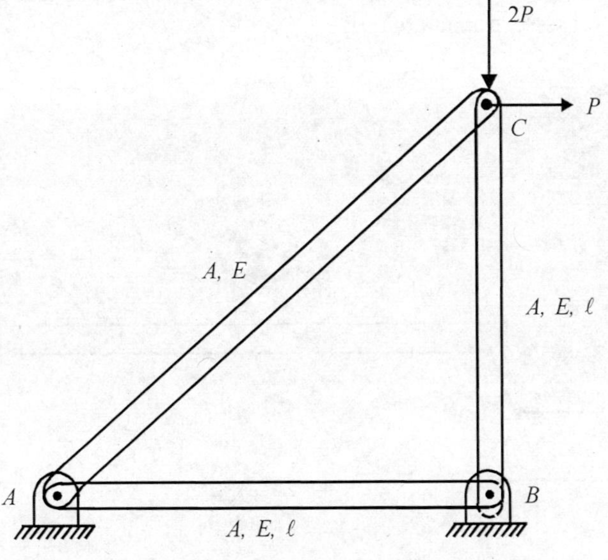

Fig. P4.5   A planar structure.

4.6   Determine the deflections for the truss structure shown in Figure P4.6.

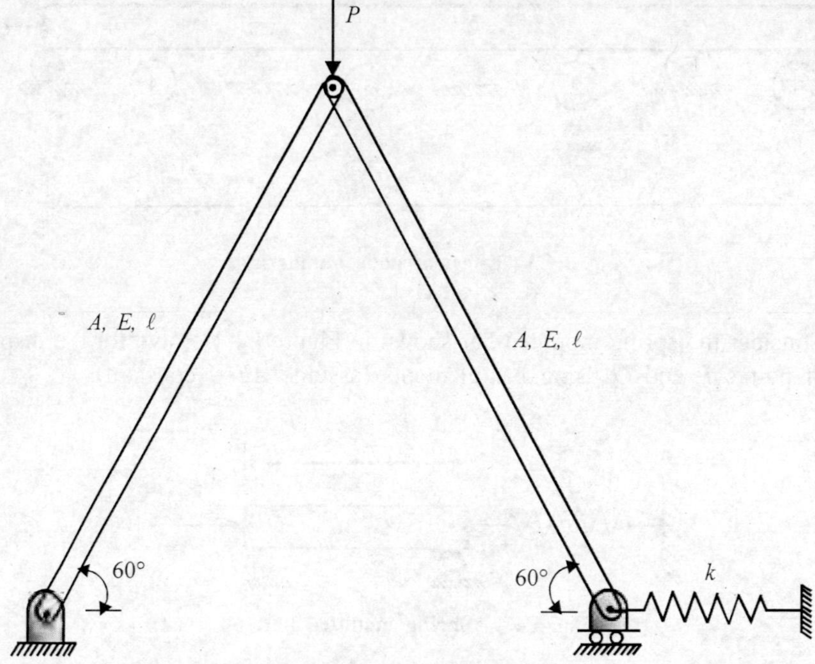

Fig. P4.6   A truss structure.

4.7   Consider a rod rotating about a fixed axis as shown in Figure P4.7 at a constant speed ω rad/s. Derive the element nodal force vector due to the centrifugal force field. Consider both linear and quadratic bar elements. Let the density be ρ and uniform cross-sectional area A.

Fig. P4.7   Rotating rod.

4.8 Consider a truss element arbitrarily oriented in space. If the coordinates of the two ends of an element are given by (x_1, y_1, z_1) and (x_2, y_2, z_2), we can get the direction cosines of the line vector as

$$\ell = \frac{\Delta x}{\sqrt{(\Delta x)^2 + (\Delta y)^2 + (\Delta z)^2}}, \qquad m = \frac{\Delta y}{\sqrt{(\Delta x)^2 + (\Delta y)^2 + (\Delta z)^2}}$$

$$n = \frac{\Delta z}{\sqrt{(\Delta x)^2 + (\Delta y)^2 + (\Delta z)^2}}$$

where $\Delta x = x_2 - x_1$; $\Delta y = y_2 - y_1$; $\Delta z = z_2 - z_1$. Show that the transformation matrix $[T]$ relating the nodal displacements in local and global coordinate frames is given by

$$[T] = \begin{bmatrix} \ell & m & n & 0 & 0 & 0 \\ 0 & 0 & 0 & \ell & m & n \end{bmatrix}$$

Derive the element stiffness matrix in the global frame XYZ.

4.9 For a uniform circular c/s bar (diameter d, length L, modulus of rigidity G) under twist, determine the stiffness equations, i.e.

$$\begin{Bmatrix} T_1 \\ T_2 \end{Bmatrix} = \begin{bmatrix} k_{11} & k_{12} \\ k_{21} & k_{22} \end{bmatrix} \begin{Bmatrix} \phi_1 \\ \phi_2 \end{Bmatrix}$$

Use this to solve for the angle of twist ϕ at points A and B in Figure P4.9

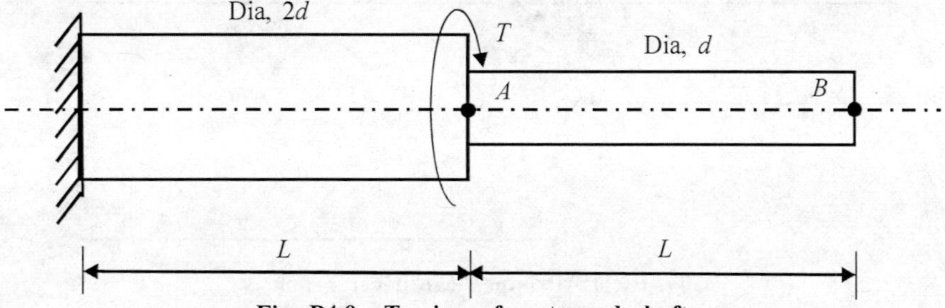

Fig. P4.9   Torsion of a stepped shaft.

4.10 A crude approximation of a dam is shown in Figure P4.10. Using one beam element and consistent nodal load vector, find the tip deflection. Using this result find the deflection of the point P.

4.11 For the beam problem shown in Figure P4.11 determine the tip deflection and slope at the roller support.

4.12 A beam element is loaded by a moment at the mid-span. Find the consistent nodal force vector. How would the result change if the moment were applied at one-third the span rather than at the centre.

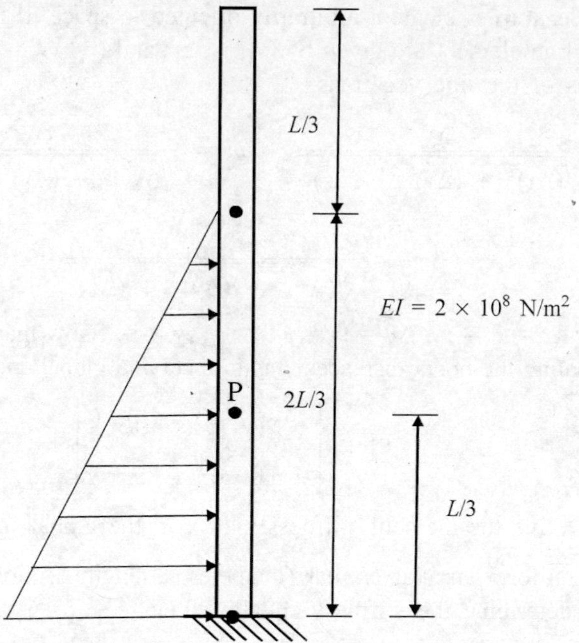

Fig. P4.10   A cantilever beam under hydrostatic load.

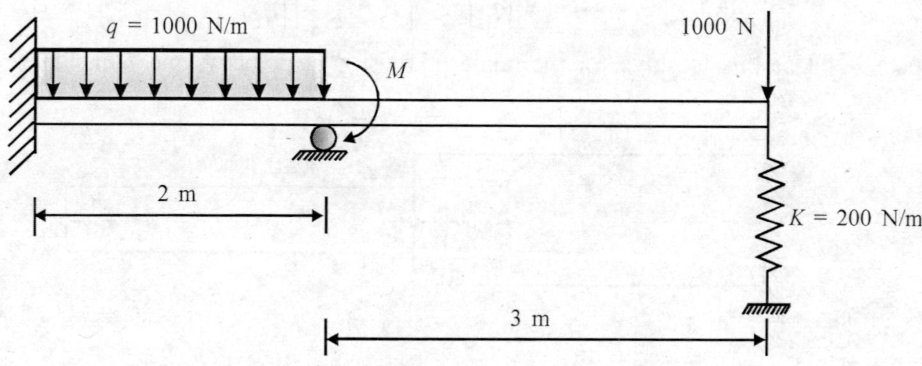

Fig. P4.11   Propped cantilever beam.

4.13 Describe how you would implement the boundary condition for the beam shown in Figure P4.13.

4.14 For the discretisation of beam elements shown in Figure P4.14, number the nodes so as to minimise the bandwidth of the assembled stiffness matrix $[K]$.

4.15 The elements of a row or column of the stiffness matrix of a bar element sum up to zero, but not so for a beam element. Explain why this is so.

4.16 The plane wall shown in Figure P4.16 is 0.5 m thick. The left surface of the wall is maintained at a constant temperature of 200°C, and the right surface is insulated.

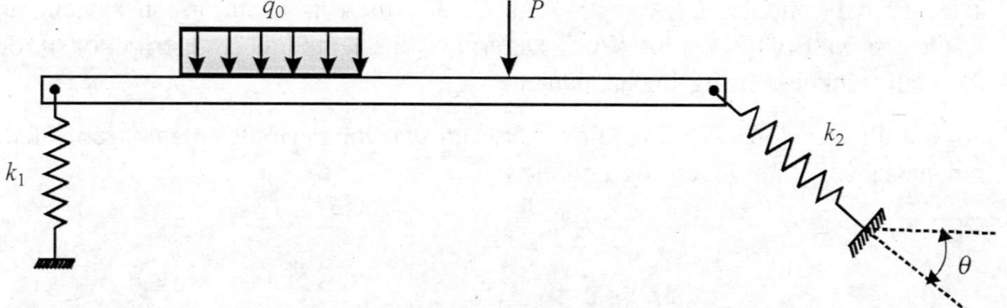

Fig. P4.13   Beam on spring supports.

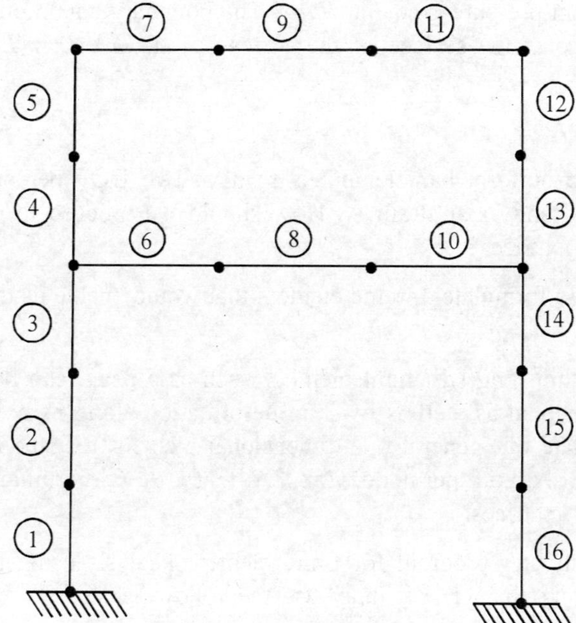

Fig. P4.14   Node numbering.

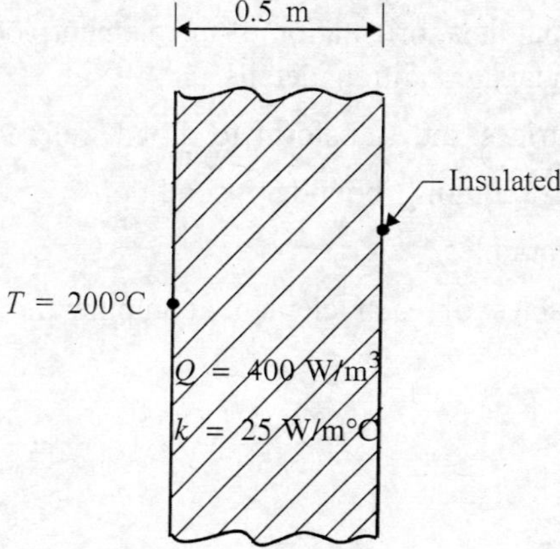

Fig. P4.16   Heat conduction in a slab.

The thermal conductivity $k = 25$ W/m°C, and there is a uniform heat generation inside the wall of $Q = 400$ W/m^3. Determine the temperature distribution through the wall thickness using linear elements.

4.17 For a fully developed laminar flow between two long parallel plates (reparated by a distance 2L), the governing equation is

$$\mu \frac{d^2u}{dy^2} = \frac{dP}{dx}$$

where μ is the viscosity of the fluid.

Derive the finite element form and hence determine the velocity distribution u_y for a given constant pressure gradient dP/dx. The boundary conditions are that the fluid velocity is zero at the surface of the plates i.e. $u(-L) = 0$; $u(L) = 0$.

Problems to Investigate

4.18 Imagine that a uniform bar element is to have two DOF per node, namely, axial displacement u and axial strain ε_x. How should we proceed to derive the element characteristic matrices?

4.19 How would you formulate a frame element that would enable us to model a buckling problem?

4.20 In general, a thin long structural member such as a beam can bend in two planes, undergo extensional as well as twisting deformation along its axis. Thus we would like to formulate our general one-dimensional element for structural analysis with two nodes and six d.o.f. per node, viz., u, v, w, θ_x, θ_y, θ_z. Formulate such an element and derive its matrices.

4.21 Develop a computer program for finite element analysis using the finite elements discussed in this chapter. You may use the following steps:

- Allow the user to choose the type of analysis (structural/heat transfer, etc.) and the type of element (truss/beam/frame, etc., linear/quadratic, and so on).

- Accept appropriate input data, orientation of the element, local and global node number correspondence, boundary condition details, etc.

- Generate element matrices and transform to global reference frame.

- Perform assembly incorporating boundary conditions.

- Solve for unknown variables.

- From the nodal variables of each element, compute the quantities of interest, e.g. stresses, heat flux.

CHAPTER 5

Two-dimensional Finite Element Analysis

5.1   Introduction—Dimensionality of a Problem

In Chapter 4 we discussed one-dimensional finite element analysis in general and applied the technique to problems of structural mechanics and heat transfer. We formulated the truss, torsional bar and beam elements and then combined all of them into a general 1-d element for structural analysis. We discussed the general one-dimensional heat transfer element with conduction as well as convection. While we solved many example problems, we did not emphasise how a real-life problem gets itself posed as a one-dimensional problem in the first place. We now briefly discuss this important issue.

Let us consider a space frame or space truss structure with arbitrary configuration in 3-d space. We consider this to be a typical one-dimensional problem. The nodal d.o.f. vary only along one direction, viz., along the length of the element (say 'x'). In view of the fact that the members are long and slender, we can make suitable assumptions on the variation of displacements in the cross-sectional plane. For example, for a slender bar of uniform cross-section, we can make a reasonable assumption that the deflection varies in a known way across the cross-section (e.g. uniform, linear). Many a time, for complex real-life structures such as a gear tooth or a thick pressure vessel or a crank shaft, the unknown field variables vary **independently** along two or even three directions and we cannot make any suitable assumptions about their variation. For a block of material of comparable dimensions, for example, the heat transfer is more closely approximated as 3-d rather than as 1-d or 2-d. Thus we seek to determine the entire field of variation of the unknown variable using the Finite element method, and such problems are said to be two- or three-dimensional problems. While every real-life problem can be modelled as a three-dimensional problem, it may be computationally very expensive to perform such an analysis. In a design situation, typically, several iterations are carried out before a final design is frozen. Thus, repeated 3-d analysis can be very cumbersome and we look for simpler but reasonably accurate one/two-dimensional models. The detailed discussion of three-dimensional finite element analysis is beyond the scope of this text. We will discuss in detail several two-dimensional elements and related issues in this chapter.

A structural mechanics problem can be considered to be two dimensional if we are able to make reasonable assumption about the variation of unknown parameters in the third dimension. Three practically important situations of this type, arising in structural mechanics, are shown in Figure 5.1. In the first case [Fig. 5.1a], the structure is very thin as compared

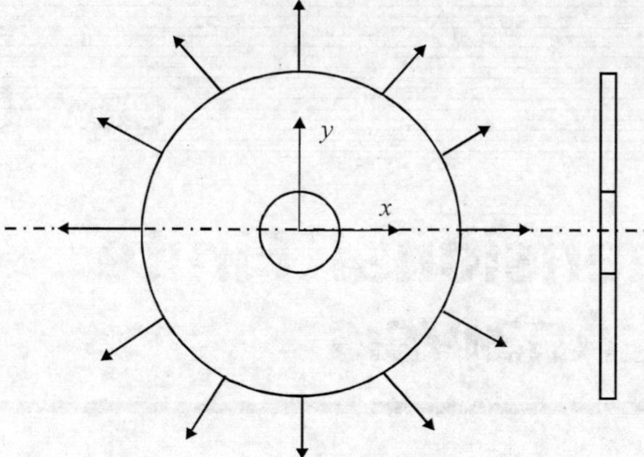

(a) Thin disk under plane stress

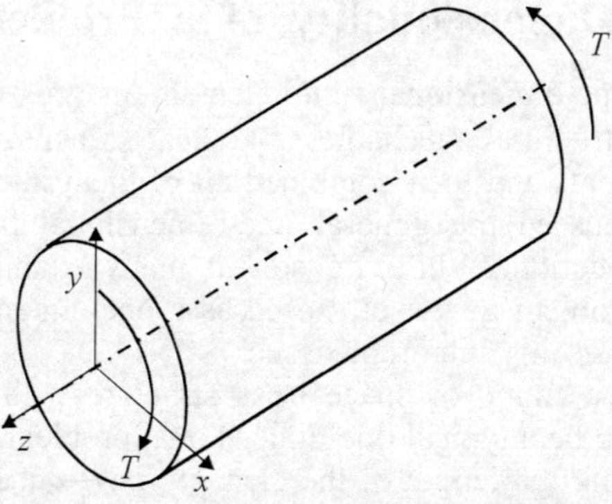

(b) Long, prismatic shaft under plane strain

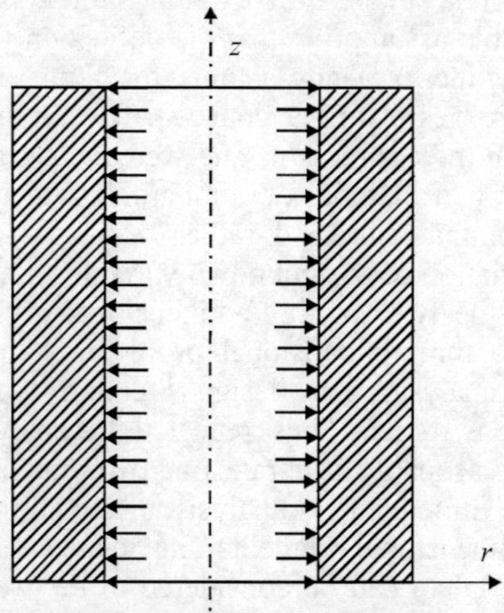

(c) Thick cylinder under axisymmetric loads

Fig. 5.1   Typical 2-d problems in stress analysis.

to its other two dimensions and is loaded in its own plane, e.g. a rotating impeller wheel. Thus the top and bottom faces are free from any normal stress σ_z and since the thickness is small, we assume σ_z is zero throughout. So also the shearing stresses τ_{xz}, τ_{yz}. Thus the only nonzero stresses are σ_x, σ_y, τ_{xy}. The stress-strain constitutive relation for a linear, elastic, isotropic material under such plane-stress condition is given by

$$\begin{Bmatrix} \sigma_x \\ \sigma_y \\ \tau_{xy} \end{Bmatrix} = \frac{E}{1-v^2} \begin{bmatrix} 1 & v & 0 \\ v & 1 & 0 \\ 0 & 0 & \dfrac{1-v}{2} \end{bmatrix} \left(\begin{Bmatrix} \varepsilon_x \\ \varepsilon_y \\ \gamma_{xy} \end{Bmatrix} - \{\varepsilon\}^0 \right) \tag{5.1}$$

where $\{\varepsilon\}^0$ represents an initial strain.

The problem situation depicted in the second case (Fig. 5.1b) is the other extreme of the first case, where the length of the member is so long that it is reasonable to assume the longitudinal strain ε_z to be zero, e.g. torsion of a long uniform shaft. So also the shearing strains γ_{xz}, γ_{yz}. Thus the only nonzero strains are ε_x, ε_y, γ_{xy}. The stress-strain constitutive relation for a linear, elastic, isotropic material under such plane-strain condition is given by

$$\begin{Bmatrix} \sigma_x \\ \sigma_y \\ \tau_{xy} \end{Bmatrix} = \frac{E}{(1+v)(1-2v)} \begin{bmatrix} 1-v & v & 0 \\ v & 1-v & 0 \\ 0 & 0 & \dfrac{1-2v}{2} \end{bmatrix} \left(\begin{Bmatrix} \varepsilon_x \\ \varepsilon_y \\ \gamma_{xy} \end{Bmatrix} - \{\varepsilon\}^0 \right) \tag{5.2}$$

It is to be observed that the normal stress σ_z is nonzero but can be readily obtained from the condition that the longitudinal strain ε_z be zero.

The problem situation depicted in the third case (Fig. 5.1c) represents an important class of problems, commonly described as *axisymmetric problems*, e.g. pressure vessels. If the *geometry of the structure, material properties, support conditions and loading are all axisymmetric*, then it is reasonable to assume that the resulting deformation will also be axisymmetric, i.e., it does not vary along the circumferential direction. Thus, it is sufficient to determine the variation of the unknown field variable (e.g. deflection, temperature) in any one plane as shown in the figure. In practical pressure vessels, there may be minor geometrical features which cause deviation from perfect axial symmetry such as helical screw threads on an end-cap. It may be worthwhile to ignore the small helix angle and model the problem as purely 2-d axisymmetric. However, sometimes a perfect axisymmetric structure may be subjected to nonaxisymmetric loading such as earthquake loading. It may still be useful to modify a two-dimensional axisymmetric analysis rather than going for a full-fledged three-dimensional analysis. The non-axisymmetric loading, expressed as a function of circumferential coordinate θ, has a periodicity 2π. It can be expressed as a Fourier series. To each harmonic of load, the resulting deflection also varies as a sine or cosine function and it can still be analysed as a two-dimensional problem. The Fourier harmonics of deflection are superposed to find the total response. This technique is known as semi-analytical FEM. The nonzero stresses and strains in a perfect axisymmetric situation are related as follows:

$$\begin{Bmatrix} \sigma_r \\ \sigma_\theta \\ \sigma_z \\ \tau_{rz} \end{Bmatrix} = \frac{(1-v)E}{(1+v)(1-2v)} \begin{bmatrix} 1 & \dfrac{v}{1-v} & \dfrac{v}{1-v} & 0 \\ & 1 & \dfrac{v}{1-v} & 0 \\ & & 1 & 0 \\ \text{Sysmmetric} & & & \dfrac{1-2v}{2(1-v)} \end{bmatrix} \left(\begin{Bmatrix} \varepsilon_r \\ \varepsilon_\theta \\ \varepsilon_z \\ \gamma_{rz} \end{Bmatrix} - \{\varepsilon\}^0 \right) \quad (5.3)$$

Sometimes, a one-dimensional problem may have to be modelled as a two-dimensional problem, e.g. a beam with a transverse load, and we wish to determine the stresses under the point of application of load. Thus, if we are interested in detailed modelling of local effects such as concentrated loads, and sudden changes in geometry (such as a stepped sheft), a one-dimensional system may have to be modelled using two-dimensional finite elements.

5.2 Approximation of Geometry and Field Variable

A typical two-dimensional problem domain is shown in Figure 5.2, where we depict a bracket with an internal hole subjected to loads. We immediately observe that unlike in one-dimensional

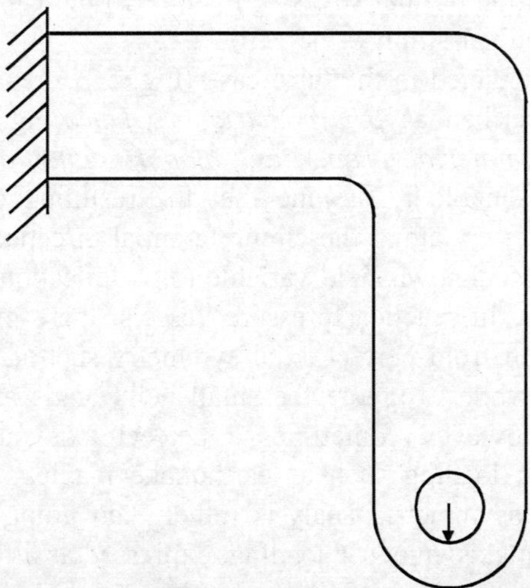

Fig. 5.2 Typical 2-d problems: Bracket with an internal hole.

finite element analysis given in Chapter 4, here we need to model the complex two-dimensional geometry of the system as well as the variation of the unknown field variable. To represent the two-dimensional geometry, we have to formulate two-dimensional elements, and most common elements consist of triangles and quadrilaterals as shown in Figure 5.3. Each node is capable of independent motion along the x and y axes and, typically, we have u and v as the independent nodal degree of freedom (d.o.f.) for plane elasticity problems. However, there is only one temperature d.o.f. for 2-d heat transfer problems. Simple lower order elements have straight sides and nodes only at the vertices. More complex higher order elements permit curved edges (so as to model curved geometric feature in the domain such as a circular hole) and mid-side or even internal nodes.

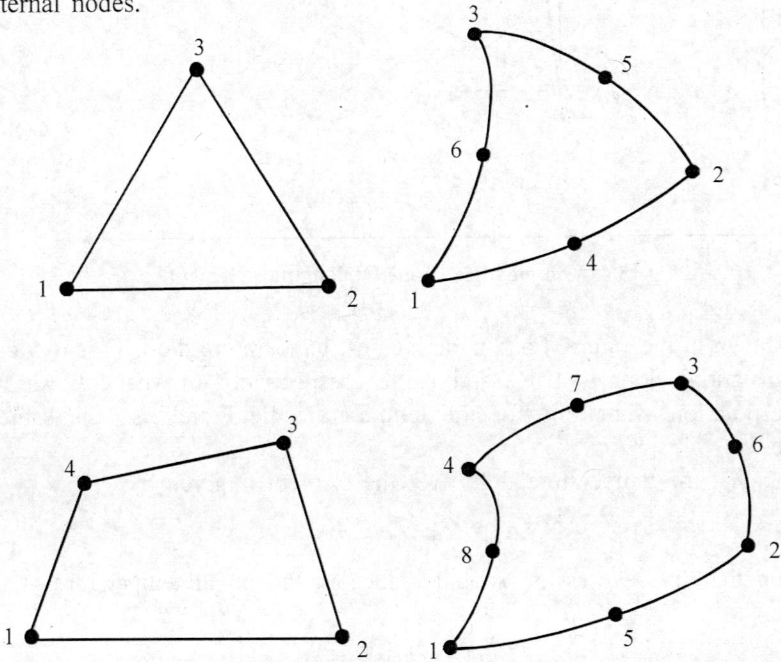

Fig. 5.3   Typical 2-d finite elements.

We will now begin our discussion of two-dimensional elements. We will first discuss a few typical elements and derive their shape functions. Subsequently, we will discuss the formulation of 2-d structural mechanics and heat transfer/fluid flow problems using these elements. We will now begin our discussion with the simplest element, viz., the three-noded triangular element. We will keep in mind that each node of the elements has just one d.o.f. (viz., temperature T) when applied to heat transfer problems and two displacements, u, v along the coordinate axes for structural mechanics problems. The flow of an ideal fluid in 2-d can be represented in terms of a scalar potential function and so we have only one d.o.f. per node.

5.2.1   Simple Three-noded Triangular Element

A typical element is shown in Figure 5.4. It has straight sides and three nodes, one at each vertex. The nodes have coordinates (x_1, y_1), (x_2, y_2), (x_3, y_3) in the global Cartesian coordinate

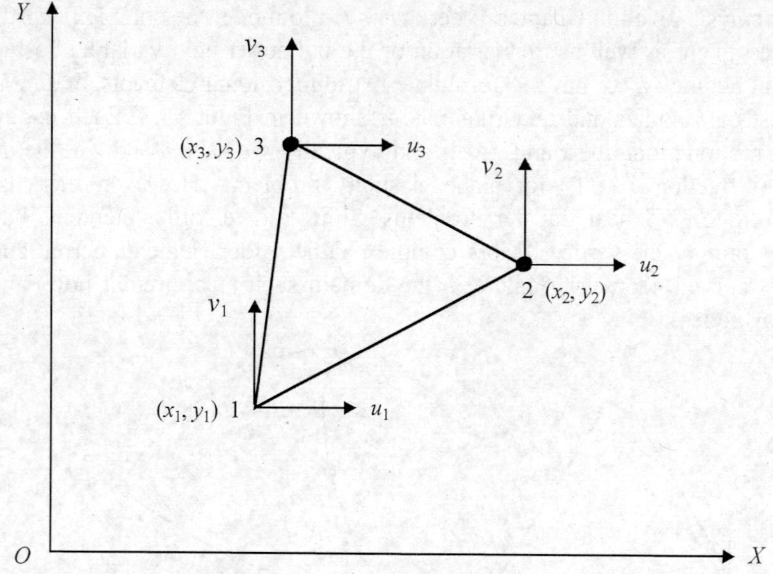

Fig. 5.4   Simple three-noded triangular element.

frame OXY as shown in the figure. Each node has just temperature d.o.f. T or two d.o.f., viz., u and v, the translations along global X and Y axes, respectively. In what follows, we derive the shape (interpolation) functions for the temperature field and use the same for the displacement field.

We assume that the temperature field over the element is given by

$$T(x, y) = c_0 + c_1 x + c_2 y \tag{5.4}$$

Considering that this expression should reduce to the nodal temperature at the nodal points, we have

$$\left. \begin{aligned} T_1 &= c_0 + c_1 x_1 + c_2 y_1 \\ T_2 &= c_0 + c_1 x_2 + c_2 y_2 \\ T_3 &= c_0 + c_1 x_3 + c_2 y_3 \end{aligned} \right\} \tag{5.5}$$

Solving for c_0, c_1, c_2, we get

$$\begin{Bmatrix} c_0 \\ c_1 \\ c_2 \end{Bmatrix} = \begin{bmatrix} 1 & x_1 & y_1 \\ 1 & x_2 & y_2 \\ 1 & x_3 & y_3 \end{bmatrix}^{-1} \begin{Bmatrix} T_1 \\ T_2 \\ T_3 \end{Bmatrix} \tag{5.6}$$

Substituting these in Eq. (5.5) and rewriting, we finally obtain

$$T(x, y) = \left(\frac{\alpha_1 + \beta_1 x + \gamma_1 y}{2\Delta} \right) T_1 + \left(\frac{\alpha_2 + \beta_2 x + \gamma_2 y}{2\Delta} \right) T_2 + \left(\frac{\alpha_3 + \beta_3 x + \gamma_3 y}{2\Delta} \right) T_3 \tag{5.7}$$

where

$$\alpha_1 = x_2 y_3 - x_3 y_2, \qquad \beta_1 = y_2 - y_3, \qquad \gamma_1 = x_3 - x_2 \qquad (5.8)$$

Other coefficients $(\alpha_2,\ \beta_2,\ \gamma_2)$ and $(\alpha_3,\ \beta_3,\ \gamma_3)$ can be obtained by a simple cyclic permutation of subscripts 1, 2, 3. Also,

$$2\Delta = \begin{vmatrix} 1 & x_1 & y_1 \\ 1 & x_2 & y_2 \\ 1 & x_3 & y_3 \end{vmatrix} = 2 \text{ (Area of triangle 123)} \qquad (5.9)$$

In our standard finite element notation, we write this as

$$T(x, y) = N_1 T_1 + N_2 T_2 + N_3 T_3 = [N]\{T\}^e \qquad (5.10)$$

Thus we have the shape functions N_i as follows:

$$\boxed{N_i = \frac{1}{2\Delta}\{\alpha_i + \beta_i x + \gamma_i y\}, \qquad i = 1, 2, 3} \qquad (5.11)$$

If this element is used to model structural mechanics problems, each node will have two d.o.f., viz., u and v, and we can write the displacements at any interior point in terms of the nodal displacements using the same shape functions as

$$u(x, y) = N_1 u_1 + N_2 u_2 + N_3 u_3 \qquad (5.12)$$

$$v(x, y) = N_1 v_1 + N_2 v_2 + N_3 v_3 \qquad (5.13)$$

In our standard finite element notation, the displacement field at an interior point in the element is written in terms of the nodal d.o.f. as

$$\begin{Bmatrix} u \\ v \end{Bmatrix} = \begin{bmatrix} N_1 & 0 & N_2 & 0 & N_3 & 0 \\ 0 & N_1 & 0 & N_2 & 0 & N_3 \end{bmatrix} \begin{Bmatrix} u_1 \\ v_1 \\ u_2 \\ v_2 \\ u_3 \\ v_3 \end{Bmatrix} = [N]\{\delta\}^e \qquad (5.14)$$

We observe that this element may only permit a linear variation of the unknown field variable within the element and thus derivatives such as strains and heat flux will be constant throughout the element. In structural mechanics, this element is therefore popularly known as the *constant strain triangle* (*CST*). We now discuss a typical rectangular element.

5.2.2   Four-noded Rectangular Element

A typical element is shown in Figure 5.5. It is rectangular and has four nodes, one at each vertex. The nodes have coordinates (X_1, Y_1), (X_2, Y_2), (X_3, Y_3) and (X_4, Y_4) in the global Cartesian coordinate frame (OXY) as shown in the figure. For ease of derivation, we define

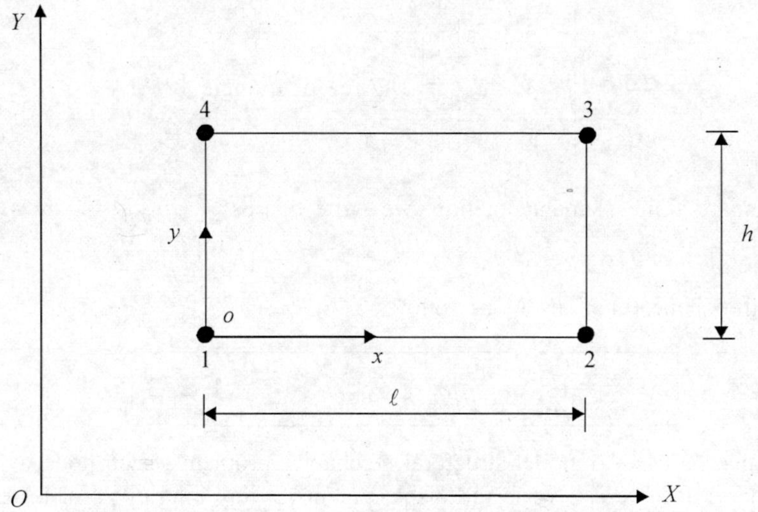

Fig. 5.5   A four-noded rectangular element.

a *local* coordinate frame (oxy). Each node has temperature d.o.f. T or two d.o.f., viz., u and v, the translations along the global X and Y axes, respectively. In what follows, we will derive the shape (interpolation) functions for the temperature field and use the same for the displacement field.

We assume that the temperature field over the element is given by

$$T(x, y) = c_0 + c_1 x + c_2 y + c_3 xy \qquad (5.15)$$

We observe that as we have four nodes now, we are able to take a fourth term in our polynomial field. Considering that this expression should reduce to the nodal temperatures at the nodal points we have, for a rectangular element of size $(\ell \times h)$ as shown in Figure 5.5,

$$T_1 = c_0, \qquad T_2 = c_0 + c_1\ell, \qquad (5.16)$$

$$T_3 = c_0 + c_1\ell + c_2 h + c_3\ell h, \qquad T_4 = c_0 + c_2 h$$

Solving for c_0, c_1, c_2 and c_3, we get

$$c_0 = T_1, \qquad c_1 = \frac{T_2 - T_1}{\ell}, \qquad c_2 = \frac{T_4 - T_1}{h}, \qquad c_3 = \frac{T_3 + T_1 - T_2 - T_4}{\ell h} \quad (5.17)$$

Substituting these in Eq. (5.15) and rewriting, we obtain

$$T(x, y) = \left(1 - \frac{x}{\ell} - \frac{y}{h} + \frac{xy}{\ell h}\right)T_1 + \left(\frac{x}{\ell} - \frac{xy}{\ell h}\right)T_2 + \left(\frac{xy}{\ell h}\right)T_3 + \left(\frac{y}{h} - \frac{xy}{\ell h}\right)T_4 \quad (5.18)$$

In our standard finite element notation, we write this as

$$
T(x, y) = \begin{bmatrix} N_1 & N_2 & N_3 & N_4 \end{bmatrix} \begin{Bmatrix} T_1 \\ T_2 \\ T_3 \\ T_4 \end{Bmatrix}
\tag{5.19}
$$

Thus we obtain the shape functions N_i:

$$
\begin{aligned}
N_1 &= 1 - \frac{x}{\ell} - \frac{y}{h} + \frac{xy}{\ell h}, & N_2 &= \frac{x}{\ell} - \frac{xy}{\ell h} \\[2ex]
N_3 &= \frac{xy}{\ell h}, & N_4 &= \frac{y}{h} - \frac{xy}{\ell h}
\end{aligned}
\tag{5.20}
$$

If this element is used to model structural mechanics problems, each node will have two d.o.f., viz., u and v, and we can write the displacements at any interior point in terms of the nodal displacements using the same shape functions as follows:

$$
\begin{Bmatrix} u \\ v \end{Bmatrix} =
\begin{bmatrix}
N_1 & 0 & N_2 & 0 & N_3 & 0 & N_4 & 0 \\
0 & N_1 & 0 & N_2 & 0 & N_3 & 0 & N_4
\end{bmatrix}
\begin{Bmatrix} u_1 \\ v_1 \\ u_2 \\ v_2 \\ u_3 \\ v_3 \\ u_4 \\ v_4 \end{Bmatrix}
= [N]\{\delta\}^e
\tag{5.21}
$$

We observe that this element permits a linear variation of unknown field variable along $x = $ constant or $y = $ constant lines, and thus it is known as the *bilinear element*. Strain and heat flux are not necessarily constant within the element.

We will now discuss a typical higher order element.

5.2.3 Six-noded Triangular Element

A typical higher order element is shown in Figure 5.6. It has six nodes, three at the vertices and three on the edges. While usually the edge nodes are located at the mid-point, it is not necessary. By convention, we number the nodes as shown in the figure. The nodes have

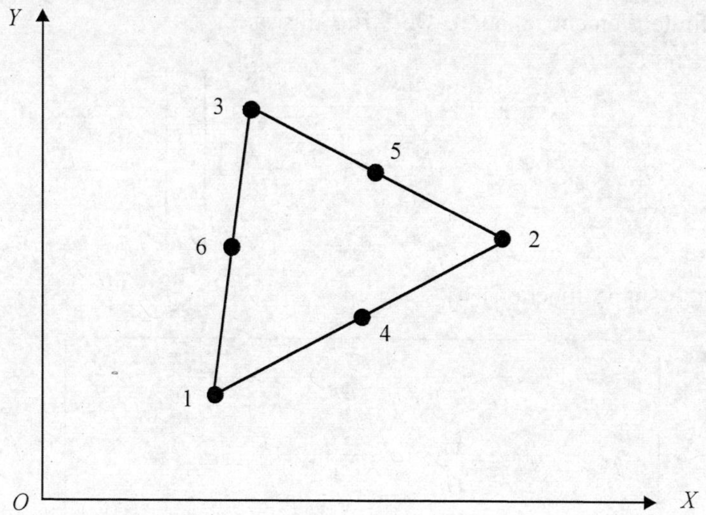

Fig. 5.6 A six-noded triangular element.

coordinates (x_1, y_1), (x_2, y_2), ..., (x_6, y_6) in the global Cartesian coordinate frame OXY as shown in the figure. Each node has just temperature d.o.f. T or two d.o.f., viz., u and v, the translations along global X and Y axes, respectively.

Since we have six nodes, we can fit a higher degree polynomial for the variation of the unknown field variable. To be precise, following our procedure, we can determine up to six coefficients $c_i (i = 0, 1, ..., 5)$. Thus we assume that the temperature field over the element is given by the complete quadratic polynomial

$$T(x, y) = c_0 + c_1 x + c_2 y + c_3 x^2 + c_4 xy + c_5 y^2 \tag{5.22}$$

Considering that this expression should reduce to the nodal temperature at the nodal points, we have

$$\left.\begin{aligned}
T_1 &= c_0 + c_1 x_1 + c_2 y_1 + c_3 x_1^2 + c_4 x_1 y_1 + c_5 y_1^2 \\
T_2 &= c_0 + c_1 x_2 + c_2 y_2 + c_3 x_2^2 + c_4 x_2 y_2 + c_5 y_2^2 \\
T_3 &= c_0 + c_1 x_3 + c_2 y_3 + c_3 x_3^2 + c_4 x_3 y_3 + c_5 y_3^2 \\
T_4 &= c_0 + c_1 x_4 + c_2 y_4 + c_3 x_4^2 + c_4 x_4 y_4 + c_5 y_4^2 \\
T_5 &= c_0 + c_1 x_5 + c_2 y_5 + c_3 x_5^2 + c_4 x_5 y_5 + c_5 y_5^2 \\
T_6 &= c_0 + c_1 x_6 + c_2 y_6 + c_3 x_6^2 + c_4 x_6 y_6 + c_5 y_6^2
\end{aligned}\right\} \tag{5.23}$$

We can, in principle, solve for coefficients $c_0 - c_5$ in terms of nodal coordinates (x_1, y_1), (x_2, y_2), ..., (x_6, y_6), and nodal temperatures $T_1 - T_6$ and substitute in Eq. (5.22) above and rearrange the terms to get the desired shape functions. However, it would be very tedious to get explicit expressions for the shape functions in this manner. Further computations, in particular evaluation of integrals, involved in deriving the element matrices, would also turn out to be quite intensive.

Our elements so far were limited to only straight-sided triangular and rectangular shape. To model typical structural geometries, we would like our elements to possess, in general, curved edges. We would also like to have alternate methods of deriving the shape functions and a way by which the computations can be performed inside a computer program through methods such as numerical integration.

In order to achieve these goals, we now introduce dimensionless natural coordinates, viz., ξ-η for typical quadrilateral elements. We propose that we have a parent rectangular element in natural coordinates, which is transformed to the child element in the real x-y coordinates as shown in Figure 5.7. The natural coordinates vary from -1 to $+1$, and the size of the parent element will be 2×2 units, irrespective of the actual element size in real Cartesian coordinates x-y. Depending on the transformation used between the $(\xi$-$\eta)$ and the $(x$-$y)$ frames, we observe that a straight-sided rectangular parent element can be transformed into an element with curved edges in the x-y coordinates. We observe that even when we use natural

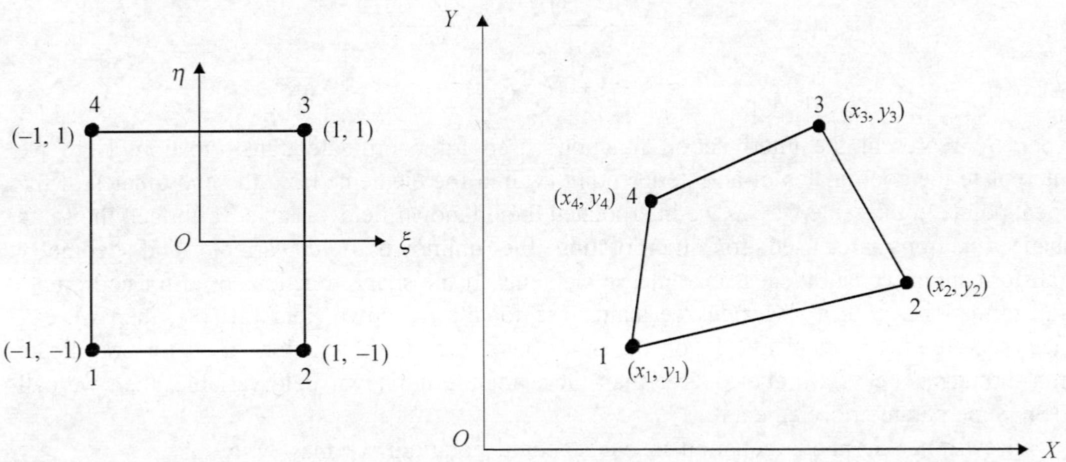

Fig. 5.7   A general four-noded quadrilateral child element with the 'parent' square element.

coordinates for the formulation, the nodal d.o.f. (such as displacements) are along the physical x-y coordinates and not along $(\xi$-$\eta)$. We will now discuss the formulation of the finite elements using natural coordinates. We will see that the formulation now involves coordinate transformation and seem to be formidable but, in fact, permits us to develop a powerful generic structure for finite element development. We will begin our discussion with an introduction to the natural coordinates in one dimension/two dimensions, and then take up the formulation of typical finite elements.

5.3   Natural Coordinates and Coordinate Transformation

We will designate the natural coordinate used for one-dimensional case by ξ. Consider a child line shown in Figure 5.8 with two end points defined by their coordinates $A(x_1)$ and $B(x_2)$. The

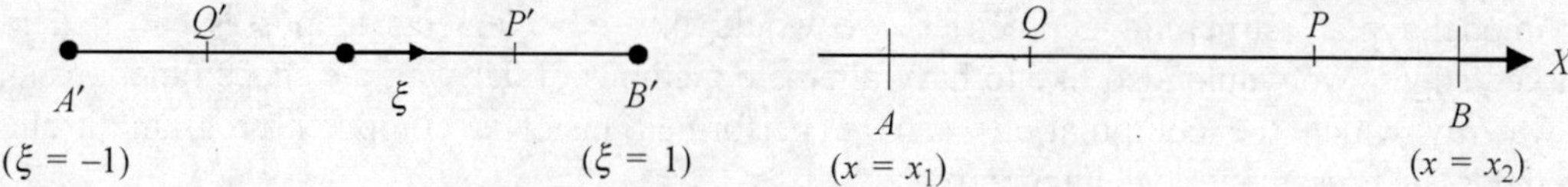

Fig. 5.8   Natural coordinates in 1-d.

parent line indicating ξ varying from $A'(-1)$ to $B'(+1)$ is also shown in the figure. An appropriate linear transformation between x and ξ is given by

$$
\begin{aligned}
x(\xi) &= \left(\frac{x_1 + x_2}{2}\right) + \left(\frac{x_2 - x_1}{2}\right)\xi \\[2mm]
&= \underbrace{\left(\frac{1 - \xi}{2}\right)}_{N_1}x_1 + \underbrace{\left(\frac{1 + \xi}{2}\right)}_{N_2}x_2 \\[2mm]
&= N_1 x_1 + N_2 x_2
\end{aligned}
\tag{5.24}
$$

where N_i represent the interpolation functions used for coordinate transformation. Here we interpolate the coordinates of an interior point (within the element) from the coordinates of the nodal points in the same way as we interpolated the unknown field variable. If, in fact, the same shape functions are used for interpolating the unknown field variable and geometry transformation, we call them isoparametric elements. If the shape functions used for coordinate transformation are of a lower degree than those for the unknown field variable, then we call them sub-parametric elements. If, on the other hand, the shape functions used for coordinate transformation are of a higher degree than those for the unknown field variable, then we call them Superparametric elements.

Rewriting the above equation in our standard notation, we may write

$$
x = \begin{bmatrix} N_1 & N_2 \end{bmatrix} \begin{Bmatrix} x_1 \\ x_2 \end{Bmatrix}
\tag{5.25}
$$

For a point P', $\xi = 0.5$, correspondingly,

$$
x = \frac{x_1 + 3x_2}{4}
$$

which is the point P.

For a point Q,

$$
x = x_1 + \frac{x_2 + x_1}{4}
$$

correspondingly, $\xi = -0.5$ which is the point Q'.

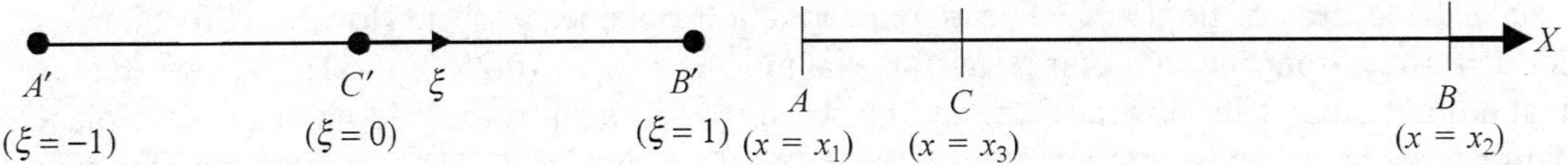

Fig. 5.9   Quadratic transformation between x and ξ.

In order to fit a quadratic transformation between x and ξ, we need one more point C as shown in Figure 5.9. The corresponding parent line ($A'\ C'\ B'$) is also shown in the figure. We observe that C' is always located at $\xi = 0$ while the corresponding point C can be located at any desired position on the line AB in the Cartesian frame.

We can write the transformation as

$$x(\xi) = \sum N_i x_i = N_1 x_1 + N_2 x_2 + N_3 x_3 \tag{5.26}$$

We will now discuss different ways of deriving the shape functions N_i in natural coordinates.

5.3.1   Alternate Methods of Deriving Shape Functions

Serendipity approach

An easy way of writing down the necessary shape or interpolation functions by inspection is as follows: If we observe the characteristic of shape functions (e.g. ref. Eq. (5.24)), we see that each N_i takes a value of unity at the point i and goes to zero at all other points $j \neq i$. This has to be so in order to satisfy the relation $x = \sum N_i x_i$, …. We can use this fact to write down the shape functions very easily. For example, we can now write the shape function N_1 in Eq. (5.26) based on the above observation that N_1 should be unity at point 1 and zero at points 2 and 3. Since the coordinates of points 2 and 3 are $\xi = 1$ and 0, respectively, the functions which become zero at the points 2 and 3 are simply $(\xi - 0)$ and $(\xi - 1)$. Thus we can write the required shape function as:

$$N_1 = (\text{SF})(\xi - 0)(\xi - 1) \tag{5.27}$$

where the scale factor (SF) is determined from the requirement that N_1 be unity at point 1, i.e. ($\xi = -1$). We readily find that SF = 0.5. Thus the required shape function is

$$N_1 = 0.5(\xi - 0)\,(\xi - 1) \tag{5.28}$$

On similar lines, we can directly write the other two shape functions as

$$N_2 = 0.5(\xi - 0)(\xi + 1) \tag{5.29}$$

$$N_3 = -(\xi + 1)(\xi - 1) \tag{5.30}$$

In this way, we can also write down very easily the shape functions for any complex finite element. Considering the example of quadratic coordinate transformation once again, we get, using Eqs. (5.28)–(5.30) in Eq. (5.26), the required transformation

$$x(\xi) = N_1 x_1 + N_2 x_2 + N_3 x_3 = 0.5(\xi)(\xi - 1)x_1 + 0.5(\xi)(\xi + 1)x_2 - (\xi + 1)(\xi - 1)x_3 \tag{5.31}$$

If point 3 in the physical space x-y were at the middle of the line, i.e. $x_3 = (0.5)(x_1 + x_2)$, then we observe from the above equation that the quadratic transformation just obtained reduces to the simple linear transformation obtained earlier. However, if point 3 is located away from the mid-point, say for example, $x_3 = x_1 + (0.25)(x_2 - x_1)$, then we require that a quarter point in the real Cartesian plane be transformed to the mid-point in the ξ plane. This cannot be achieved through a linear transformation and Eq. (5.31) becomes (for simplicity, let $x_1 = 0$, $x_2 = 4$, $x_3 = 1$),

$$x(\xi) = 0.5(\xi)(\xi - 1)(0) + 0.5(\xi)(\xi + 1)(4) - (\xi + 1)(\xi - 1)(1)$$

$$= (1 + \xi)^2 \tag{5.32}$$

Thus we can fit a nonlinear transformation between x and ξ. Such a transformation can be used to formulate finite elements which possess curved edges so that we can model curved structural geometry.

Lagrange's interpolation

If we observe our process of interpolation of either the coordinates or the field variable from the nodal values, we notice that it is simply to find a polynomial curve fit passing through the prescribed function values at specified points (nodes). It is possible to achieve this through the classical Lagrange's interpolation formula. If we are given, in general, a set of data points $x_1, x_2, \ldots, x_n$ and the corresponding function values at these data points as $f_1, f_2, \ldots, f_n$, we can then write

$$f(x) \approx L_1 f_1 + L_2 f_2 + \cdots + L_n f_n \tag{5.33}$$

where L_i are the Lagrange polynomials given by

$$L_i = \frac{(x_1 - x)(x_2 - x) \cdots (x_{i-1} - x)(x_{i+1} - x) \cdots (x_n - x)}{(x_1 - x_i)(x_2 - x_i) \cdots (x_{i-1} - x_i)(x_{i+1} - x_i) \cdots (x_n - x_i)} \tag{5.34}$$

We observe that the L_i given above are simply the required shape functions N_i. Thus we can write the shape functions for the quadratic coordinate transformation above as follows: Given the data points $(\xi_1 = -1)$, $(\xi_2 = 1)$ and $(\xi_3 = 0)$ and the corresponding function values $x_1, x_2,$ and x_3, we can write the required Lagrange interpolation polynomials as

$$N_1 = \frac{(\xi_2 - \xi)(\xi_3 - \xi)}{(\xi_2 - \xi_1)(\xi_3 - \xi_1)} = \frac{(1 - \xi)(0 - \xi)}{(1 + 1)(0 + 1)} = \frac{\xi(\xi - 1)}{2} \tag{5.35}$$

$$N_2 = \frac{(\xi_1 - \xi)(\xi_3 - \xi)}{(\xi_1 - \xi_2)(\xi_3 - \xi_2)} = \frac{(-1 - \xi)(0 - \xi)}{(-1 - 1)(0 - 1)} = \frac{\xi(1 + \xi)}{2} \tag{5.36}$$

$$N_3 = \frac{(\xi_1 - \xi)(\xi_2 - \xi)}{(\xi_1 - \xi_3)(\xi_2 - \xi_3)} = \frac{(-1 - \xi)(1 - \xi)}{(-1)(1)} = 1 - \xi^2 \tag{5.37}$$

We observe that, in general, the shape functions derived following either approach, viz., Serendipity or Lagrange's, will be different. Since the shape function is the most fundamental property of the finite elements, elements formulated based on these shape functions also exhibit different characteristics. Elements formulated based on Serendipity (Lagrange) shape functions are popularly known as Serendipity (Lagrange) elements. In general, Lagrange elements tend to have more internal nodes and admit more, higher degree terms into the polynomial shape function.

5.3.2 Natural Coordinates—Quadrilateral Elements

We will designate the natural coordinates used for two-dimensional domains by ξ and η. Thus we are looking for transformation between $(x\text{-}y)$ and $(\xi\text{-}\eta)$. Consider the general quadrilateral *ABCD* shown in Figure 5.10. The parent square $A'B'C'D'$ indicating ξ, η varying from -1 to $+1$ is also shown in the figure. If the element in the $(x\text{-}y)$ frame were rectangular (size $\ell \times h$),

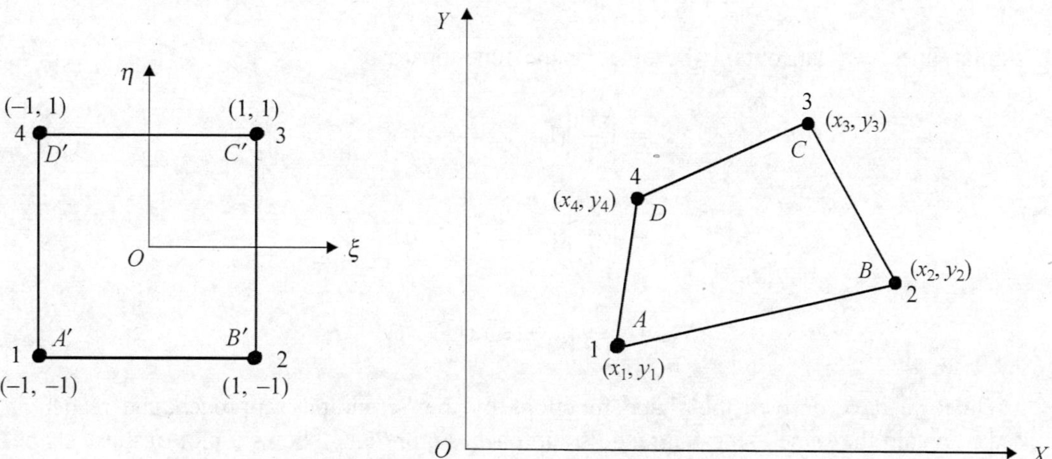

Fig. 5.10   A general four-noded quadrilateral element.

it is easy to see that the coordinate transformation would be simply a scale factor and is given by

$$\xi = \frac{2x}{\ell} \tag{5.38}$$

$$\eta = \frac{2y}{h} \tag{5.39}$$

For a general quadrilateral element, we have to obtain the transformation

$$(x, y) \Leftrightarrow f(\xi, \eta) \tag{5.40}$$

This transformation, which yields the coordinates of a point $P(x, y)$ within the element in terms of the nodal coordinates (just as we interpolate the displacement/temperature field), can be written as

$$x_P = \sum N_i x_i, \qquad y_P = \sum N_i y_i \tag{5.41}$$

where $i = 1, 2, \ldots,$ NNOEL which stands for the number of nodes per element ($= 4$ here). Considering that N_i must be unity at node i and zero at all other nodes, the shape functions N_i can be obtained as

$$N_1 = (\text{SF}) (\text{Equation of line } 2\text{--}3) (\text{Equation of line } 3\text{--}4)$$

$$= (\text{SF}) (1 - \xi)(1 - \eta) \tag{5.42}$$

where the scale factor (SF) is to be chosen such that N_1 is unity at node 1 and the product of equation of lines 2–3 and 3–4 has been taken to ensure that N_1 is zero at nodes 2, 3 and 4.

At node 1, $\xi = \eta = -1$ and, therefore, SF = 0.25. Thus our shape function is given by

$$N_1 = \left(\frac{1}{4}\right)(1 - \xi)(1 - \eta) \tag{5.43}$$

On similar lines, we can obtain the other shape functions as

$$N_2 = \left(\frac{1}{4}\right)(1 + \xi)(1 - \eta) \tag{5.44}$$

$$N_3 = \left(\frac{1}{4}\right)(1 + \xi)(1 + \eta) \tag{5.45}$$

$$N_4 = \left(\frac{1}{4}\right)(1 - \xi)(1 + \eta) \tag{5.46}$$

While we have derived the shape functions by the Serendipity approach, the reader is advised to obtain the same using Lagrange's approach. Figure 5.11 shows a plot of these shape functions.

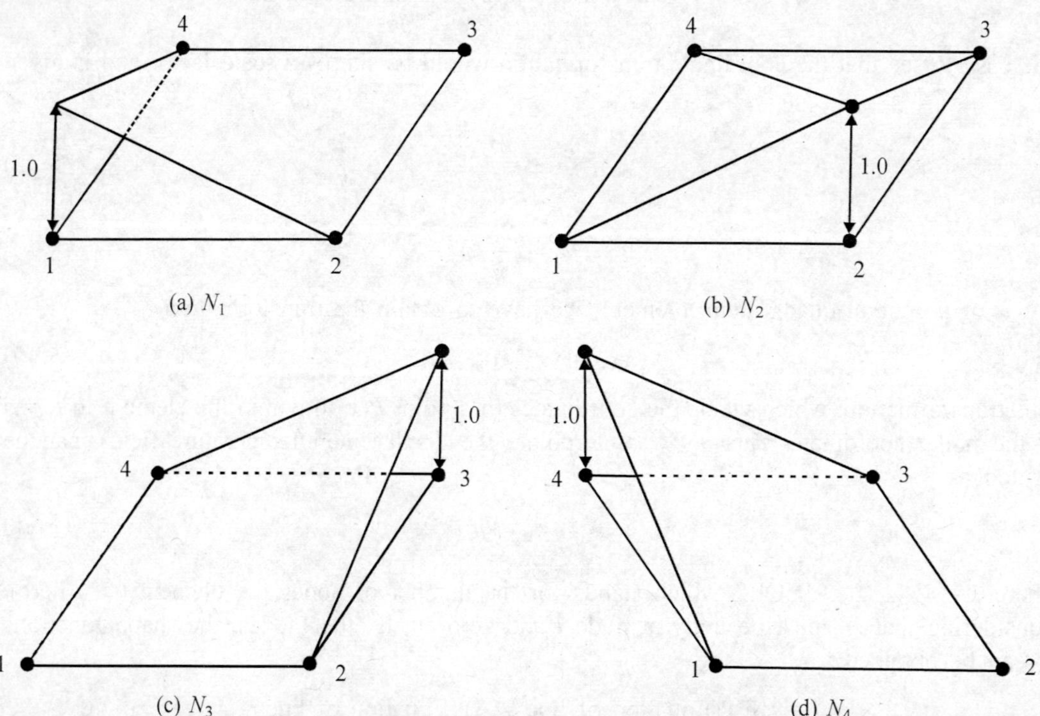

Fig. 5.11   Shape functions N_i (i = 1, 2, 3, 4) for Quad4 element.

Using this coordinate transformation, we will be able to map a parent square element in $(\xi\text{-}\eta)$ space into a general quadrilateral element in physical $(x\text{-}y)$ coordinate space. Consider, for example, the general quadrilateral shown in Figure 5.12. Using the nodal coordinate data given, we can write

$$x = \Sigma\, N_i x_i = -N_1 + N_3 \tag{5.47}$$

$$y = \Sigma\, N_i y_i = -5N_2 + 5N_4 \tag{5.48}$$

Thus we have

$$x = \frac{1}{4}\big[(1 + \xi)(1 + \eta) - (1 - \xi)(1 - \eta)\big] \tag{5.49}$$

$$y = \frac{5}{4}\big[(1 - \xi)(1 + \eta) - (1 + \xi)(1 - \eta)\big] \tag{5.50}$$

Using this transformation relation, we can make certain interesting observations. If we consider the line $\xi = 0$ in the $(\xi\text{-}\eta)$ space, its mapping onto the $(x\text{-}y)$ space is given by

$$x = (1/2)\eta, \qquad y = (5/2)\eta \tag{5.51}$$

Therefore, the equation of the line in the $(x\text{-}y)$ space is $y = 5x$.

If we consider the line $\eta = 0$ in the $(\xi\text{-}\eta)$ space, its mapping onto the $(x\text{-}y)$ space is given by

$$x = (1/2)\xi, \qquad y = -(5/2)\xi \tag{5.52}$$

Thus the equation of the line in the $(x\text{-}y)$ space is $y = -5x$. These two lines are indicated in Figure 5.12. We observe that the $\xi\text{-}\eta$ axis lines, when transformed into the physical $(x\text{-}y)$ space, need not be parallel to the x and y axes. The transformed lines need not even be orthogonal in the $x\text{-}y$ space. We reiterate the fact that our nodal d.o.f. (such as displacements) are along real, physical $x\text{-}y$ axes, and not along $\xi\text{-}\eta$.

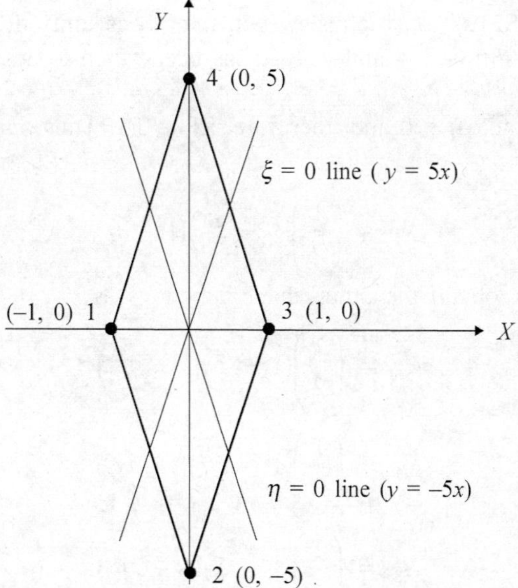

Fig. 5.12   Natural and Cartesian coordinates for a general quadrilateral element.

For the four-noded element, with the above linear shape functions, we have been able to transform a parent square element in $(\xi\text{-}\eta)$ space into a general quadrilateral child element in the physical $(x\text{-}y)$ space. If we need general curved edge element, we require a higher order transformation and we will now discuss the eight-noded element for this purpose.

Consider the general quadrilateral element shown in Figure 5.13 with eight nodes. The parent eight-noded element in $(\xi\text{-}\eta)$ space has also been shown in the figure. The attention of

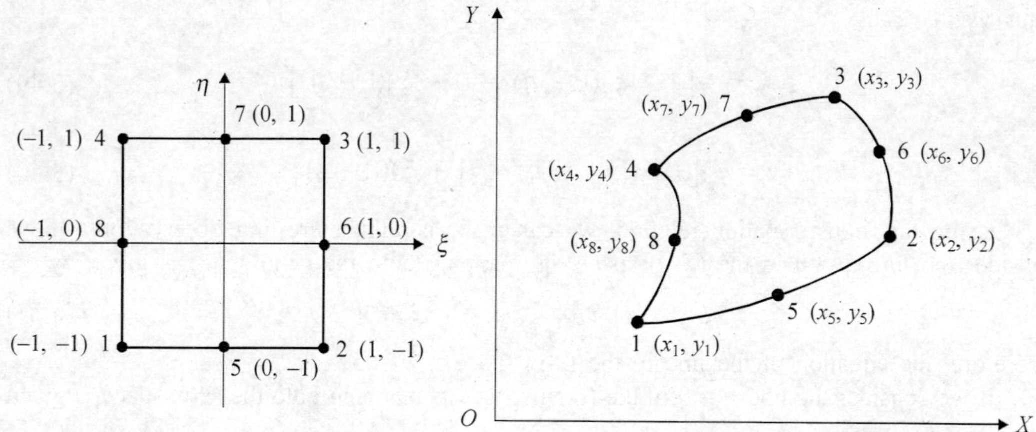

Fig. 5.13 A general eight-noded quadrilateral child element with its parent.

the reader is drawn to the convention followed in numbering the nodes. The shape functions N_i can be obtained as

$$N_5 = (\text{SF})(\text{Eq. of line } 2\text{–}6\text{–}3)(\text{Eq. of line } 3\text{–}7\text{–}4)(\text{Eq. of line } 4\text{–}8\text{–}1)$$

$$= (\text{SF})(1 - \xi)(1 - \eta)(1 + \xi) \tag{5.53}$$

where the scale factor (SF) is to be chosen such that N_5 is unity at node 5 and the product of equation of lines 2–6–3, 3–7–4 and 4–8–1 has been taken to ensure that N_5 is zero at all other nodes.

At node 5, $\xi = -1$ and $\eta = 0$ and, therefore, SF = 0.5. Thus our shape function is given by

$$N_5 = \left(\frac{1}{2}\right)\left(1 - \xi^2\right)(1 - \eta) \tag{5.54}$$

On similar lines, we can obtain the other shape functions as

$$N_6 = \left(\frac{1}{2}\right)(1 + \xi)\left(1 - \eta^2\right) \tag{5.55}$$

$$N_7 = \left(\frac{1}{2}\right)\left(1 - \xi^2\right)(1 + \eta) \tag{5.56}$$

$$N_8 = \left(\frac{1}{2}\right)(1 - \xi)\left(1 - \eta^2\right) \tag{5.57}$$

We observe, for example, that N_5 is quadratic in ξ but linear in η. Similarly, N_8 is quadratic in η but linear in ξ.

Referring to the plot of N_1 for the four-noded element (Fig. 5.11(a)), we can infer that we can readily modify that shape function to obtain the shape function N_1 for the present eight-noded element by making it vanish at nodes 5 and 8 also. Thus,

$$N_1\big|_{8\ Node} = N_1\big|_{4\ Node} - \frac{1}{2}N_5 - \frac{1}{2}N_8$$

$$= \frac{1}{4}(1 - \xi)(1 - \eta) - \frac{1}{4}(1 - \xi)(1 - \eta)(2 + \xi + \eta) \qquad (5.58)$$

$$= \frac{1}{4}(1 - \xi)(1 - \eta)(-1 - \xi - \eta)$$

On similar lines we can obtain the shape functions for other nodes as

$$N_2\big|_{8\ Node} = N_2\big|_{4\ Node} - \frac{1}{2}N_5 - \frac{1}{2}N_6$$

$$\qquad (5.59)$$

$$= \frac{1}{4}(1 + \xi)(1 - \eta)(-1 + \xi - \eta)$$

$$N_3\big|_{8\ Node} = N_3\big|_{4\ Node} - \frac{1}{2}N_5 - \frac{1}{2}N_7$$

$$= \frac{1}{4}(1 + \xi)(1 + \eta)(-1 + \xi + \eta) \qquad (5.60)$$

$$N_4\big|_{8\ Node} = N_4\big|_{4\ Node} - \frac{1}{2}N_7 - \frac{1}{2}N_8$$

$$= \frac{1}{4}(1 - \xi)(1 + \eta)(-1 - \xi + \eta) \qquad (5.61)$$

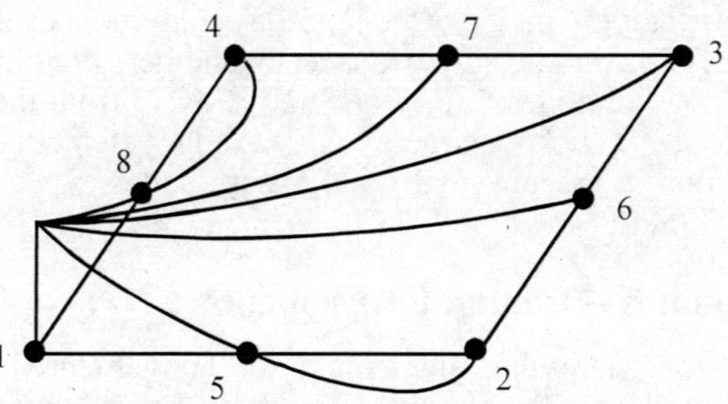

Fig. 5.14 N_1 **function of Quad-8 element.**

As an illustration, shape function N_1 is plotted in Figure 5.14. Using these shape functions, it is possible to map a straight line in the $(\xi-\eta)$ space to a curve in the physical $(x-y)$ space. Consider edge 1–5–2 ($\eta = -1$). On this edge, we have $N_3 = N_4 = N_6 = N_7 = N_8 = 0$. Thus we have, for this edge, the relations

$$x = N_1 x_1 + N_2 x_2 + N_5 x_5 \tag{5.62}$$

$$y = N_1 y_1 + N_2 y_2 + N_5 y_5 \tag{5.63}$$

Thus,

$$x = \frac{-\xi(1 - \xi)}{2} x_1 + \frac{\xi(1 + \xi)}{2} x_2 + (1 - \xi^2) x_5 \tag{5.64}$$

$$y = \frac{-\xi(1 - \xi)}{2} y_1 + \frac{\xi(1 + \xi)}{2} y_2 + (1 - \xi^2) y_5 \tag{5.65}$$

For the general values of nodal coordinates (x_1, y_1), (x_2, y_2) and (x_5, y_5), this would represent a nonlinear transformation. For example, if the nodal coordinates are given by $(0, 0)$, $(3, 9)$ and $(2, 4)$, then we have

$$x = \frac{3}{2}(1 + \xi) + \frac{1}{2}(1 - \xi^2) \tag{5.66}$$

$$y = \frac{9}{2}(1 + \xi) - \frac{1}{2}(1 - \xi^2) \tag{5.67}$$

Therefore, the equation representing the edge 1–5–2 in the Cartesian space is (eliminating ξ from Eqs. (5.66)–(5.67))

$$x^2 + y^2 + 2xy + 42x - 30y = 0 \tag{5.68}$$

Thus it is possible, in general, to map a straight line in the $(\xi-\eta)$ space to a curve in the physical $(x-y)$ space, and we can thus develop elements with curved edges so as to be able to model curved domains. In the special case, when node 5 is located at the mid-point of edge 1–2 in the physical space, i.e., if

$$x_5 = \frac{x_1 + x_2}{2} \tag{5.69}$$

$$y_5 = \frac{y_1 + y_2}{2} \tag{5.70}$$

then the transformation (as given in Eq. (5.64))(5.65) and reduces to the simple case of mapping a straight edge 1–5–2 in the $(\xi-\eta)$ space to another straight edge in the physical $(x-y)$ space. Thus, by making the nodes 5, 6, 7 or 8 move away from the middle of their edges in the physical $(x-y)$ space, we can achieve curved edge elements. It is observed that, irrespective of their locations in the physical $(x-y)$ space, nodes, 5, 6, 7, 8 are at the middle of their edges in the $\xi-\eta$ frame.

5.3.3 Natural Coordinates—Triangular Elements

The natural coordinates for triangular elements are conveniently defined as shown in Figure 5.15. The three natural coordinates are defined by the following expressions:

$$L_1 = \frac{A_1}{A} = \frac{\text{Area of } \Delta P23}{\text{Area of } \Delta 123} \tag{5.71}$$

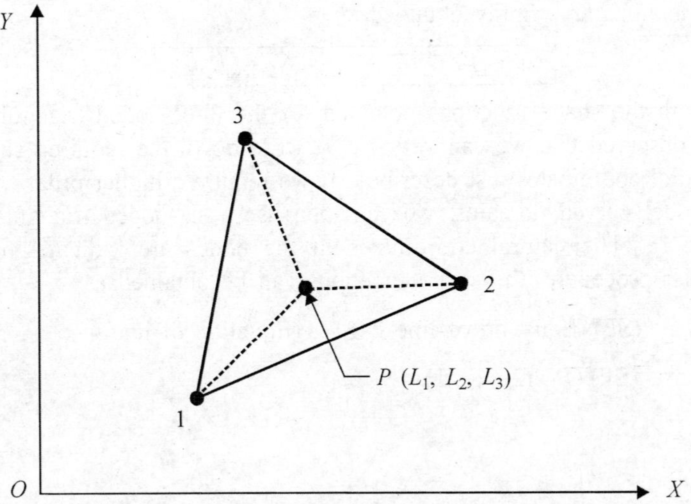

Fig. 5.15 **Natural coordinates for triangular elements.**

$$L_2 = \frac{A_2}{A} = \frac{\text{Area of } \Delta 1P3}{\text{Area of } \Delta 123} \tag{5.72}$$

$$L_3 = \frac{A_3}{A} = \frac{\text{Area of } \Delta P12}{\text{Area of } \Delta 123} \tag{5.73}$$

It is observed that in the parent equilateral triangle, each of the natural coordinates varies from 0 to 1. We observe that the three natural coordinates for a point are not independent and should sum up to unity. Thus,

$$L_1 + L_2 + L_3 = 1 \tag{5.74}$$

For a simple three-noded triangle, the required linear transformation can readily be written as

$$x = L_1 x_1 + L_2 x_2 + L_3 x_3 \tag{5.75}$$

$$y = L_1 y_1 + L_2 y_2 + L_3 y_3 \tag{5.76}$$

In our standard finite element notation, the coordinate mapping is written as

$$\left\{\begin{matrix} x \\ y \end{matrix}\right\} = \begin{bmatrix} N_1 & 0 & N_2 & 0 & N_3 & 0 \\ 0 & N_1 & 0 & N_2 & 0 & N_3 \end{bmatrix} \left\{\begin{matrix} x_1 \\ y_1 \\ x_2 \\ y_2 \\ x_3 \\ y_3 \end{matrix}\right\} \tag{5.77}$$

where the shape functions are simply obtained as

$$\boxed{N_1 = L_1, \quad N_2 = L_2, \quad N_3 = L_3} \tag{5.78}$$

Comparing with the shape functions obtained for the three-noded triangular element in section 5.2.1, it is observed that we can very easily write down the required shape functions in terms of the natural coordinates just described. If we require a higher order transformation, for example, to model curved domains, we can then use a six-noded triangular element as shown in Figure 5.16. The natural coordinates of the nodes are indicated in the figure. Following our earlier procedure, the shape functions can be obtained as

$$N_1 = \text{(SF)} \,\text{(Equation of line 2–5–3)}\,\text{(Equation of line 4–6)}$$

$$= \text{(SF)}\,(L_1)\,(2L_1 - 1) \tag{5.79}$$

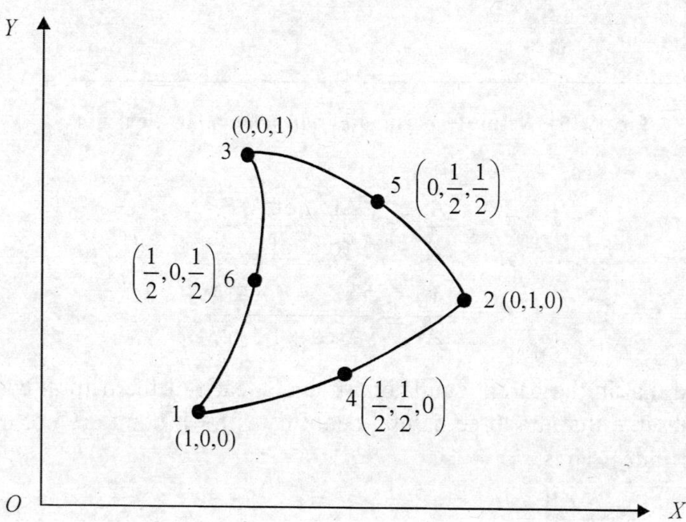

Fig. 5.16 **A six-noded triangular element.**

where the scale factor (SF) is to be chosen such that N_1 is unity at node 1 and the product of the equations of lines 2–5–3 and 4–6 has been taken to ensure that N_1 is zero at all these nodes. Equation of line 2–5–3 is $L_1 = 0$ and that of line 4–6 is $L_1 = 1/2$, i.e., $2L_1 - 1 = 0$. At node 1, $L_1 = 1$ and, therefore, SF = 1. Thus the desired shape function is readily obtained as

$$\boxed{N_1 = (L_1)\,(2L_1 - 1)} \tag{5.80}$$

Similarly, other shape functions can be obtained as

$$\boxed{\begin{aligned} N_2 &= (L_2)(2L_2 - 1), \qquad N_3 = (L_3)(2L_3 - 1) \\ N_4 &= 4L_1L_2, \qquad N_5 = 4L_2L_3, \qquad N_6 = 4L_3L_1 \end{aligned}} \tag{5.81}$$

Using these shape functions, we can write

$$
\begin{Bmatrix} x \\ y \end{Bmatrix} =
\begin{bmatrix}
N_1 & 0 & N_2 & 0 & N_3 & 0 & N_4 & 0 & N_5 & 0 & N_6 & 0 \\
0 & N_1 & 0 & N_2 & 0 & N_3 & 0 & N_4 & 0 & N_5 & 0 & N_6
\end{bmatrix}
\begin{Bmatrix} x_1 \\ y_1 \\ x_2 \\ y_2 \\ x_3 \\ y_3 \\ x_4 \\ y_4 \\ x_5 \\ y_5 \\ x_6 \\ y_6 \end{Bmatrix}
\tag{5.82}
$$

Comparing with our discussion in section 5.2.3, we readily observe that the natural coordinates as defined here for the triangular coordinates significantly simplify the derivation of shape functions. We will now discuss the formulation of the element level equations for these elements, viz., three- and six-noded triangles, four- and eight-noded quadrilaterals, etc. for both structural mechanics problems and fluid flow problems.

5.4 2-d Elements for Structural Mechanics

5.4.1 Generic Relations

We will now summarise certain generic relations which will be useful for all the 2-d finite elements for structural mechanics. For a two-dimensional structural problem, each point on the structure may have two independent displacements, viz., u and v, along the two Cartesian coordinates X and Y, respectively. The strain-displacement relations are given by

$$
\varepsilon_x = \frac{\partial u}{\partial x}
\tag{5.83}
$$

$$
\varepsilon_y = \frac{\partial v}{\partial y}
\tag{5.84}
$$

$$
\gamma_{xy} = \frac{\partial u}{\partial y} + \frac{\partial v}{\partial x}
\tag{5.85}
$$

For axisymmetric problems, we can consider the *r-z* plane (analogous to *x-y* plane for plane elasticity) as shown in Figure 5.17. We observe that, even though the deformation is

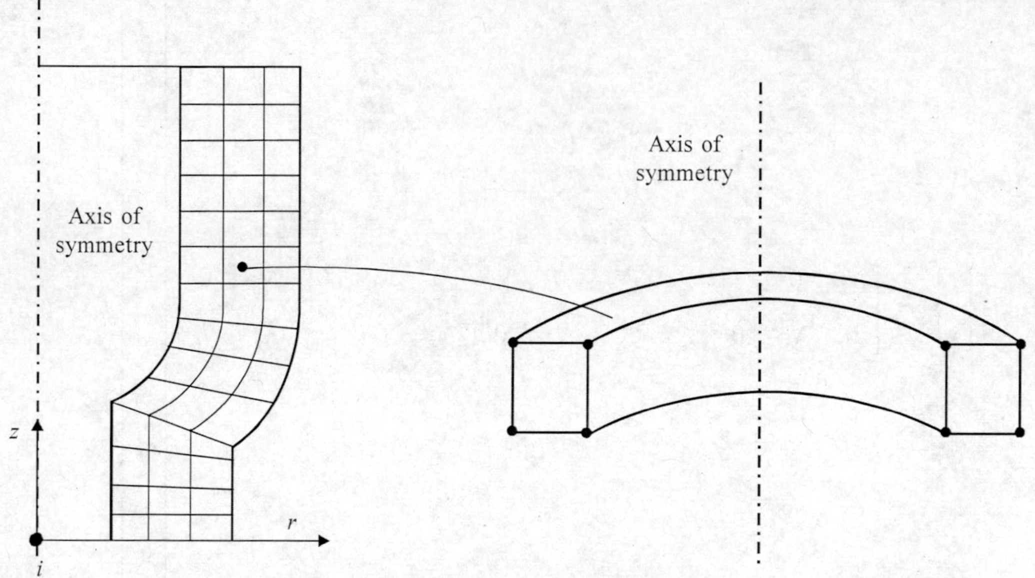

Fig. 5.17   An axisymmetric ring element.

axisymmetric, there can be four independent, nonzero strains (as a change in radius automatically leads to circumferential strain ε_0):

$$\varepsilon_r = \frac{\partial u}{\partial r} \tag{5.86}$$

$$\varepsilon_\theta = \frac{u}{r} \tag{5.87}$$

$$\varepsilon_z = \frac{\partial v}{\partial z} \tag{5.88}$$

$$\gamma_{rz} = \frac{\partial u}{\partial z} + \frac{\partial v}{\partial r} \tag{5.89}$$

In our standard finite element notation, we write the strain-displacement relations as

$$\{\varepsilon\} = [B]\{\delta\}^e \tag{5.90}$$

where the size of $[B]$ and $\{\delta\}$ will be dictated by the element and $[B]$ will contain the derivatives of the shape functions. The stress-strain relations for plane stress specialised from the 3-d Hooke's law, are given by

$$\begin{Bmatrix} \sigma_x \\ \sigma_y \\ \tau_{xy} \end{Bmatrix} = \frac{E}{1-v^2} \begin{bmatrix} 1 & v & 0 \\ v & 1 & 0 \\ 0 & 0 & 1-v^2/2 \end{bmatrix} \left(\begin{Bmatrix} \varepsilon_x \\ \varepsilon_y \\ \gamma_{xy} \end{Bmatrix} - \{\varepsilon\}^0 \right) + \{\sigma\}^0 \tag{5.91}$$

where $\{\varepsilon\}^0$ and $\{\sigma\}^0$ refer to the initial strain and stress, respectively.

The stress-strain relations for plane strain are given by

$$\begin{Bmatrix} \sigma_x \\ \sigma_y \\ \tau_{xy} \end{Bmatrix} = \frac{E}{(1+v)(1-2v)} \begin{bmatrix} 1-v & v & 0 \\ v & 1-v & 0 \\ 0 & 0 & 1-2v/2 \end{bmatrix} \left(\begin{Bmatrix} \varepsilon_x \\ \varepsilon_y \\ \gamma_{xy} \end{Bmatrix} - \{\varepsilon\}^0 \right) + \{\sigma\}^0 \qquad (5.92)$$

For axisymmetric problems, we have four independent nonzero stresses given by

$$\begin{Bmatrix} \sigma_r \\ \sigma_\theta \\ \sigma_z \\ \tau_{rz} \end{Bmatrix} = \frac{(1-v)E}{(1+v)(1-2v)} \begin{bmatrix} 1 & v/1-v & v/1-v & 0 \\ & 1 & v/1-v & 0 \\ & & 1 & 0 \\ \text{Symmetric} & & & 1-2v/2(1-v) \end{bmatrix} \left(\begin{Bmatrix} \varepsilon_r \\ \varepsilon_\theta \\ \varepsilon_z \\ \gamma_{rz} \end{Bmatrix} - \{\varepsilon\}^0 \right) + \{\sigma\}^0$$

$$(5.93)$$

In our standard finite element notation, we write the generic stress-strain relation as

$$\{\sigma\} = [D](\{\varepsilon\} - \{\varepsilon\}^0) + \{\sigma\}^0 \qquad (5.94)$$

where the elements of the $[D]$ matrix depend on plane stress/strain/axisymmetric situation and are taken from Eqs. (5.91)–(5.93).

A structure can, in general, be subjected to distributed loading in the form of body (volume) forces $\{q_v\}$ such as gravity, forces distributed over a surface $\{q_s\}$ such as a pressure or concentrated forces lumped at certain points $\{P_i\}$. It may also be subjected to initial strains and stresses (such as those due to preloading and thermal loading). Under the action of all these loads, the structure, when properly supported (so as to prevent rigid body motion), undergoes deformation and stores internal strain energy. The strain energy expression can be written as

$$U = \frac{1}{2} \int_v \{\varepsilon\}^T \{\sigma\} \, dv = \frac{1}{2} \int_v \{\varepsilon\}^T \left([D](\{\varepsilon\} - \{\varepsilon\}^0) + \{\sigma^0\} \right) dv$$

$$= \int_v \left(\frac{1}{2} \{\varepsilon\}^T [D]\{\varepsilon\} - \{\varepsilon\}^T [D]\{\varepsilon\}^0 + \{\varepsilon\}^T \{\sigma^0\} \right) dv \qquad (5.95)$$

The derivation of the above expression follows closely our discussion in Section 4.1 (Eq. (4.3)) except that we introduce the matrix notation to account for two-dimensional state of stress/strain.

From Eqs. (5.90) and (5.95), element level strain energy is given as

$$U^e = \int \frac{1}{2} \{\delta\}^{e^T} [B]^T [D][B]\{\delta\}^e \, dv - \int \{\delta\}^{e^T} [B]^T [D] \{\varepsilon\}^0 \, dv$$

$$+ \int \{\delta\}^{e^T} [B]^T \{\sigma\}^0 \, dv \qquad (5.96)$$

The potential of external forces is given by

$$V^e = -\int_v \{\delta\}^T \{q_v\} \, dv - \int_s \{\delta\}^T \{q_s\} \, ds - \Sigma \{\delta_i\}^T \{P_i\} \tag{5.97}$$

where $\{\delta\}_i$ is the displacement of the point i on which a concentrated force P_i is acting and the summation is taken over all such points. It is observed that these displacements $\{\delta\}$ and forces $\{q\}$ have both X- and Y-components. Hence, typically at a point,

$$\{\delta\} = \begin{Bmatrix} u \\ v \end{Bmatrix} \tag{5.98}$$

$$\{q\} = \begin{Bmatrix} q_x \\ q_y \end{Bmatrix} \tag{5.99}$$

From Eq. (5.97), using the notation of Eq. (5.14) and (5.21), we get

$$V^e = -\int_v \{\delta\}^{e^T} [N]^T \{q_v\} dv - \int_S \{\delta\}^{e^T} [N]^T \{Q_S\} \, ds - \Sigma\{\delta\}^{e^T} [N]_i^T \{P_i\} \tag{5.100}$$

where $[N]_i$ is the value of shape functions at the location i. Thus the total potential of an element can be written as

$$\boxed{\begin{aligned} \Pi_p^e &= \int \frac{1}{2}\{\delta\}^{e^T} [B]^T [D][B]\{\delta\}^e \, dv - \int\{\delta\}^{e^T} [B]^T [D]\{\varepsilon\}^0 dv \\ &+ \int\{\delta\}^{e^T} [B]^T \{\sigma\}^0 \, dv - \int_v \{\delta\}^{e^T} [N]^T \{q_v\} \, dv - \int_S \{\delta\}^{e^T} [N]^T \{Q_S\} \, ds - \Sigma \{\delta\}^{e^T} [N]_i^T \{P_i\} \end{aligned}} \tag{5.101}$$

Since $\{\delta\}^e$ is a vector of nodal d.o.f., it can be taken outside the integration. Thus we get

$$\Pi_p^e = \frac{1}{2}\{\delta\}^{e^T} \int_v \left([B]^T [D][B] \, dv\right)\{\delta\}^e - \{\delta\}^{e^T} \int_v [B]^T [D]\{\varepsilon\}^0 \, dv + \{\delta\}^{e^T} \int_v [B]^T \{\sigma\}^0 dv$$

$$- \{\delta\}^{e^T} \int_v [N]^T \{q_v\} \, dv - \{\delta\}^{e^T} \int_s [N]^T \{q_s\} ds - \{\delta\}^{e^T} \Sigma [N]_i^T \{P_i\} \tag{5.102}$$

Defining the element stiffness matrix $[k]^e$ and the load vector $\{f\}^e$ as

$$\boxed{[k]^e = \int_v [B]^T [D][B] \, dv} \tag{5.103}$$

$$\boxed{\{f\}^e = \int_v [B]^T [D]\{\varepsilon\}^0 dv - \int_v [B]^T \{\sigma\}^0 dv + \int_v [N]^T \{q_v\} dv + \int_s [N]^T \{q_s\} \, ds + \Sigma [N]_i^T \{P_i\}} \tag{5.104}$$

we can rewrite Eq. (5.102) as

$$\boxed{\Pi_p^e = \frac{1}{2}\{\delta\}^{e^T} [k]^e \{\delta\}^e - \{\delta\}^{e^T} \{f\}^e} \tag{5.105}$$

Once all the element matrices are 'assembled' together, we obtain the total potential of the system as

$$\Pi_p = \Sigma \Pi_p^e = \frac{1}{2}\{\delta\}^T[K]\{\delta\} - \{\delta\}^T\{F\} \qquad (5.106)$$

where the global stiffness matrix of the structure $[K]$ and the global load vector $\{F\}$ are given by

$$[K] = \sum_{n=1}^{\text{NOELEM}} [k]^e \qquad (5.107)$$

$$\{F\} = \sum_{1}^{\text{NOELEM}} \{f\}^e \qquad (5.108)$$

NOELEM refers to the number of elements and $\{\delta\}$ contains all the nodal d.o.f. variables for the entire finite element mesh. The summations indicating assembly imply that the individual element matrices have been appropriately placed in the global matrices following the standard procedure of assembly.

Using the Principle of stationary total potential we set the total potential stationary with respect to small variations in the nodal d.o.f., i.e.*

$$\frac{\partial \Pi_p}{\partial \{\delta\}^T} = 0 \qquad (5.109)$$

Thus we have the system level equations given by

$$[K]\{\delta\} = \{F\} \qquad (5.110)$$

We will now discuss each individual 2-d element and describe how the element stiffness matrix and nodal load vectors can be obtained.

5.4.2 Three-noded Triangular Element

This is one of the simplest and historically one of the first finite elements to be formulated. A typical element is shown in Figure 5.18. As discussed in section 5.2.1, the shape functions are given by (from Eq. (5.11)).

$$N_1 = \frac{1}{2\Delta}(\alpha_1 + \beta_1 x + \gamma_1 y) \qquad (5.111)$$

$$N_2 = \frac{1}{2\Delta}(\alpha_2 + \beta_2 x + \gamma_2 y) \qquad (5.112)$$

$$N_3 = \frac{1}{2\Delta}(\alpha_3 + \beta_3 x + \gamma_3 y) \qquad (5.113)$$

Since it has been possible to explicitly derive the shape functions in physical Cartesian coordinates, we shall use these directly rather than using the natural coordinates.

---

*Though this equation is written in compact matrix notation, it is to be observed that it simply states that the total potential is stationary with respect to small variations in each and every nodal d.o.f.

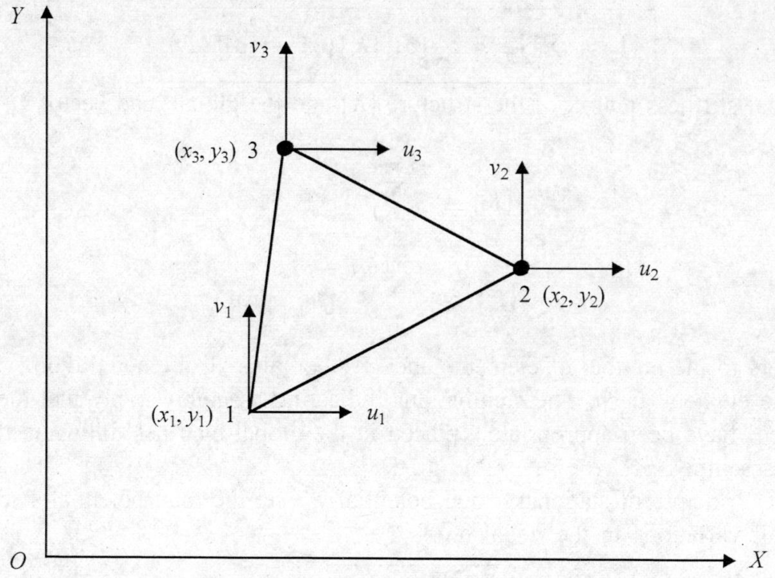

Fig. 5.18 **Simple three-noded triangular element.**

Strains and stresses

The independent nonzero strains for plane stress/strain are given by

$$
\begin{Bmatrix} \varepsilon_x \\ \varepsilon_y \\ \gamma_{xy} \end{Bmatrix} =
\begin{Bmatrix} \dfrac{\partial u}{\partial x} \\[2mm] \dfrac{\partial v}{\partial y} \\[2mm] \dfrac{\partial u}{\partial y} + \dfrac{\partial v}{\partial x} \end{Bmatrix} =
\begin{bmatrix}
\dfrac{\partial N_1}{\partial x} & 0 & \dfrac{\partial N_2}{\partial x} & 0 & \dfrac{\partial N_3}{\partial x} & 0 \\[2mm]
0 & \dfrac{\partial N_1}{\partial y} & 0 & \dfrac{\partial N_2}{\partial y} & 0 & \dfrac{\partial N_3}{\partial y} \\[2mm]
\dfrac{\partial N_1}{\partial y} & \dfrac{\partial N_1}{\partial x} & \dfrac{\partial N_2}{\partial y} & \dfrac{\partial N_2}{\partial x} & \dfrac{\partial N_3}{\partial y} & \dfrac{\partial N_3}{\partial x}
\end{bmatrix}
\begin{Bmatrix} u_1 \\ v_1 \\ u_2 \\ v_2 \\ u_3 \\ v_3 \end{Bmatrix}
\tag{5.114}
$$

Thus the strain-displacement relation matrix $[B]$ is given by

$$
[B] =
\begin{bmatrix}
\dfrac{\partial N_1}{\partial x} & 0 & \dfrac{\partial N_2}{\partial x} & 0 & \dfrac{\partial N_3}{\partial x} & 0 \\[2mm]
0 & \dfrac{\partial N_1}{\partial y} & 0 & \dfrac{\partial N_2}{\partial y} & 0 & \dfrac{\partial N_3}{\partial y} \\[2mm]
\dfrac{\partial N_1}{\partial y} & \dfrac{\partial N_1}{\partial x} & \dfrac{\partial N_2}{\partial y} & \dfrac{\partial N_2}{\partial x} & \dfrac{\partial N_3}{\partial y} & \dfrac{\partial N_3}{\partial x}
\end{bmatrix}
\tag{5.115}
$$

Thus

$$[B] = \frac{1}{2\Delta} \begin{bmatrix} \beta_1 & 0 & \beta_2 & 0 & \beta_3 & 0 \\ 0 & \gamma_1 & 0 & \gamma_2 & 0 & \gamma_3 \\ \gamma_1 & \beta_1 & \gamma_2 & \beta_2 & \gamma_3 & \beta_3 \end{bmatrix} \tag{5.116}$$

We observe that the displacement field being linear, the strains are constant (elements of $[B]$ are all constants) within an element and thus this element is known as a Constant Strain Triangle (CST). This may be considered the two-dimensional analogue of the simple linear truss element, and we expect that it will not be able to accurately model complex problems involving significant strain/stress gradients. The stress-strain relation matrix $[D]$ is given in Eq. (5.91) for plane stress and in Eq. (5.92) for plane strain cases.

Element matrices

Following Eq. (5.103), the element stiffness matrix is given by

$$[k]^e = \int_v [B]^T_{6\times 3}[D]_{3\times 3}\,[B]_{3\times 6}\,\, dv \tag{5.117}$$

The size of the stiffness matrix is (6×6) and, even after exploiting the symmetry of the stiffness matrix, we still need to evaluate 21 integrals! Since the arguments are all constants, the integration can be readily carried out and the element stiffness matrix obtained as

$$[k]^e = [B]^T[D][B](t)(A) \tag{5.118}$$

where t is the uniform thickness and A the area of the element.

For more complex, higher order elements, the elements of $[B]$ matrix will vary from point to point within an element, and explicit integration of expressions in the above equation becomes very tedious. Also, the size of the matrix itself will be larger and we need to evaluate many integrals. Thus we look for an efficient way of computing these integrals numerically within a computer program, which is readily accomplished using natural coordinates.

External loads can be concentrated forces acting at the nodes along global X and/or Y axes, and they may be distributed over the element edges as shown in Figure 5.19. Two schemes of finding equivalent nodal loads exist. In the first scheme, we generate a lumped nodal force vector by appropriately lumping the total force. In the second scheme, we compute the consistent nodal force vector which does the same amount of work as the distributed force. The element nodal force vector $\{f\}^e$ given in Eq. (5.104) is a consistent nodal force vector. We will illustrate the use of this element with a simple example.

Example 5.1.   *Illustration of the use of CST element.* We wish to determine the deflection of a thin plate subjected to extensional loads as shown in Figure 5.20. We will use two CST elements to solve this problem. From the coordinates of the nodes, we first compute α_i, β_i, γ_i, etc. as follows:

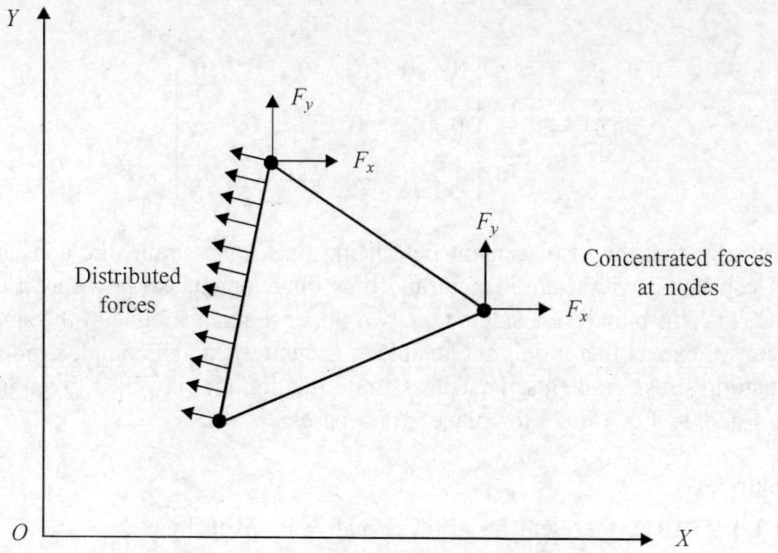

Fig. 5.19   Forces on a CST element.

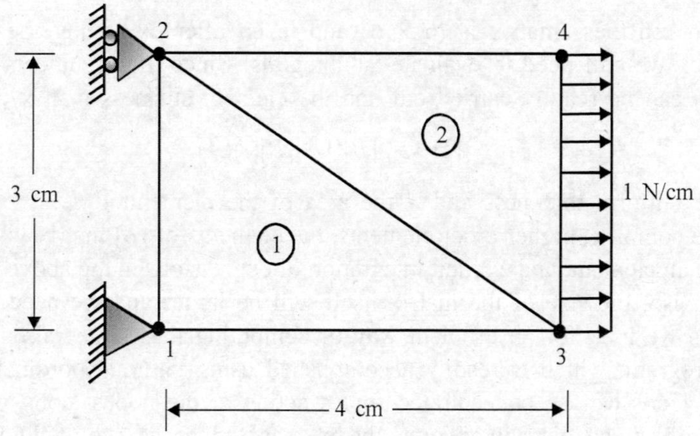

Fig. 5.20   Thin plate under uniform tension (Example 5.1).

Element 1: Shape function coefficients

$$\begin{aligned}
\alpha_i &= 12, & \beta_i &= 0, & \gamma_i &= -4 \\
\alpha_j &= 0, & \beta_j &= 3, & \gamma_j &= 0 \\
\alpha_k &= 0, & \beta_k &= -3, & \gamma_k &= 4
\end{aligned}$$

$$(5.119)$$

Element 2: Shape function coefficients

$$\begin{aligned}
\alpha_i &= 12, & \beta_i &= -3, & \gamma_i &= 0 \\
\alpha_j &= 0, & \beta_j &= 3, & \gamma_j &= -4 \\
\alpha_k &= 0, & \beta_k &= 0, & \gamma_k &= 4
\end{aligned}$$

$$(5.120)$$

We observe that the area of both the elements is 6 cm^2 and we let the thickness to be unity for plane stress. The strain-displacement matrix $[B]$ is now obtained as given below.

Element 1: Strain-displacement matrix

$$[B]^{(1)} = \begin{bmatrix} -0.25 & 0 & 0.25 & 0 & 0 & 0 \\ 0 & -0.333 & 0 & 0 & 0 & 0.333 \\ -0.333 & -0.25 & 0 & 0.25 & 0.333 & 0 \end{bmatrix} \quad (5.121)$$

Element 2: Strain-displacement matrix

$$[B]^{(2)} = \begin{bmatrix} 0.25 & 0 & -0.25 & 0 & 0 & 0 \\ 0 & 0.333 & 0 & 0 & 0 & -0.333 \\ 0.333 & 0.25 & 0 & -0.25 & -0.333 & 0 \end{bmatrix} \quad (5.122)$$

Assuming $E = 2 \times 10^7$ N/cm^2 and $v = 0.3$ for steel, we have the stress-strain relation matrix $[D]$ as

$$[D] = \begin{bmatrix} 0.22 \times 10^8 & 0.659 \times 10^7 & 0 \\ 0.659 \times 10^7 & 0.22 \times 10^8 & 0 \\ 0 & 0 & 0.769 \times 10^7 \end{bmatrix} \quad (5.123)$$

The element stiffness matrix for each element is next obtained as

$$[k]^e = [B]^T[D][B](t)(A) \quad (5.124)$$

The numerical values of $[k]^e$ for the two elements are obtained as given below.

Element 1: Stiffness matrix. The element stiffness matrix is given by

$$[k]^{(1)} = 10^7 \begin{bmatrix} 1.34 \\ 0.714 & 1.75 & & & \text{Symmetric} \\ -0.824 & -0.33 & 0.824 \\ -0.385 & -0.288 & 0 & 0.288 \\ -0.513 & -0.385 & 0 & 0.385 & 0.513 \\ -0.330 & -1.47 & 0.33 & 0 & 0 & 1.47 \end{bmatrix} \quad (5.125)$$

1	2	3	4	5	6	[Local]
1	2	5	6	3	4	[Global]

Element 2: Stiffness matrix. The element stiffness matrix is given by

$$[k]^{(2)} = 10^7 \begin{bmatrix}
1.34 & & & & & \\
0.714 & 1.75 & & & \text{Symmetric} & \\
-0.824 & -0.33 & 0.824 & & & \\
-0.385 & -0.288 & 0 & 0.288 & & \\
-0.513 & -0.385 & 0 & 0.385 & 0.513 & \\
-0.330 & -1.47 & 0.33 & 0 & 0 & 1.47
\end{bmatrix} \quad (5.126)$$

1	2	3	4	5	6	[Local]
7	8	3	4	5	6	[Global]

Since the two elements are essentially identical, we observe that our stiffness matrices are also identical.

For ease of assembly, the local and global destinations of the elements of the stiffness matrices have also been indicated above. Following the standard procedure of assembly and incorporating the boundary conditions that $u_1 = 0 = v_1 = u_2$, we get the final set of equations as follows:

$$10^7 \begin{bmatrix}
1.75 & & & & \\
0.714 & 1.34 & & \text{Symmetric} & \\
0 & 0 & 1.75 & & \\
-0.385 & -0.513 & -0.33 & 1.34 & \\
-0.288 & -0.385 & -1.47 & 0.714 & 1.75
\end{bmatrix} \begin{Bmatrix} v_2 \\ u_3 \\ v_3 \\ u_4 \\ v_4 \end{Bmatrix} = \begin{Bmatrix} 0 \\ 1.5 \\ 0 \\ 1.5 \\ 0 \end{Bmatrix} \quad (5.127)$$

The RHS force vector is equivalent nodal forces for the distributed force on the edge 3–4, viz., half the total force at each node in the X-direction. Solving these equations, we get the nodal deflections as

$$v_2 = -4.5 \times 10^{-8} \text{ cm}$$

$$u_3 = 2 \times 10^{-7} \text{ cm}$$

$$v_3 = 0 \quad (5.128)$$

$$u_4 = 2 \times 10^{-7} \text{ cm}$$

$$v_4 = -4.5 \times 10^{-8} \text{ cm}$$

From the nodal deflections, we can compute element level forces as described now:

Element 1: Forces

$$\{f\}^{(1)} = [k]^{(1)}\{\delta\}^{(1)} = 10^7 \begin{bmatrix} 1.34 & & & & & \\ 0.714 & 1.75 & & & \text{Symmetric} & \\ -0.824 & -0.33 & 0.824 & & & \\ -0.385 & -0.288 & 0 & 0.288 & & \\ -0.513 & -0.385 & 0 & 0.385 & 0.513 & \\ -0.330 & -1.47 & 0.33 & 0 & 0 & 1.47 \end{bmatrix}$$

$$\times \begin{Bmatrix} 0 \\ 0 \\ 2 \times 10^{-7} \\ 0 \\ 0 \\ -4.5 \times 10^{-8} \end{Bmatrix} = \begin{Bmatrix} -1.5 \\ 0 \\ 1.5 \\ 0 \\ 0 \\ 0 \end{Bmatrix} \qquad (5.129)$$

Element 2: Forces

$$\{f\}^{(2)} = [k]^{(2)}\{\delta\}^{(2)} = 10^7 \begin{bmatrix} 1.34 & & & & & \\ 0.714 & 1.75 & & & \text{Symmetric} & \\ -0.824 & -0.33 & 0.824 & & & \\ -0.385 & -0.288 & 0 & 0.288 & & \\ -0.513 & -0.385 & 0 & 0.385 & 0.513 & \\ -0.330 & -1.47 & 0.33 & 0 & 0 & 1.47 \end{bmatrix}$$

$$\times \begin{Bmatrix} 2 \times 10^{-7} \\ -4.5 \times 10^{-8} \\ 0 \\ -4.5 \times 10^{-8} \\ 2 \times 10^{-7} \\ 0 \end{Bmatrix} = \begin{Bmatrix} 1.5 \\ 0 \\ -1.5 \\ 0 \\ 0 \\ 0 \end{Bmatrix} \qquad (5.130)$$

These forces are along the d.o.f. From the nodal deflections, the strains in elements 1 and 2 can be computed as follows:

$$\{\varepsilon\}^e = [B]^e\{\delta\}^e \tag{5.131}$$

Element 1: Strains

$$\{\varepsilon\}^{(1)} = \begin{Bmatrix} \varepsilon_x \\ \varepsilon_Y \\ \gamma_{xy} \end{Bmatrix} = \begin{bmatrix} -0.25 & 0 & 0.25 & 0 & 0 & 0 \\ 0 & -0.333 & 0 & 0 & 0 & 0.333 \\ -0.333 & -0.25 & 0 & 0.25 & 0.333 & 0 \end{bmatrix}$$

$$\times \begin{Bmatrix} 0 \\ 0 \\ 2 \times 10^{-7} \\ 0 \\ 0 \\ -4.5 \times 10^{-8} \end{Bmatrix} = \begin{Bmatrix} 5 \times 10^{-8} \\ -1.5 \times 10^{-8} \\ 0 \end{Bmatrix} \tag{5.132}$$

Element 2: Strains

$$\{\varepsilon\}^{(2)} = \begin{Bmatrix} \varepsilon_x \\ \varepsilon_Y \\ \gamma_{xy} \end{Bmatrix} = \begin{bmatrix} 0.25 & 0 & -0.25 & 0 & 0 & 0 \\ 0 & 0.333 & 0 & 0 & 0 & -0.333 \\ -0.333 & 0.25 & 0 & -0.25 & -0.333 & 0 \end{bmatrix}$$

$$\times \begin{Bmatrix} 2 \times 10^{-7} \\ -4.5 \times 10^{-8} \\ 0 \\ -4.5 \times 10^{-8} \\ 2 \times 10^{-7} \\ -4.5 \times 10^{-8} \end{Bmatrix} = \begin{Bmatrix} 5 \times 10^{-8} \\ -1.5 \times 10^{-8} \\ 0 \end{Bmatrix} \tag{5.133}$$

It is observed that our solution tallies perfectly with the expected solution, viz., $\delta = (PL/AE)$ and ε_x and ε_y are in the ratio of v. The element stresses can be readily obtained from these strains.

While we can get excellent results for this simple problem with the CST element, in general, we cannot expect such a good performance because it is only a constant strain element. In problems involving significant gradients of stress (e.g., stress concentration near a hole or any sudden geometry change), the CST element will not give good results.

5.4.3 Four-noded Rectangular Element

This is the simplest element of the family of general quadrilateral elements and is depicted in Figure 5.21. It has four nodes, straight sides, and each node admits two d.o.f., viz., u and v translations along global the X and Y axes. Our formulation of the element follows standard procedure and is described now.

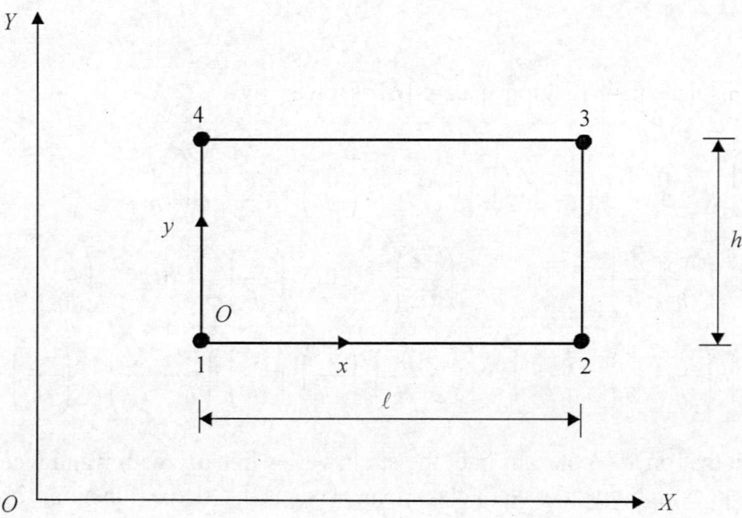

Fig. 5.21   A four-noded rectangular element.

(i) Displacement field and shape functions

From Eq. (5.20), we have the following shape functions:

$$N_1 = 1 - \frac{x}{\ell} - \frac{y}{h} + \frac{xy}{\ell h}$$

$$N_2 = \frac{x}{\ell} - \frac{xy}{\ell h}$$

$$N_3 = \frac{xy}{\ell h} \tag{5.134}$$

$$N_4 = \frac{y}{h} - \frac{xy}{\ell h}$$

(ii) Strains and stresses

The independent nonzero strains for plane stress/strain are given by

$$\{\varepsilon\}^e = \begin{Bmatrix} \varepsilon_x \\ \varepsilon_y \\ \gamma_{xy} \end{Bmatrix} = \begin{Bmatrix} \dfrac{\partial u}{\partial x} \\[2mm] \dfrac{\partial v}{\partial y} \\[2mm] \dfrac{\partial u}{\partial y} + \dfrac{\partial v}{\partial x} \end{Bmatrix} = \begin{bmatrix} \dfrac{\partial N_1}{\partial x} & 0 & \dfrac{\partial N_2}{\partial x} & 0 & \dfrac{\partial N_3}{\partial x} & 0 & \dfrac{\partial N_4}{\partial x} & 0 \\[2mm] 0 & \dfrac{\partial N_1}{\partial y} & 0 & \dfrac{\partial N_2}{\partial y} & 0 & \dfrac{\partial N_3}{\partial y} & 0 & \dfrac{\partial N_4}{\partial y} \\[2mm] \dfrac{\partial N_1}{\partial y} & \dfrac{\partial N_1}{\partial x} & \dfrac{\partial N_2}{\partial y} & \dfrac{\partial N_2}{\partial x} & \dfrac{\partial N_3}{\partial y} & \dfrac{\partial N_3}{\partial x} & \dfrac{\partial N_4}{\partial y} & \dfrac{\partial N_4}{\partial x} \end{bmatrix}_{3\times8} \begin{Bmatrix} u_1 \\ v_1 \\ u_2 \\ v_2 \\ u_3 \\ v_3 \\ u_4 \\ v_4 \end{Bmatrix}_{8\times1}$$

$$(5.135)$$

Thus the strain-displacement relation matrix $[B]$ is given by

$$[B] = \begin{bmatrix} \left(\dfrac{y}{\ell h} - \dfrac{1}{\ell}\right) & 0 & \left(\dfrac{1}{\ell} - \dfrac{y}{\ell h}\right) & 0 & \left(\dfrac{y}{\ell h}\right) & 0 & \left(\dfrac{-y}{\ell h}\right) & 0 \\[3mm] 0 & \left(\dfrac{x}{\ell h} - \dfrac{1}{h}\right) & 0 & \left(\dfrac{-x}{\ell h}\right) & 0 & \left(\dfrac{x}{\ell h}\right) & 0 & \left(\dfrac{1}{h} - \dfrac{x}{\ell h}\right) \\[3mm] \left(\dfrac{x}{\ell h} - \dfrac{1}{h}\right) & \left(\dfrac{y}{\ell h} - \dfrac{1}{\ell}\right) & \left(\dfrac{-x}{\ell h}\right) & \left(\dfrac{1}{\ell} - \dfrac{y}{\ell h}\right) & \left(\dfrac{x}{\ell h}\right) & \left(\dfrac{y}{\ell h}\right) & \left(\dfrac{1}{h} - \dfrac{x}{\ell h}\right) & \left(\dfrac{-y}{\ell h}\right) \end{bmatrix} \quad (5.136)$$

We observe that on $x = $ constant line, the strain varies linearly with y and vice versa. This element is known as a bilinear rectangular element since its shape functions involve linear polynomials in both the x and y directions. The stress-strain relation matrix $[D]$ is given in Eq. (5.91) for plane stress and in Eq. (5.92) for plane strain.

(iii) Element matrices

The element stiffness matrix, following Eq. (5.103), is given by

$$[k]^e = \int [B]^T [D][B] \, dv \qquad (5.137)$$

The size of the stiffness matrix is (8×8) since there are four nodes and two d.o.f. per node and even after exploiting the symmetry of the stiffness matrix, we still need to evaluate 36 integrals! Also, the elements of $[B]$ matrix now vary from point to point within an element and explicit integration of expressions in the above equation becomes very tedious.

External loads can be concentrated forces acting at the nodes along the global X and/or Y axes (or) they may be distributed over the element edges as shown in Figure 5.22. Two schemes of finding equivalent nodal loads exist. In the first scheme, we generate a lumped nodal force vector by appropriately lumping the total force. In the second scheme, we compute the consistent nodal force vector which does the same amount of work as the distributed force. The element nodal force vector $\{f\}^e$ given in Eq. (5.104) is a consistent nodal force vector.

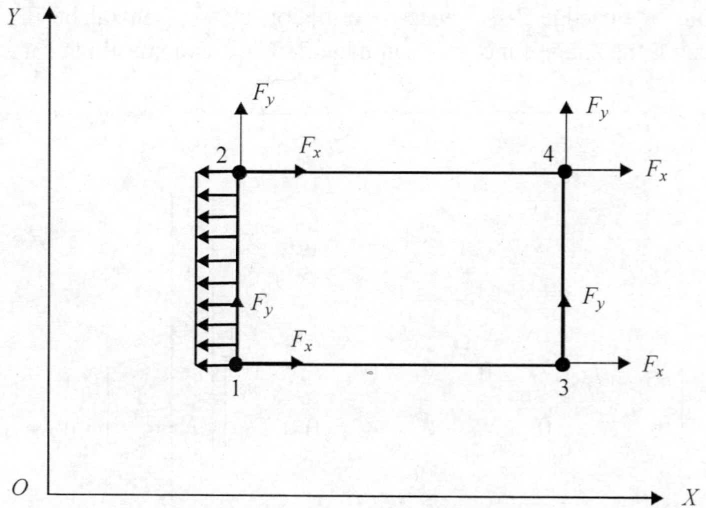

Fig. 5.22   Forces on RECT4 element.

5.4.4   Compatibility of Displacements

In the one-dimensional elements described in Chapter 4, elements shared only common nodes and the nodal d.o.f. were such that across elements, displacements were compatible, i.e., in a deformed position, there would be no openings or kinks. However, in the two-dimensional case, the elements share common edges and we need to ensure that the deformation of a point lying on such a common edge would be the same whether it is assumed to belong to one element or the other. Let us consider the bilinear rectangular element just described. Consider two adjacent elements as shown in Figure 5.23. We see that they share common nodes and displacements are therefore the same at these nodal points for both elements. Now consider

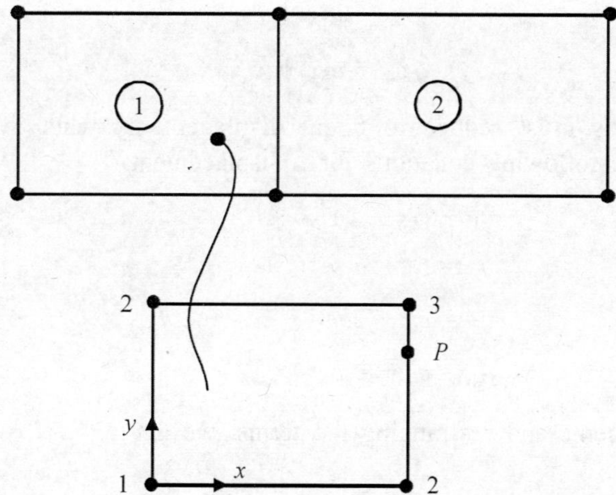

Fig. 5.23   Continuity of displacement across two elements.

an intermediate point P on edge 2–3 whose deformation can in general be interpolated from nodal deflections using the shape functions. On edge 2–3, we can substitute for $x = \ell$ and have

$$N_1 = 0, \qquad N_2 = 1 - \frac{y}{h}, \qquad N_3 = 1 - \frac{y}{h}, \qquad N_4 = 0 \tag{5.138}$$

$$\begin{Bmatrix} u_P \\ v_P \end{Bmatrix} = \begin{bmatrix} N_1 & 0 & N_2 & 0 & N_3 & 0 & N_4 & 0 \\ 0 & N_1 & 0 & N_2 & 0 & N_3 & 0 & N_4 \end{bmatrix} \begin{Bmatrix} u_1 \\ v_1 \\ u_2 \\ v_2 \\ u_3 \\ v_3 \\ u_4 \\ v_4 \end{Bmatrix} = \begin{Bmatrix} N_2 u_2 + N_3 u_3 \\ N_2 v_2 + N_3 v_3 \end{Bmatrix} \tag{5.139}$$

Thus we see that u_P and v_P, the deflections at the interior point P, are functions of only the nodes on that edge, viz., 2–3. Since nodal deflections are common for both the elements, it can be concluded that the deflection of an interior point on the edge will also be the same for both the elements. Thus we have compatibility of displacements along the entire edge. One of the major limitations of the bilinear element as formulated above is that it can only be rectangular in shape. We will now illustrate that once this element becomes a general quadrilateral, we cannot guarantee compatibility of displacements across elements. Let us consider a four-node quadrilateral element as shown in Figure 5.24 and let the displacements be given as

$$u(x, y) = c_0 + c_1 x + c_2 y + c_3 xy \tag{5.140}$$

$$v(x, y) = c_4 + c_5 x + c_6 y + c_7 xy \tag{5.141}$$

Since these expressions must reduce to nodal displacements when we substitute nodal coordinates, we get the following equations for u-displacement:

$$u_1 = c_0$$

$$u_2 = c_0 + c_1 a$$

$$u_3 = c_0 + 2ac_1 + bc_2 + 2abc_3 \tag{5.142}$$

$$u_4 = c_0 + c_2 b$$

Solving for the coefficients and rearranging the terms, we get

$$u(x, y) = N_1 u_1 + N_2 u_2 + N_3 u_3 + N_4 u_4 \tag{5.143}$$

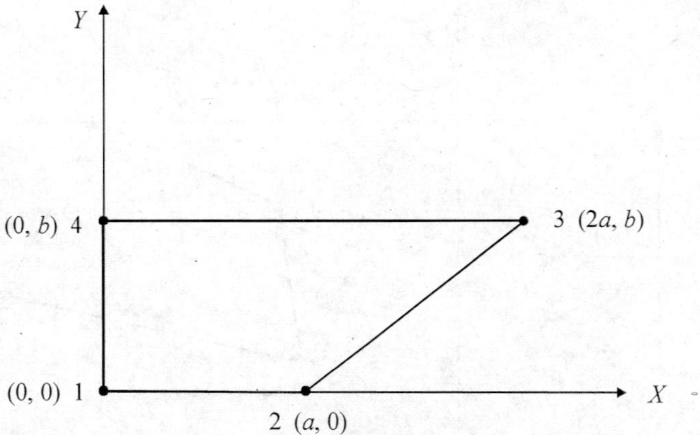

Fig. 5.24   A quadrilateral element.

where

$$N_1 = 1 - \frac{x}{a} - \frac{y}{b} + \frac{xy}{ab}, \qquad N_2 = \frac{x}{a} - \frac{xy}{ab} \tag{5.144}$$

$$N_3 = \frac{xy}{2ab}, \qquad N_4 = \frac{y}{b} - \frac{xy}{2ab} \tag{5.145}$$

At node 2, $N_2 = 1$ and $N_1 = N_3 = N_4 = 0$. Similarly, at node 3, $N_3 = 1$, $N_1 = N_2 = N_4$

$= 0$. However, for any intermediate point on the edge 2–3, $\left(\dfrac{y}{b} = \dfrac{x}{a} - 1\right)$ we see that the shape functions N_1 and N_4 do not vanish. So, the displacement of a point on edge 2–3, in general, is given as

$$u = N_1 u_1 + N_2 u_2 + N_3 u_3 + N_4 u_4 \tag{5.146}$$

Since this is not dependent only on u_2 and u_3 (the common nodes between two adjacent elements sharing this edge), we cannot guarantee that on edge 2–3 every interior point will have the same u-displacement for both the elements. While some incompatible finite elements are used in practice, we will limit our discussion in this text only to those elements where displacements are compatible across inter-element boundaries. We will now formulate the four-node quadrilateral element using the natural coordinates and show that it is possible to achieve full compatibility.

5.4.5   Four-node Quadrilateral Element

Consider the parent four-node square element in ξ–η frame as well as the general quadrilateral element in the physical x-y space as shown in Figure 5.25. The coordinates of any point $P(x, y)$ are interpolated from nodal coordinates as follows:

$$x = \sum_{i=1}^{4} N_i x_i \tag{5.147}$$

$$y = \sum_{i=1}^{4} N_i y_i \tag{5.148}$$

where N_i are as given in Eqs. (5.43)–(5.46).

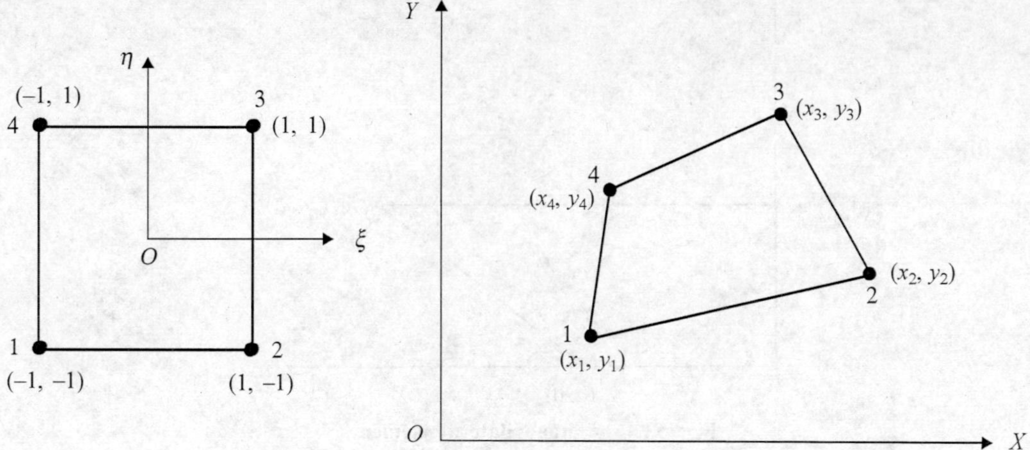

Fig. 5.25   **A general four-noded quadrilateral element.**

The displacements of the interior point P are also interpolated from nodal deflections using the same shape functions for this element, which is therefore called an "isoparametric element". The displacements can be written as

$$\begin{Bmatrix} u \\ v \end{Bmatrix} = \begin{bmatrix} N_1 & 0 & N_2 & 0 & N_3 & 0 & N_4 & 0 \\ 0 & N_1 & 0 & N_2 & 0 & N_3 & 0 & N_4 \end{bmatrix} \begin{Bmatrix} u_1 \\ v_1 \\ u_2 \\ v_2 \\ u_3 \\ v_3 \\ u_4 \\ v_4 \end{Bmatrix} \tag{5.149}$$

Let us now show that the displacements in this element are compatible across inter-element boundaries. Along edge 2–3 ($\xi = 1$), for example, the shape functions $N_1 = 0 = N_4$. Thus, irrespective of the shape of the element in the physical x-y space (i.e. even for a general quadrilateral), the displacements of any point on edge 2–3 are given as

$$u = N_2 u_2 + N_3 u_3$$

$$v = N_2 v_2 + N_3 v_3 \tag{5.150}$$

Therefore, we observe that even for a general quadrilateral shape, the isoparametric element guarantees the continuity of displacements across inter-element boundary edges.

Recall our observation that the nodal d.o.f. are in X-Y directions and not along ξ–η. Also the ξ–η axes, as transformed into real physical x–y space, may not be orthogonal. Thus we write the strain displacement relations in an orthogonal, Cartesian frame of reference, as

$$\{\varepsilon\} = \begin{Bmatrix} \varepsilon_x \\ \varepsilon_y \\ \gamma_{xy} \end{Bmatrix} = \begin{Bmatrix} \dfrac{\partial u}{\partial x} \\[2ex] \dfrac{\partial v}{\partial y} \\[2ex] \dfrac{\partial u}{\partial y} + \dfrac{\partial v}{\partial x} \end{Bmatrix} \tag{5.151}$$

We observe that u and v are given in terms of shape functions N_i which are expressed in the ξ–η coordinates rather than x–y. For a general function f, using the chain rule of differentiation, we obtain the equations

$$\frac{\partial f}{\partial x} = \frac{\partial f}{\partial \xi}\frac{\partial \xi}{\partial x} + \frac{\partial f}{\partial \eta}\frac{\partial \eta}{\partial x}$$

$$\frac{\partial f}{\partial y} = \frac{\partial f}{\partial \xi}\frac{\partial \xi}{\partial y} + \frac{\partial f}{\partial \eta}\frac{\partial \eta}{\partial y} \tag{5.152}$$

However, from Eqs. (5.147)–(5.148), it is easier to write down derivatives of $(x$–$y)$ in terms of $(\xi$–$\eta)$ rather than the other way round. Thus we write

$$\frac{\partial f}{\partial \xi} = \frac{\partial f}{\partial x}\frac{\partial x}{\partial \xi} + \frac{\partial f}{\partial y}\frac{\partial y}{\partial \xi}$$

$$\frac{\partial f}{\partial \eta} = \frac{\partial f}{\partial x}\frac{\partial x}{\partial \eta} + \frac{\partial f}{\partial y}\frac{\partial y}{\partial \eta} \tag{5.153}$$

Thus, in general, we can write

$$\begin{Bmatrix} \dfrac{\partial}{\partial \xi} \\[2ex] \dfrac{\partial}{\partial \eta} \end{Bmatrix} = \begin{bmatrix} \dfrac{\partial x}{\partial \xi} & \dfrac{\partial y}{\partial \xi} \\[2ex] \dfrac{\partial x}{\partial \eta} & \dfrac{\partial y}{\partial \eta} \end{bmatrix} \begin{Bmatrix} \dfrac{\partial}{\partial x} \\[2ex] \dfrac{\partial}{\partial y} \end{Bmatrix} \tag{5.154}$$

i.e.

$$\boxed{\begin{Bmatrix} \dfrac{\partial}{\partial \xi} \\[2ex] \dfrac{\partial}{\partial \eta} \end{Bmatrix} = [J] \begin{Bmatrix} \dfrac{\partial}{\partial x} \\[2ex] \dfrac{\partial}{\partial y} \end{Bmatrix}} \tag{5.155}$$

where

$$[J] = \begin{bmatrix} \dfrac{\partial x}{\partial \xi} & \dfrac{\partial y}{\partial \xi} \\[2mm] \dfrac{\partial x}{\partial \eta} & \dfrac{\partial y}{\partial \eta} \end{bmatrix} \tag{5.156}$$

From Eqs. (5.147)–(5.148),

$$[J] = \begin{bmatrix} \displaystyle\sum \dfrac{\partial N_i}{\partial \xi} x_i & \displaystyle\sum \dfrac{\partial N_i}{\partial \xi} y_i \\[4mm] \displaystyle\sum \dfrac{\partial N_i}{\partial \eta} x_i & \displaystyle\sum \dfrac{\partial N_i}{\partial \eta} y_i \end{bmatrix} \tag{5.157}$$

Here, $[J]$ is known as the Jacobian matrix relating the derivatives in two coordinate frames. This generic expression for $[J]$ is true for all 2-d elements, but the actual coefficients in the matrix will depend on the shape functions being employed in a given element and the nodal coordinates.

For this particular element, for example, the Jacobian is

$$[J] = \begin{bmatrix} \left(\dfrac{1-\eta}{4}\right)(x_2 - x_1) + \left(\dfrac{1+\eta}{4}\right)(x_3 - x_4) & \left(\dfrac{1-\eta}{4}\right)(y_2 - y_1) + \left(\dfrac{1+\eta}{4}\right)(y_3 - y_4) \\[4mm] \left(\dfrac{1-\xi}{4}\right)(x_4 - x_1) + \left(\dfrac{1+\xi}{4}\right)(x_3 - x_2) & \left(\dfrac{1-\xi}{4}\right)(y_4 - y_1) + \left(\dfrac{1+\xi}{4}\right)(y_3 - y_2) \end{bmatrix} \tag{5.158}$$

From Eq. (5.155),

$$\begin{Bmatrix} \dfrac{\partial}{\partial x} \\[3mm] \dfrac{\partial}{\partial y} \end{Bmatrix} = [J]^{-1} \begin{Bmatrix} \dfrac{\partial}{\partial \xi} \\[3mm] \dfrac{\partial}{\partial \eta} \end{Bmatrix} = \dfrac{1}{|J|} \begin{bmatrix} J_{22} & -J_{12} \\[2mm] -J_{21} & J_{11} \end{bmatrix} \begin{Bmatrix} \dfrac{\partial}{\partial \xi} \\[3mm] \dfrac{\partial}{\partial \eta} \end{Bmatrix} \tag{5.159}$$

We can now obtain the required strain displacement relations.

$$\{\varepsilon\} = \begin{Bmatrix} \dfrac{\partial u}{\partial x} \\[3mm] \dfrac{\partial v}{\partial y} \\[3mm] \dfrac{\partial u}{\partial y} + \dfrac{\partial v}{\partial x} \end{Bmatrix} = \underbrace{\begin{bmatrix} 1 & 0 & 0 & 0 \\ 0 & 0 & 0 & 1 \\ 0 & 1 & 1 & 0 \end{bmatrix}}_{[B_1]} \begin{Bmatrix} \dfrac{\partial u}{\partial x} \\[3mm] \dfrac{\partial u}{\partial y} \\[3mm] \dfrac{\partial v}{\partial x} \\[3mm] \dfrac{\partial v}{\partial y} \end{Bmatrix} \tag{5.160}$$

However, from Eq. (5.159),

$$
\begin{Bmatrix} \dfrac{\partial u}{\partial x} \\[2ex] \dfrac{\partial u}{\partial y} \\[2ex] \dfrac{\partial v}{\partial x} \\[2ex] \dfrac{\partial v}{\partial y} \end{Bmatrix}
=
\underbrace{\begin{bmatrix}
\dfrac{J_{22}}{|J|} & -\dfrac{J_{12}}{|J|} & 0 & 0 \\[2ex]
-\dfrac{J_{21}}{|J|} & \dfrac{J_{11}}{|J|} & 0 & 0 \\[2ex]
0 & 0 & \dfrac{J_{22}}{|J|} & -\dfrac{J_{12}}{|J|} \\[2ex]
0 & 0 & \dfrac{-J_{21}}{|J|} & \dfrac{J_{11}}{|J|}
\end{bmatrix}}_{[B_2]}
\begin{Bmatrix} \dfrac{\partial u}{\partial \xi} \\[2ex] \dfrac{\partial u}{\partial \eta} \\[2ex] \dfrac{\partial v}{\partial \xi} \\[2ex] \dfrac{\partial u}{\partial \eta} \end{Bmatrix}
\qquad (5.161)
$$

From Eq. (5.149),

$$
\begin{Bmatrix} \dfrac{\partial u}{\partial \xi} \\[2ex] \dfrac{\partial u}{\partial \eta} \\[2ex] \dfrac{\partial v}{\partial \xi} \\[2ex] \dfrac{\partial v}{\partial \eta} \end{Bmatrix}
=
\begin{bmatrix}
\dfrac{\partial N_1}{\partial \xi} & 0 & \dfrac{\partial N_2}{\partial \xi} & 0 & \dfrac{\partial N_3}{\partial \xi} & 0 & \dfrac{\partial N_4}{\partial \xi} & 0 \\[2ex]
\dfrac{\partial N_1}{\partial \eta} & 0 & \dfrac{\partial N_2}{\partial \eta} & 0 & \dfrac{\partial N_3}{\partial \eta} & 0 & \dfrac{\partial N_4}{\partial \eta} & 0 \\[2ex]
0 & \dfrac{\partial N_1}{\partial \xi} & 0 & \dfrac{\partial N_2}{\partial \xi} & 0 & \dfrac{\partial N_3}{\partial \xi} & 0 & \dfrac{\partial N_4}{\partial \xi} \\[2ex]
0 & \dfrac{\partial N_1}{\partial \eta} & 0 & \dfrac{\partial N_2}{\partial \eta} & 0 & \dfrac{\partial N_3}{\partial \eta} & 0 & \dfrac{\partial N_4}{\partial \eta}
\end{bmatrix}
\begin{Bmatrix} u_1 \\ v_1 \\ u_2 \\ v_2 \\ u_3 \\ v_3 \\ u_4 \\ v_4 \end{Bmatrix}
\qquad (5.162)
$$

$$
= [B_3]\,\{\delta\}^e
$$

For this element, for example, the shape function derivatives in the ξ-η frame are as follows:

$$
\begin{aligned}
\frac{\partial N_1}{\partial \xi} &= \frac{-(1-\eta)}{4}, & \frac{\partial N_1}{\partial \xi} &= \frac{-(1-\eta)}{4} \\[2ex]
\frac{\partial N_2}{\partial \xi} &= \frac{1-\eta}{4}, & \frac{\partial N_2}{\partial \eta} &= \frac{-(1+\xi)}{4} \\[2ex]
\frac{\partial N_3}{\partial \xi} &= \frac{1+\eta}{4}, & \frac{\partial N_3}{\partial \eta} &= \frac{(1+\xi)}{4} \\[2ex]
\frac{\partial N_4}{\partial \xi} &= \frac{-(1+\eta)}{4}, & \frac{\partial N_4}{\partial \eta} &= \frac{(1-\xi)}{4}
\end{aligned}
\qquad (5.163)
$$

Thus, finally, combining Eqs. (5.160)–(5.162), the strain-displacement relations can be written as

$$\{\varepsilon\} = \begin{Bmatrix} \varepsilon_x \\ \varepsilon_y \\ \varepsilon_{xy} \end{Bmatrix} = [B]\{\delta\}^e \tag{5.164}$$

where $[B] = [B_1][B_2][B_3]$.

The stress-strain relation matrix $[D]$ remains as given earlier (see, e.g. Eq. (5.91)). Thus we have the element stiffness matrix

$$[k]^e_{8\times8} = \int_v [B]^T[D][B]\,dv = \iint [B]^T[D][B]\,t\,dx\,dy$$

$$= \int_{-1}^{1}\int_{-1}^{1} [B]^T[D][B]t\,|J|\,d\xi\,d\eta \tag{5.165}$$

where the coordinate transformation has been taken into account and the integrals need now be evaluated over the parent square element in natural coordinates. Since $[B]$ in general varies from point to point within the element, we look for ways of evaluating these integrals numerically within a computer program rather than attempting to derive them explicitly. We will discuss numerical integration schemes in Section 5.5.

5.4.6   Eight-node Quadrilateral Element

Consider the parent eight-node rectangular element in ξ-η frame as well as the general quadrilateral element in the physical x-y space as shown in Figure 5.26. The coordinates of any point $P(x, y)$ are interpolated from nodal coordinates as follows:

$$x = \sum N_i x_i, \qquad y = \sum N_i y_i \tag{5.166}$$

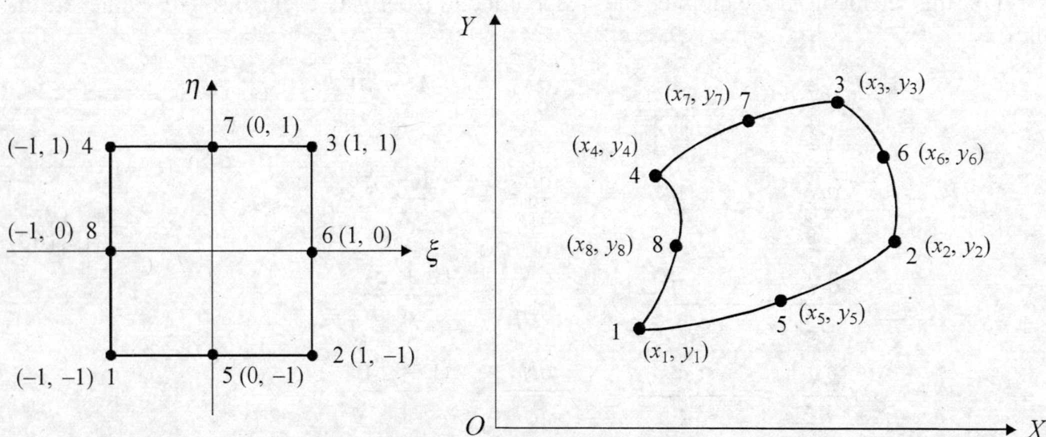

Fig. 5.26 A general eight-noded quadrilateral element.

where we may take coordinates of just the four vertices and follow a simple linear interpolation for coordinate transformation (or) use the full eight nodes to permit curved edges. The displacements of the interior point P are also interpolated from nodal deflections using the same shape functions. Three possibilities arise as shown below:

Coordinate Interpolation	**Displacement Interpolation**
Linear (i.e. four vertex nodes)	Quadratic (i.e. all eight nodes)
Quadratic	Quadratic
Quadratic	Linear

The first element belongs to the category of sub-parametric elements and is useful when the structural geometry is simple polygonal, but unknown field variation may involve sharp variations. The second element is an isoparametric element and permits curved edge modelling as well as quadratic variations in displacements. If the geometry has certain curved features but is in a low stress region, we can use the third element belonging to the superparametric category. The displacements can, in general, be written as

$$
\begin{Bmatrix} u \\ v \end{Bmatrix}_{2\times1} = \begin{bmatrix} N_1 & 0 & N_2 & 0 & \cdots \\ 0 & N_1 & 0 & N_2 & \cdots \end{bmatrix}_{2\times16} \begin{Bmatrix} u_1 \\ v_1 \\ u_2 \\ v_2 \\ \vdots \end{Bmatrix}_{16\times1}
\tag{5.167}
$$

where the shape functions are appropriately taken from Eqs. (5.54)–(5.61) and sizes of the matrices shown are valid for the sub- and isoparametric elements.

The corresponding strain-displacement relation, following Eq. (5.164), is written as

$$
\{\varepsilon\}_{3\times1} = \underbrace{[B_1]_{3\times4}[B_2]_{4\times4}[B_3]_{4\times16}}\ \{\delta\}^e_{16\times1}
$$
$$
= [B]_{3\times16}\{\delta\}^e_{16\times1}
\tag{5.168}
$$

For the eight-noded isoparametric element, for example, the Jacobian can be obtained as

$$
[J] = \begin{bmatrix} \sum_{i=1}^{8} \dfrac{\partial N_i}{\partial \xi} x_i & \sum_{i=1}^{8} \dfrac{\partial N_i}{\partial \xi} y_i \\[2ex] \sum_{i=1}^{8} \dfrac{\partial N_i}{\partial \eta} x_i & \sum_{i=1}^{8} \dfrac{\partial N_i}{\partial \eta} y_i \end{bmatrix}
\tag{5.169}
$$

The necessary shape function derivatives in the ξ-η frame are summarised now from Eqs. (5.54)–(5.61):

$$\frac{\partial N_1}{\partial \xi} = \frac{(1 - \eta)(\eta + 2\xi)}{4}, \qquad \frac{\partial N_1}{\partial \eta} = \frac{(1 - \xi)(\xi + 2\eta)}{4}$$

$$\frac{\partial N_2}{\partial \xi} = \frac{(1 - \eta)(2\xi - \eta)}{4}, \qquad \frac{\partial N_2}{\partial \eta} = \frac{(1 + \xi)(2\eta - \xi)}{4}$$

$$\frac{\partial N_3}{\partial \xi} = \frac{(1 + \eta)(2\xi + \eta)}{4}, \qquad \frac{\partial N_3}{\partial \eta} = \frac{(1 + \xi)(2\eta + \xi)}{4}$$

$$\frac{\partial N_4}{\partial \xi} = \frac{(1 + \eta)(2\xi - \eta)}{4}, \qquad \frac{\partial N_4}{\partial \eta} = \frac{(1 - \xi)(2\eta - \xi)}{4}$$

$$\frac{\partial N_5}{\partial \xi} = -\xi(1 - \eta), \qquad \frac{\partial N_5}{\partial \eta} = -\frac{(1 - \xi^2)}{2} \tag{5.170}$$

$$\frac{\partial N_6}{\partial \xi} = \frac{1 - \eta^2}{2}, \qquad \frac{\partial N_6}{\partial \eta} = -\eta(1 + \xi)$$

$$\frac{\partial N_7}{\partial \xi} = -\xi(1 + \eta), \qquad \frac{\partial N_7}{\partial \eta} = \frac{(1 - \xi^2)}{2}$$

$$\frac{\partial N_8}{\partial \xi} = \frac{-(1 - \eta^2)}{2}, \qquad \frac{\partial N_8}{\partial \eta} = -\eta(1 - \xi)$$

We can write the expression for the element stiffness matrix as

$$
\begin{aligned}
\left[k\right]^e &= \int_v [B]^T [D][B]\, dv \\
&= \iint [B]^T [D][B]\, t\, dx\, dy \\
&= \int_{-1}^{1}\int_{-1}^{1} [B]^T [D][B]\, t\, |J|\, d\xi\, d\eta
\end{aligned}
\tag{5.171}
$$

which can be evaluated by using suitable numerical integration schemes.

5.4.7 Nine-node Quadrilateral Element

This element, as shown in Figure 5.27, is essentially similar to the eight-node element discussed in section 5.4.6 except that we have added one more node in the centre ($\xi = \eta = 0$) of the element. The shape functions for the eight-node element were derived based on the serendipity approach. The basic idea behind adding the centre node here is that we can use the product of one-dimensional Lagrange polynomials. For node 1, the quadratic, one-dimensional Lagrange polynomial in ξ can be written as

$$L_1(\xi) = \frac{(\xi - \xi_5)(\xi - \xi_2)}{(\xi_1 - \xi_5)(\xi_1 - \xi_2)} = \frac{(\xi - 0)(\xi - 1)}{(-1-0)(-1 - 1)} = \frac{1}{2}\xi(\xi - 1) \tag{5.172}$$

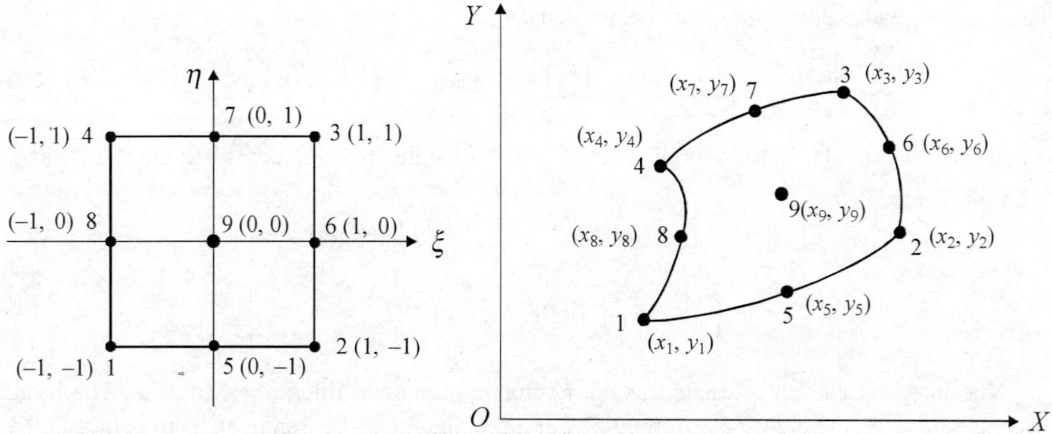

Fig. 5.27 A general nine-noded quadrilateral element.

and in η it can be written as

$$L_1(\eta) = \frac{(\eta - \eta_8)(\eta - \eta_4)}{(\eta_1 - \eta_8)(\eta_1 - \eta_4)} = \frac{(\eta - 0)(\eta - 1)}{(-1 - 0)(-1 - 1)} = \frac{1}{2}\eta(\eta - 1) \qquad (5.173)$$

The complete shape function for node 1 can now be obtained as product of the above polynomials, in the form

$$N_1 = L_1(\xi)L_1(\eta) = \frac{1}{4}\xi\eta(\xi - 1)(\eta - 1) = \frac{\xi\eta}{4}(1-\xi)(1-\eta) \qquad (5.174)$$

The other shape functions can similarly be obtained as

$$N_2 = \frac{-\xi\eta}{4}(1 + \xi)(1 - \eta), \qquad N_3 = \frac{\xi\eta}{4}(1 + \xi)(1 + \eta), \qquad N_4 = \frac{-\xi\eta}{4}(1 + \eta)(1 - \xi)$$

$$N_5 = \frac{-\eta}{2}(1 - \eta)(1 - \xi^2), \qquad N_6 = \frac{\xi}{2}(1 + \xi)(1 - \eta^2), \qquad N_7 = \frac{\eta}{2}(1 + \eta)(1 - \xi^2)$$

$$N_8 = \frac{-\xi}{2}(1-\xi)(1-\eta^2), \qquad N_9 = (1 - \xi^2)(1 - \eta^2) \qquad (5.175)$$

The Jacobian matrix of coordinate transformation is given by

$$[J] = \begin{bmatrix} \displaystyle\sum_{i=1}^{9} \frac{\partial N_i}{\partial \xi} x_i & \displaystyle\sum_{i=1}^{9} \frac{\partial N_i}{\partial \xi} y_i \\[2em] \displaystyle\sum_{i=1}^{9} \frac{\partial N_i}{\partial \eta} x_i & \displaystyle\sum_{i=1}^{9} \frac{\partial N_i}{\partial \eta} y_i \end{bmatrix} \qquad (5.176)$$

The strain-displacement matrix can be written as

$$\{\varepsilon\}_{3\times 1} = \underbrace{[B_1]_{3\times 4}[B_2]_{4\times 4}[B_3]_{4\times 18}}\ \{\delta\}^e_{18\times 1} = [B]_{3\times 18}\{\delta\}^e_{18\times 1} \tag{5.177}$$

Using the standard stress-strain matrix, we can write the element stiffness matrix as

$$\begin{aligned}
[k]^e &= \int_v [B]^T[D][B]\ dv = \iint [B]^T[D][B]t\ dx\ dy \\
&= \int_{-1}^{1}\int_{-1}^{1} [B]^T[D][B]t\,|J|\,d\xi\ d\eta
\end{aligned} \tag{5.178}$$

The integrals can be evaluated using a suitable numerical integration scheme. The basic difference between the eight-node serendipity and the nine-node Lagrange element comes in the shape functions because of the addition of a node. It is observed that the serendipity shape functions involves the following eight polynomial terms:

$$1,\ \xi,\ \eta,\ \xi^2,\ \xi\eta,\ \eta^2,\ \xi^2\eta,\ \xi\eta^2 \tag{5.179}$$

The Lagrange shape functions for the nine-node element include an additional ninth polynomial term, viz.

$$\xi^2\eta^2 \tag{5.180}$$

Thus, both the elements adopt more terms than necessary for the complete quadratic polynomial in $\xi - \eta$, viz., $(1,\ \xi,\ \eta,\ \xi^2,\ \xi\eta,\ \eta^2)$. In general, serendipity elements have fewer extra terms. These two elements are commonly used in two-dimensional analysis and have certain crucial variations in their behaviour, a detailed discussion of which is beyond the scope of the present discussion.

5.4.8 Six-node Triangular Element

A typical six-node triangular element is shown in Figure 5.28. The quadratic shape functions in natural coordinates are obtained from Eqs. (5.80) and (5.81) as follows:

$$\begin{aligned}
N_1 &= L_1(2L_1 - 1), \qquad N_2 = L_2(2L_2 - 1), \qquad N_3 = L_3(2L_3 - 1) \\
N_4 &= 4L_1L_2, \qquad N_5 = 4L_2L_3, \qquad N_6 = 4L_3L_1
\end{aligned} \tag{5.181}$$

We recall that L_1, L_2, L_3 are not independent:

$$L_1 + L_2 + L_3 = 1$$

Hence,

$$L_3 = 1 - L_1 - L_2 \tag{5.182}$$

So, in terms of L_1, L_2, we rewrite the shape functions as

$$N_3 = (1 - L_1 - L_2)(1 - 2L_1 - 2L_2), \quad N_5 = 4L_2(1 - L_1 - L_2), \quad N_6 = 4L_1(1 - L_1 - L_2) \tag{5.183}$$

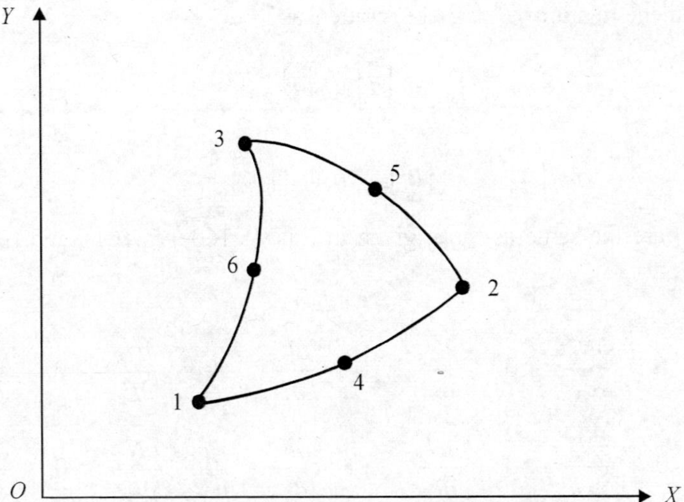

Fig. 5.28   A six-noded triangular element.

The derivatives of the shape function are given by

$$\frac{\partial N_1}{\partial L_1} = 4L_1 - 1, \qquad \frac{\partial N_1}{\partial L_2} = 0$$

$$\frac{\partial N_2}{\partial L_1} = 0, \qquad \frac{\partial N_2}{\partial L_2} = 4L_2 - 1$$

$$\frac{\partial N_3}{\partial L_1} = -3 + 4L_1 + 4L_2, \qquad \frac{\partial N_3}{\partial L_2} = -3 + 4L_1 + 4L_2$$

$$\frac{\partial N_4}{\partial L_1} = 4L_2, \qquad \frac{\partial N_4}{\partial L_2} = 4L_1 \qquad (5.184)$$

$$\frac{\partial N_5}{\partial L_1} = -4L_2, \qquad \frac{\partial N_5}{\partial L_2} = 4 - 4L_1 - 8L_2$$

$$\frac{\partial N_6}{\partial L_1} = 4 - 4L_2 - 8L_1, \qquad \frac{\partial N_6}{\partial L_2} = -4L_1$$

The Jacobian for the isoparametric element is given by

$$[J] = \begin{bmatrix} \displaystyle\sum_{i=1}^{6} \frac{\partial N_i}{\partial L_1} x_i & \displaystyle\sum_{i=1}^{6} \frac{\partial N_i}{\partial L_1} y_i \\[2em] \displaystyle\sum_{i=1}^{6} \frac{\partial N_i}{\partial L_2} x_i & \displaystyle\sum_{i=1}^{6} \frac{\partial N_i}{\partial L_2} y_i \end{bmatrix} \qquad (5.185)$$

The strain-displacement relation $[B]$ can be written as

$$\{\varepsilon\}_{3\times1} = [B]_{3\times12}\{\delta\}^e_{12\times1} \tag{5.186}$$

where

$$[B]_{3\times12} = [B_1]_{3\times4}[B_2]_{4\times4}[B_3]_{4\times12} \tag{5.187}$$

where $[B_1]$ and $[B_2]$ are the same as those given in Eqs. (5.160)–(5.161), and $[B_3]$ is obtained as

$$[B_3]_{3\times12} = \begin{bmatrix} \dfrac{\partial N_1}{\partial L_1} & 0 & \dfrac{\partial N_2}{\partial L_1} & 0 & \dfrac{\partial N_3}{\partial L_1} & 0 & \dfrac{\partial N_4}{\partial L_1} & 0 & \dfrac{\partial N_5}{\partial L_1} & 0 & \dfrac{\partial N_6}{\partial L_1} & 0 \\[2ex] \dfrac{\partial N_1}{\partial L_2} & 0 & \dfrac{\partial N_2}{\partial L_2} & 0 & \dfrac{\partial N_3}{\partial L_2} & 0 & \dfrac{\partial N_4}{\partial L_2} & 0 & \dfrac{\partial N_5}{\partial L_2} & 0 & \dfrac{\partial N_6}{\partial L_2} & 0 \\[2ex] 0 & \dfrac{\partial N_1}{\partial L_1} & 0 & \dfrac{\partial N_2}{\partial L_1} & 0 & \dfrac{\partial N_3}{\partial L_1} & 0 & \dfrac{\partial N_4}{\partial L_1} & 0 & \dfrac{\partial N_5}{\partial L_1} & 0 & \dfrac{\partial N_6}{\partial L_1} \\[2ex] 0 & \dfrac{\partial N_1}{\partial L_2} & 0 & \dfrac{\partial N_2}{\partial L_2} & 0 & \dfrac{\partial N_3}{\partial L_2} & 0 & \dfrac{\partial N_4}{\partial L_2} & 0 & \dfrac{\partial N_5}{\partial L_2} & 0 & \dfrac{\partial N_6}{\partial L_2} \end{bmatrix}_{4\times12} \tag{5.188}$$

The element stiffness matrix can be expressed as

$$[k]^e = \int_v [B]^T[D][B]\, dv = \int_0^1 \int_{-0}^{1-L_1} [B]^T[D][B]t\,|J|\,dL_1\,dL_2 \tag{5.189}$$

where again, the evaluation of the integrals can be done using suitable integration rule in triangular coordinates. We observe that L_1, L_2 are similar to ξ, η except that range of L_1 is $(0, 1)$ rather than $(-1, 1)$ and, in view of the triangular domain, L_2 varies between 0 and $(1 - L_1)$.

We have discussed the formulation of several two-dimensional elements for structural analysis purpose. Using natural coordinates and isoparametric formulation, the procedure for derivation of element characteristic matrices is fairly systematic and amenable to computerisation. While we have discussed linear and quadratic elements that are most commonly used in practice, we can extend the same procedure to formulate cubic or quartic or other higher order elements. Explicit derivation of element level matrices is quite tedious and we look for appropriate schemes for numerical integration, which will now be discussed.

5.5   Numerical Integration

We will first discuss numerical integration in one dimension and then extend the scheme to two dimensions. Consider the evaluation of the following integral:

$$I = \int_{x_1}^{x_2} f(x)\, dx \tag{5.190}$$

The geometrical interpretation of the integration is that we are trying to determine the area under the curve $f(x)$ between the limits x_1 and x_2. Let the function be known at discrete points x_i in this interval. Two classical schemes of numerical integration are the Trapezoidal rule and the Simpson's rule. These are now summarised.

5.5.1 Trapezoidal Rule

Let a discrete set of uniformly spaced points x_i and the corresponding values of the function f_i be given. In this scheme, we assume that the function is varying linearly within each segment as shown in Figure 5.29. We know that the area of the trapezium $ABCD$ is given by

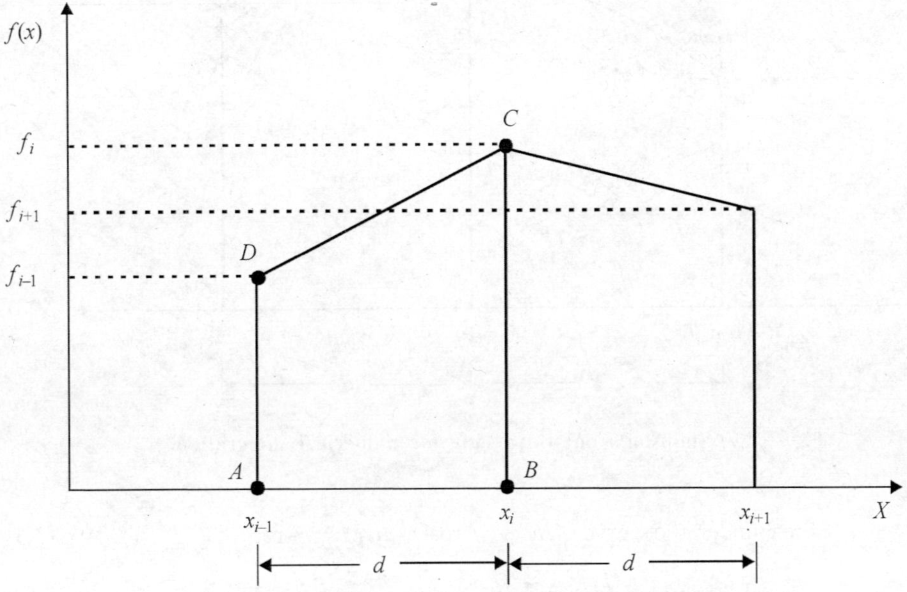

Fig. 5.29 **Trapezoidal rule for numerical integration.**

$$\text{Area } ABCD = \frac{f_i + f_{i-1}}{2} d \tag{5.191}$$

The area under the curve is then approximated as the sum of the areas of the trapezia as indicated in the figure. Thus we have

$$\int f(x)\, dx \approx \sum_{i=1}^{n} \frac{f_i + f_{i-1}}{2} d = d\left[\frac{f_1 + f_n}{2} + (f_2 + f_3 + \cdots + f_{n-1})\right] \tag{5.192}$$

For later use, we now rewrite this formula as

$$\int f(x)\, dx \approx \left(\frac{d}{2}\right) f_1 + (d)f_2 + (d)f_2 + \cdots + (d)f_{n-1} + \left(\frac{d}{2}\right) f_n = \sum_{i=1}^{n} W_i f_i \tag{5.193}$$

where W_i are the weights corresponding to f_i.

5.5.2   Simpson's 1/3 Rule

In this scheme, instead of linear approximation, we use a parabolic approximation for the function within each segment as shown in Figure 5.30. We can find the area under the curve in one such segment as follows:

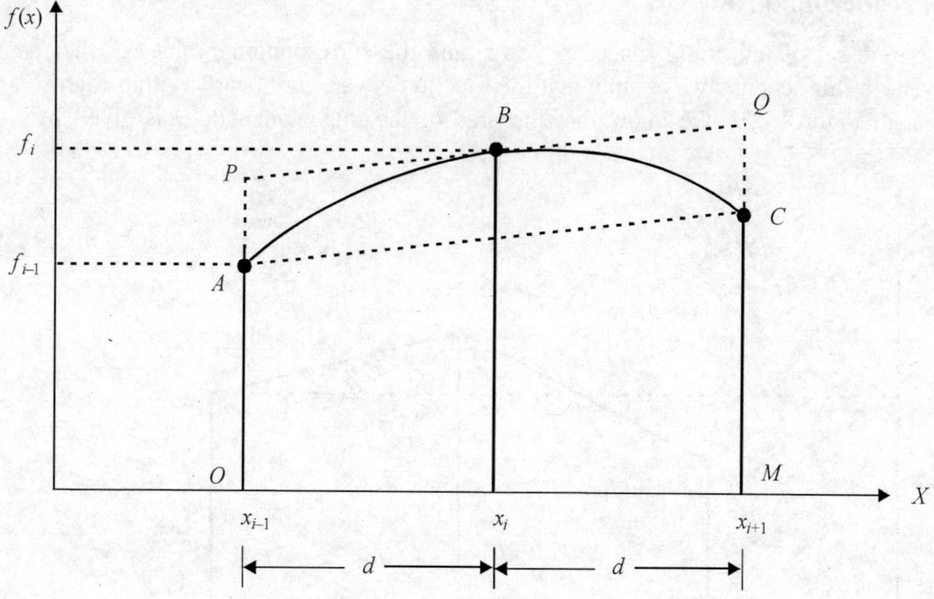

Fig. 5.30   Simpson's one-third rule for numerical integration.

$$\text{Area under the curve } ABC = \text{Area } OACM + \text{Area } ABCA \tag{5.194}$$

$$\text{Area } OACM = \left(\frac{f_{i-1} + f_{i+1}}{2}\right)(2d) = d(f_{i-1} + f_{i+1}) \tag{5.195}$$

$$\text{Area } ABCA = \frac{2}{3}\ (\text{area } APBQCA) = \frac{2}{3}\left(f_i - \frac{f_{i-1} + f_{i+1}}{2}\right)(2d) \tag{5.196}$$

Thus

$$\text{Area of one segment} = \frac{d}{3}(f_{i-1} + 4f_i + f_{i+1}) \tag{5.197}$$

$$\text{Total area under the curve} = \sum_{i=2}^{n}\frac{d}{3}(f_{i-1} + 4f_i + f_{i+1})$$

$$= \left(\frac{d}{3}\right)f_1 + \left(\frac{5d}{3}\right)f_2 + \left(\frac{6d}{3}\right)f_3 + \cdots + \left(\frac{5d}{3}\right)f_{n-1} + \left(\frac{d}{3}\right)f_n \tag{5.198}$$

Thus our procedure for numerical integration consists of two simple steps, viz., first do a curve fit (straight line, parabola, etc.) for the function within each segment and next evaluate the total area, and express the result as a generic formula of the type

$$\int f(x)\, dx = \sum_{i=1}^{n} W_i f_i \tag{5.199}$$

The numerical integration scheme popularly used in finite element computations is known as the *Gauss quadrature*, which is essentially based on similar lines. We will first discuss the Newton–Cotes formula and then present the Gauss quadrature rule, followed by a set of examples. In view of the intended application in finite element computations using natural coordinates, we shall take, the limits of integration to be -1 and 1.

5.5.3   Newton–Cotes Formula

Let us consider the evaluation of the integral

$$I(x) = \int_{-1}^{1} f(x)\, dx \tag{5.200}$$

where we assume that the function values f_i at some arbitrarily spaced $x_i (i = 1, 2, \ldots, n)$ are known. We then approximate the function $f(x)$ using a polynomial $g(x)$ as

$$f(x) \approx g(x) = c_0 + c_1 x + c_2 x^2 + \cdots + c_{n-1} x^{n-1} \tag{5.201}$$

Considering that at $x = x_i$, $g(x_i) = f_i$, we get equations

$$
\begin{aligned}
c_0 + c_1 x_1 + c_2 x_1^2 + \cdots + c_{n-1} x_1^{n-1} &= f_1 \\
\vdots \qquad \vdots \qquad \vdots \qquad\qquad \vdots \qquad\quad \vdots & \\
c_0 + c_1 x_n + c_2 x_n^2 + \cdots + c_{n-1} x_n^{n-1} &= f_n
\end{aligned}
\tag{5.202}
$$

If $x_1, x_2, \ldots, x_n$ are given, then Eqs. (5.202) are simple linear algebraic equations which can be readily solved using, for example, the Gauss elimination. We thus have n equations in n unknowns, which can be used to solve for the coefficients $c_0, c_1, \ldots, c_{n-1}$ in terms of the function values f_i. Assuming that these coefficients have been determined, we can evaluate the integral as

$$I = \int_{-1}^{1} f(x)\, dx \approx \int_{-1}^{1} g(x)\, dx \tag{5.203}$$

However,

$$\int_{-1}^{1} g(x)\, dx = \int_{-1}^{1} (c_0 + c_1 x + c_2 x^2 + \cdots + c_{n-1} x^{n-1})\, dx$$

$$= \left[c_0 x + c_1 \frac{x^2}{2} + c_2 \frac{x^3}{3} + \cdots + c_{n-1} \frac{x^n}{n} \right]_{-1}^{1}$$

$$= 2c_0 + 0 + \frac{2}{3} c_2 + 0 + \frac{2}{5} c_4 + \cdots \tag{5.204}$$

If we substitute the values of the coefficients in this expression, we obtain a formula of the type

$$I \approx W_1 f_1 + W_2 f_2 + \cdots + W_n f_n \tag{5.205}$$

Using this formula, called the Newton–Cotes formula, our evaluation of the integral will be exact if the original function $f(x)$ were any polynomial up to degree $n-1$. Here we have assumed that at given locations (sampling points) x_i, the function values f_i are known. How do we know which are the "good" locations for sampling the function? The Gauss quadrature formula attempts to find the locations of the sampling points x_i as well as the weight factors W_i. We will now discuss the details of this scheme which is commonly used in finite element computations.

5.5.4   Gauss Quadrature Formula

Let the function be approximated by a polynomial of degree p, which is also to be found, i.e., let

$$f(x) \approx g(x) = c_0 + c_1 x + c_2 x^2 + \cdots + c_p x^p \tag{5.206}$$

The evaluation of the integral can be done as follows:

$$I = \int_{-1}^{1} f(x)\, dx \approx \int_{-1}^{1} g(x)\, dx = 2c_0 + \frac{2}{3} c_2 + \frac{2}{5} c_4 + \cdots + \frac{c_p}{p+1}[1 - (-1)^{p+1}] \tag{5.207}$$

As per our generic formula, we attempt to express the integral in the following way:

$$I \approx W_1 f_1 + W_2 f_2 + \cdots + W_n f_n \tag{5.208}$$

Using Eq. (5.206), we can write

$$\begin{aligned}
I \approx\ & W_1(c_0 + c_1 x_1 + c_2 x_1^2 + \cdots + c_p x_1^p) \\
& + W_2(c_0 + c_1 x_2 + c_2 x_2^2 + \cdots + c_p x_2^p) \\
& + \cdots \\
& + \cdots \\
& + W_n(c_0 + c_1 x_n + c_2 x_n^2 + \cdots + c_p x_n^p)
\end{aligned} \tag{5.209}$$

Comparing the coefficients of c_0, c_1, c_2, ... in Eqs. (5.207) and (5.209), we can write

$$\begin{aligned}
W_1 + W_2 + \cdots + W_n &= 2 \\
W_1 x_1 + W_2 x_2 + \cdots + W_n x_n &= 0 \\
W_1 x_1^2 + W_2 x_2^2 + \cdots + W_n x_n^2 &= \frac{2}{3} \\
&\ \ \vdots \\
W_1 x_1^p + W_2 x_2^p + \cdots + W_n x_n^p &= \frac{1}{p+1}[1 - (-1)^{p+1}]
\end{aligned} \tag{5.210}$$

These equations can be solved for the locations of the sampling points $x_i(i = 1, 2, ..., n)$ and the corresponding weight factors $W_i(i = 1, 2, ..., n)$. It is easy to see that our number of unknowns (viz., $2n$) and equations available ($p + 1$) match exactly when the polynomial degree p is such that

$$p + 1 = 2n$$

or

$$\boxed{p = 2n - 1} \tag{5.211}$$

In other words, using n Gauss sampling points as determined above, we can evaluate an integral exactly if the integrand function $f(x)$ is a polynomial of degree less than or equal to $(2n - 1)$. Thus, with one-point rule, we can integrate a linear polynomial exactly, while with two-point rule we can integrate up to cubic polynomial exactly. In practical use, we observe the degree of the highest terms in the integrand and then decide the order of Gauss quadrature rule. As compared to the Newton–Cotes formula, we observe that, for a given number of sampling points, the Gauss quadrature rule can exactly integrate much higher degree polynomial.

We now list the sampling points and the weight factors for Gauss quadrature for one-dimensional case in Table 5.1.

Table 5.1   Sampling Points and Weights for Gauss Quadrature

n	Location of sampling point x_i	Weight factor W_i
1	0	2
2	$+\dfrac{1}{\sqrt{3}}$	1
	$-\dfrac{1}{\sqrt{3}}$	1
3	$+\sqrt{0.6}$	5/9
	0	8/9
	$-\sqrt{0.6}$	5/9
4	$+0.8611363$	0.347854845
	$+0.3399810$	0.652145155
	-0.3399810	0.652145155
	-0.8611363	0.347854845

It is easy to observe that the exact numerical values of the sampling points and the weight factors, obtained as a solution of Eq. (5.210), depend on the integration interval. Thus the formula is standardised for the interval -1 to 1, keeping in mind the finite element computations using natural coordinates. If the interval is different, we need to perform an appropriate change of coordinates such that the limits tally with -1 and 1. We will now illustrate the use of the formula through the following examples.

Example 5.2. *Evaluate the integral*

$$I = \int_{-1}^{1} \left(2 + x + x^2\right) dx \tag{5.212}$$

It is easy to verify that we need at least a two-point integration rule since the integrand contains a quadratic term. We will obtain our approximation to the integral with one and two-point Gauss quadrature rules and show that the two-point rule result matches with the exact solution.

For a one-point rule, from Table 5.1, we have

$$x_1 = 0, \qquad W_1 = 2$$

Hence,

$$I \approx 2(2 + 0 + 0) = 4 \tag{5.213}$$

For a two-point rule, from Table 5.1 we have

$$x_1 = + \frac{1}{\sqrt{3}}, \qquad W_1 = 1$$

$$x_2 = - \frac{1}{\sqrt{3}}, \qquad W_2 = 1$$

Therefore,

$$I \approx (1)\left(2 + \frac{1}{\sqrt{3}} + \frac{1}{3} \right) + (1)\left(2 - \frac{1}{\sqrt{3}} + \frac{1}{3} \right) = 4.6667 \tag{5.214}$$

The exact solution is easily verified to be

$$I = \int_{-1}^{1} (2 + x + x^2)\, dx = \left(2x + x^2 + x^3/3 \right)_{-1}^{1} = 4.6667 \tag{5.215}$$

Example 5.3.   *Evaluate the integral*

$$I = \int_{-1}^{1} \cos \frac{\pi x}{2}\, dx \tag{5.216}$$

The exact solution can be readily obtained as follows:

$$I = \left(\sin \frac{\pi x}{2} \bigg/ \frac{\pi}{2} \right)_{-1}^{1} = 1.273239544 \tag{5.217}$$

For the one-point Gauss rule,

$$I \approx 2 \cos \left(\frac{(\pi)(0)}{2} \right) = 2 \tag{5.218}$$

For the two-point Gauss rule,

$$I \approx 1 \cos \left(\frac{\pi}{2} \frac{1}{\sqrt{3}} \right) + 1 \cos \left(\frac{\pi}{2} \frac{-1}{\sqrt{3}} \right) \tag{5.219}$$

$$= 2 \cos \left(\frac{\pi}{2\sqrt{3}} \right) = 1.232381 \tag{5.220}$$

For the three-point Gauss rule,

$$I \approx \frac{5}{9} \cos\left(\frac{\pi}{2}\sqrt{0.6}\right) + \frac{8}{9} \cos\left(\frac{\pi}{2} 0\right) + \frac{5}{9} \cos\left(\frac{\pi}{2}\left(-\sqrt{0.6}\right)\right)$$

$$= 1.274123754 \tag{5.221}$$

With a four-point Gauss rule, we get

$$I \approx (0.347854845) \cos\left(\frac{\pi}{2}(0.8611363)\right)$$

$$+ (0.652145155) \cos\left(\frac{\pi}{2}(0.3399810)\right)$$

$$+ (0.652145155) \cos\left(\frac{\pi}{2}(-0.3399810)\right)$$

$$+ (0.347854845) \cos\left(\frac{\pi}{2}(-0.8611363)\right)$$

$$= 1.273229508 \tag{5.222}$$

Since the integrand is not an exact polynomial, our numerical scheme can never give the exact solution but approximates fairly well when sufficient number of points are taken.

5.5.4   Gauss Quadrature in Two Dimensions

We shall first discuss two-dimensional rules appropriate for quadrilateral finite elements and then present integration rules for triangular elements.

For two-dimensional integration, we can write

$$I = \int_{-1}^{1} \int_{-1}^{1} f(\xi,\eta)\ d\xi\, d\eta$$

$$\approx \int_{-1}^{1} \left[\sum_i W_i\, f(\xi_i,\eta)\right] d\eta$$

$$\tag{5.223}$$

$$\approx \sum_j W_j \left[\sum_i W_i\, f(\xi_i,\eta_j)\right] \approx \sum_i \sum_j W_i W_j\, f(\xi_i,\eta_j)$$

Thus we can obtain the necessary Gauss quadrature rule simply by taking the product of appropriate one-dimensional rules. Thus, a (2×2) Gauss rule would be as shown in Figure 5.31.

$$I \approx (1)(1)f\left(\frac{-1}{\sqrt{3}},\frac{-1}{\sqrt{3}}\right) + (1)(1)f\left(\frac{1}{\sqrt{3}},\frac{-1}{\sqrt{3}}\right) + (1)(1)f\left(\frac{1}{\sqrt{3}},\frac{1}{\sqrt{3}}\right)$$

$$+ (1)(1)f\left(\frac{-1}{\sqrt{3}},\frac{1}{\sqrt{3}}\right) \tag{5.224}$$

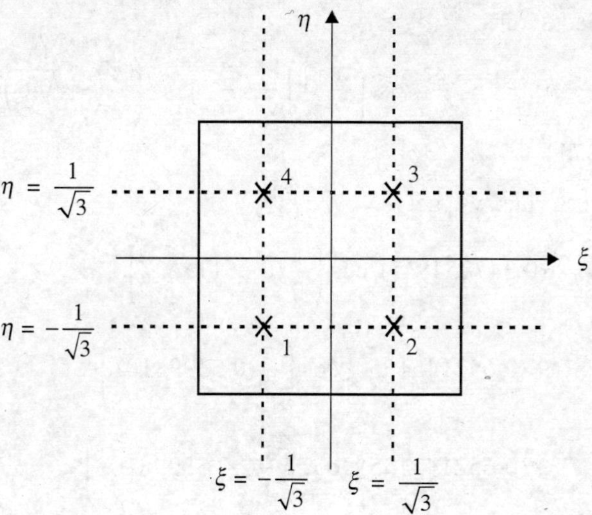

Fig. 5.31 Sampling points for a 2×2 Gauss quadrature rule.

Similarly, a (3×3) rule would be as shown in Figure 5.32. Thus,

$$I \approx \left(\frac{5}{9}\right)\left(\frac{5}{9}\right)f_1 + \left(\frac{8}{9}\right)\left(\frac{5}{9}\right)f_2 + \left(\frac{5}{9}\right)\left(\frac{5}{9}\right)f_3$$
$$+ \left(\frac{5}{9}\right)\left(\frac{8}{9}\right)f_4 + \left(\frac{8}{9}\right)\left(\frac{8}{9}\right)f_5 + \left(\frac{5}{9}\right)\left(\frac{8}{9}\right)f_6 \qquad (5.225)$$
$$+ \left(\frac{5}{9}\right)\left(\frac{5}{9}\right)f_7 + \left(\frac{8}{9}\right)\left(\frac{5}{9}\right)f_8 + \left(\frac{5}{9}\right)\left(\frac{5}{9}\right)f_9$$

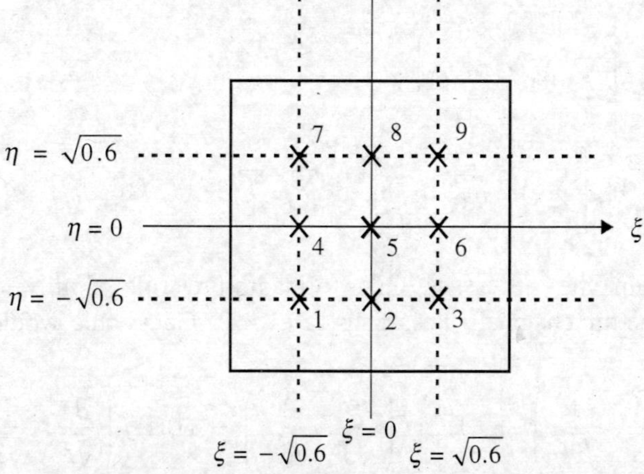

Fig. 5.32 Sampling points for a 3×3 Gauss quadrature rule.

It is not necessary that we take the same number of points along both ξ and η directions. For example, a (2×3) rule would be as shown in Figure 5.33.

$$I \approx (1)\left(\frac{5}{9}\right)f_1 + (1)\left(\frac{5}{9}\right)f_2 + (1)\left(\frac{8}{9}\right)f_3 + (1)\left(\frac{8}{9}\right)f_4 + (1)\left(\frac{5}{9}\right)f_5 + (1)\left(\frac{5}{9}\right)f_6 \quad (5.226)$$

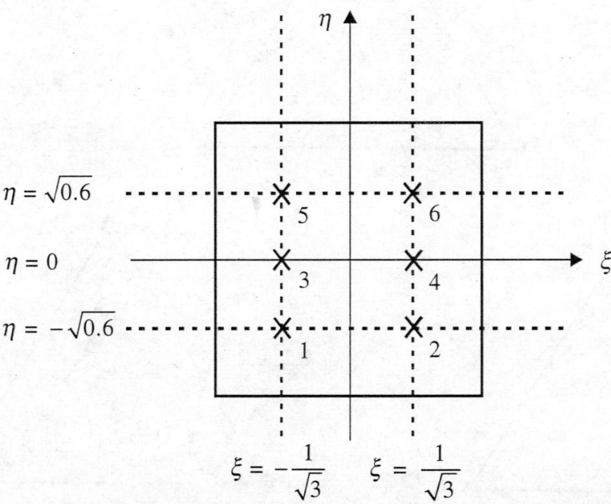

Fig. 5.33   **Sampling points for a 2×3 Gauss quadrature rule.**

The Gauss quadrature rules can be effectively used for finite element computations. Consider that we need to evaluate the following:

$$[k]^e = \int_{-1}^{1} \int_{-1}^{1} [B]^T [D][B]\, t\, |J|\, d\xi\, d\eta \quad (5.227)$$

Using Gauss quadrature rule we rewrite this equation as

$$[k]^e = \sum_i \sum_j \left([B]^T [D][B]\, t\, |J|\right) W_i W_j \quad (5.228)$$

where all the terms within the parentheses are evaluated at each $(\xi_i,\, \eta_j)$. Once we select the appropriate number of Gauss points needed, based on the highest polynomial term expected in the integrand, we can readily evaluate the integral. The integrals corresponding to element nodal force vector are evaluated in a similar manner. Numerical integration for triangular domains follows similar pattern. However, we do not take the product of Gauss quadrature rule in two directions. Instead, we write

$$I = \int_0^1 \int_{-1}^{1-L_1} f(L_1,\, L_2)\, dL_1\, dL_2 \approx \sum W_i f_i \quad (5.229)$$

The locations of the sampling points and the corresponding weights are given in Table 5.2 and are depicted in Figure 5.34.

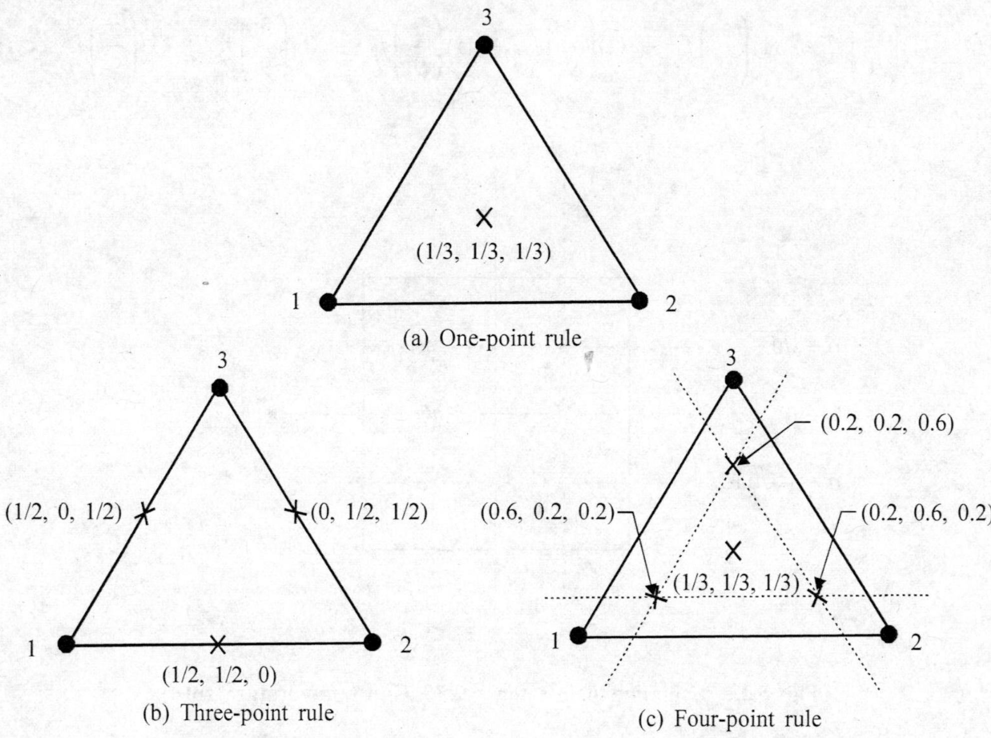

Fig. 5.34   Sampling points for numerical integration over triangular domain (Table 5.2).

Table 5.2   Gauss Quadrature Points and Weights for Triangular Domains

n, No. of points	Location of sampling points	Weight factor W
1	$L_1 = 1/3$ $L_2 = 1/3$ $L_3 = 1/3$	1
3	(1/2, 1/2, 0)	1/3
	(1/2, 0, 1/2)	1/3
	(0, 1/2, 1/2)	1/3
4	(1/3, 1/3, 1/3)	−27/48
	(0.6, 0.2, 0.2)	25/48
	(0.2, 0.6, 0.2)	25/48
	(0.2, 0.2, 0.6)	25/48

We have so far discussed, in detail, the formulation of element level equations and evaluation of element characteristic matrices through numerical integration. These element matrices can be assembled together to form the global system level matrices, in a manner similar to the assembly procedure used for one-dimensional problems. Before we can actually use these elements to solve some example problems, we need to discuss two more aspects, viz., incorporating the prescribed boundary conditions and, secondly, the solution of system level equations. In the sections that follow, we will discuss these two issues.

5.6 Incorporation of Boundary Conditions

Let the system level assembled equations be given as

$$[A]_{n \times n} \{x\}_{n \times 1} = \{b\}_{n \times 1} \qquad (5.230)$$

In general, a d.o.f. x_i may be prescribed to be zero (e.g. a fixed support on a structure) or nonzero (e.g. a prescribed wall temperature). An effective way of implementing either boundary condition is now described. Retaining all other coefficients of $[A]$ and $\{b\}$ the same, we modify a_{ii} and b_i as

$$\boxed{\overline{a}_{ii} = a_{ii} + a_p} \qquad (5.231)$$

$$\boxed{\overline{b}_i = a_p \overline{x}_i} \qquad (5.232)$$

where a_p is called a penalty number and $\overline{x}_i$ is the prescribed value for x_i ($\overline{x}_i$ may or may not be zero). Thus our method is to simply replace existing b_i by $a_p \overline{x}_i$ and add a_p to the existing diagonal coefficient a_{ii}. If $a_p \gg a_{ij}$ ($j = 1, 2, ..., n$), this equation will essentially give $x_i \approx \overline{x}_i$. The larger the value of a_p, the closer our enforcement of the constraint $x_i = \overline{x}_i$. Typically, $a_p = 10^3 - 10^6 \, a_{ii}$. This method is easy to implement in a computer program; it can take care of prescribed zero/nonzero constraints; does not involve any rearrangement of coefficients in $[A]$ or $\{b\}$. However, a large a_p may lead to numerical difficulties in solution. A simpler alternative to take care of prescribed zero boundary condition is to incorporate it at the time of assembly itself, i.e., if a d.o.f. is prescribed to be zero, do not assemble that equation itself (i.e. row and column in the matrices)! A more cumbersome way of implementing the general nonzero prescribed boundary condition is as follows. Consider three equations in three unknowns as

$$a_{11}x_1 + a_{12}x_2 + a_{13}x_3 = b_1$$

$$a_{21}x_1 + a_{22}x_2 + a_{23}x_3 = b_2$$

$$a_{31}x_1 + a_{32}x_2 + a_{33}x_3 = b_3$$

and let $x_1 = \overline{x}_1$. We can rewrite these equations as

$$x_1 = \overline{x}_1$$

$$a_{22}x_2 + a_{23}x_3 = b_2 - a_{21}\overline{x}_1$$

$$a_{32}x_2 + a_{33}x_3 = b_3 - a_{31}\overline{x}_1$$

Thus, our general strategy is to make $a_{ii} = 1$ and all other coefficients in that row and column to be zero. Also, the RHS $\{b\}$ is modified as

$$\left\{ (b_j - a_{ji}\overline{x}_i) \right\}, \qquad j = 1, 2, \ldots, n \tag{5.233}$$

The ith element in $\{b\}$ is simply taken as $\overline{x}_i$. The key idea of this method is that the ith equation is made to read simply as $x_i = \overline{x}_i$; the value $\overline{x}_i$ is plugged in all other equations and taken to the RHS. The enforcement of the constraint $x_i = \overline{x}_i$ is exact and not approximate as in the penalty method.

5.7 Solution of Static Equilibrium Equations

The general form of the finite element equations in linear static analysis is given by

$$[A]_{n \times n} \{x\}_{n \times 1} = \{b\}_{n \times 1} \tag{5.234}$$

In typical finite element models of complex, real-life problems, a large number of elements are required. Hence the size of the matrices (viz. n) is usually very large. The finite element models with several thousand d.o.f. are routinely used. Thus, computationally efficient solution algorithms are a crucial part of a total finite element software package. Two approaches to the solution of Eq. (5.234) are possible—direct method and the iterative methods. The direct solution methods are essentially based on traditional Gauss elimination strategy. Among the iterative methods, the Gauss–Siedel and the conjugate gradient method (with preconditioning) are popular. We will discuss here only Gauss elimination based direct methods. It may be recalled that, in the basic Gauss elimination process, through a series of operations on the matrices $[A]$ and $\{b\}$, we aim to transform Eq. (5.234) into the following form:

$$
\begin{bmatrix}
\overline{a}_{11} & a_{11} & \cdots & \cdots & \cdots & a_{1n} \\
0 & \overline{a}_{22} & \cdots & \cdots & \cdots & \overline{a}_{2n} \\
0 & 0 & \overline{a}_{33} & \cdots & \cdots & \overline{a}_{3n} \\
& & & \ddots & & \\
& & & & \overline{a}_{n-1,n-1} & \overline{a}_{n-1,n} \\
0 & 0 & \cdots & \cdots & \cdots & \overline{a}_{nn}
\end{bmatrix}
\left\{ x \right\} = \left\{ \overline{b} \right\} \tag{5.235}
$$

The coefficient matrix in this form is a simple upper triangular matrix. The last row in Eq. (5.235) reads

$$\overline{a}_{nn} x_n = \overline{b}_n \tag{5.236}$$

Thus,

$$x_n = \overline{b}_n / \overline{a}_{nn} \tag{5.237}$$

From the penultimate row, we have

$$\overline{a}_{n-1,n-1} x_{n-1} + \overline{a}_{n-1,n} x_n = \overline{b}_{n-1} \tag{5.238}$$

Substituting for x_n from Eq. (5.237) into Eq. (5.238), we can find x_{n-1}. Using x_n and x_{n-1}, we can find x_{n-2}, and so on through the process of backward substitution.

With respect to the finite element equations, we observe that the nonzero elements in $[A]$ are clustered around the principal diagonal as shown below.

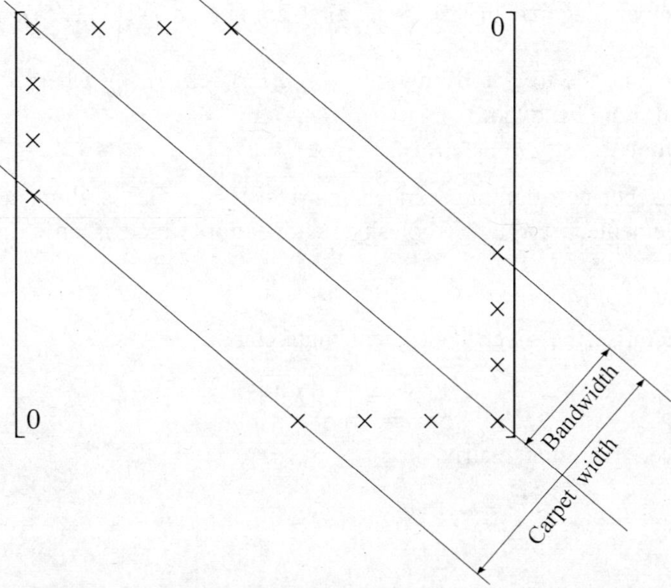

The elements of $[A]$ beyond the carpet width are zero and remain zero throughout the Gauss elimination solution process. Thus it is unnecessary to store and operate on these elements.

A further characteristic of the finite element coefficient matrices in typical structural mechanics applications is symmetry and positive definiteness (i.e. all the eigenvalues of $[A]$ are positive). For a symmetric, banded $[A]$, we need only store and operate on the nonzero elements contained within the bandwidth, as shown below.

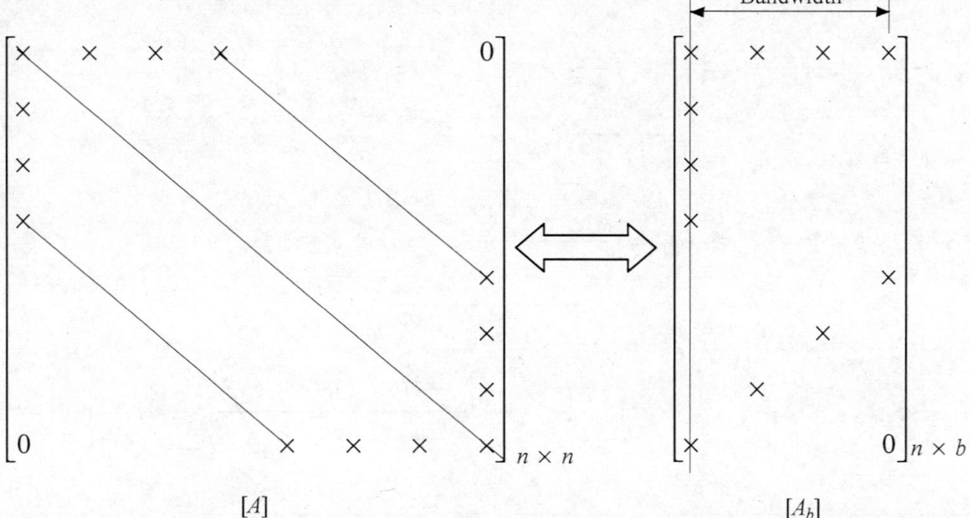

Since $b \ll n$, the saving in computer storage space and also further computations could be tremendous. However, this calls for a (usually small) computer program to keep track of the correspondence of elements in the original $[A]$ and $[A_b]$ matrices. The key ideas in such a program are:

> ➤ The principal diagonal in $[A]$ is the first column in $[A_b]$, i.e. $a(i, i) = a_b(i, 1)$ for $i = 1, n$.
> ➤ Similarly, for $j = i$ to $(i + \text{bandwidth} -1)$, $a(i, j) = a_b (i, \ell)$, where ℓ is an index that goes from 1 to bandwidth.
> ➤ From symmetry, $a(j,i) = a(i,j)$.

For symmetric, positive definite, banded matrix $[A]$, the Gauss elimination solution can be effectively implemented through Cholesky factorisation which is now discussed.

Cholesky factorisation

Here we aim to factorise the given coefficient matrix as

$$\boxed{[A] = [L][L]^T} \tag{5.239}$$

where $[L]$ is a lower triangular matrix, i.e.,

$$[L] = \begin{bmatrix} \ell_{11} & & & \\ \ell_{12} & \ell_{22} & 0 & \\ \vdots & & \ddots & \\ \ell_{1n} & & \cdots & \ell_{nn} \end{bmatrix} \tag{5.240}$$

If we perform the multiplication $[L] [L]^T$ and compare coefficient by coefficient with elements in $[A]$, we get the necessary equations to find ℓ_{ij}.

$$\ell_{11}^2 = a_{11} \Rightarrow \ell_{11} = \sqrt{a_{11}}$$

$$\ell_{11}\ell_{1j} = a_{1j} \Rightarrow \ell_{1j} = \frac{a_{1j}}{\ell_{11}}, \qquad j = 2, n$$

$$\ell_{12}^2 + \ell_{22}^2 = a_{22} \Rightarrow \ell_{22} = \sqrt{a_{22} - \ell_{12}^2}$$

$$\ell_{12}\ell_{1j} + \ell_{22}\ell_{2j} = a_{2j} \Rightarrow \ell_{2j} = \frac{a_{2j} - \ell_{12}\ell_{1j}}{\ell_{22}}, \qquad j = 3, n$$

$$\ell_{13}^2 + \ell_{23}^2 + \ell_{33}^2 = a_{33} \Rightarrow \ell_{33} = \sqrt{a_{33} - (\ell_{13}^2 + \ell_{23}^2)}$$

$$\ell_{13}\ell_{1j} + \ell_{23}\ell_{2j} + \ell_{33}\ell_{3j} = a_{3j} \Rightarrow \ell_{3j} = \frac{a_{3j} - (\ell_{13}\ell_{1j} + \ell_{23}\ell_{2j})}{\ell_{33}}, \qquad j = 4, n \tag{5.241}$$

Thus, all the coefficients in $[L]$ can be determined. We have

$$[A]\{x\} = \{b\} \tag{5.242}$$

or

$$[L][L]^T\{x\} = \{b\} \tag{5.243}$$

let

$$[L]^T\{x\} = \{y\} \tag{5.244}$$

Hence

$$[L]\{y\} = \{b\} \tag{5.245}$$

i.e.

$$\begin{bmatrix} \ell_{11} & & & \\ \ell_{12} & \ell_{22} & 0 & \\ \vdots & & \ddots & \\ \ell_{1n} & & \cdots & \ell_{nn} \end{bmatrix} \begin{Bmatrix} y_1 \\ y_2 \\ \vdots \\ y_n \end{Bmatrix} = \begin{Bmatrix} b_1 \\ b_2 \\ \vdots \\ b_n \end{Bmatrix} \tag{5.246}$$

The first row gives

$$\ell_{11} y_1 = b_1 \Rightarrow y_1 = \frac{b_1}{\ell_{11}} \tag{5.247}$$

and the second row yields

$$\ell_{12} y_1 + \ell_{22} y_2 = b_2 \tag{5.248}$$

Substituting for y_1 from Eq. (5.247), we can find y_2 and so on till y_n through the process of "forward substitution". Once $\{y\}$ is found, we use (from Eq. (5.244)).

$$\begin{bmatrix} \ell_{11} & \ell_{12} & & \ell_{1n} \\ & \ell_{22} & & \\ 0 & & \ddots & \\ & & & \ell_{nn} \end{bmatrix} \begin{Bmatrix} x_1 \\ x_2 \\ \vdots \\ x_n \end{Bmatrix} = \begin{Bmatrix} y_1 \\ y_2 \\ \vdots \\ y_n \end{Bmatrix} \tag{5.249}$$

Equation (5.249) gives x_n and we find x_{n-1}, x_{n-2} ..., x_1 through backward substitution. Thus the Cholesky scheme of solution is in three steps:

Step 1: Cholesky factorisation $[A] = [L][L]^T$.

Step 2: Forward substitution $[L]\{y\} = \{b\}$.

Step 3: Backward substitution $[L]^T\{x\} = \{y\}$.

In the actual implementation on a banded matrix, the coefficients (ℓ_{ij}) outside the bandwidth are zero and are therefore never computed. For a matrix $[A_b]_{n \times b}$ size, $[L]$ is also of size $(n \times b)$. It can be shown that the total number of floating point operations involved in the above three steps is approximately $\left[(nb^2/2) + 2nb\right]$, where the Cholesky factorisation takes $(nb^2/2)$ flops and each of steps 2 and 3 takes (nb) flops. For a given mesh (i.e. n is given), the node numbering scheme dictates bandwidth b and the computational effort depends significantly on b. Hence an efficient solution strategy will require that bandwidth "b" be as small as possible. Thus efficient node numbering schemes, which optimize the resulting bandwidth, are an integral part of any pre-processor in a typical finite element software package. We will now present a few example problems.

Example 5.4. *Illustration of use of bilinear rectangular element.* We wish to redo the problem of Example 5.1, but this time, using a single rectangular bilinear element as shown in Figure 5.35. The computations presented here are based on a 2×2 Gauss quadrature. At each integration point, the Jacobian and $[B]$ are given as follows:

Integration Point 1:

$$\xi = -1/\sqrt{3}, \qquad \eta = -1/\sqrt{3}$$

$$[J] = \begin{bmatrix} 2 & 0 \\ 0 & 1.5 \end{bmatrix}, \qquad |J| = 3.0$$

$$[B] = \begin{bmatrix} -0.197 & 0 & 0.197 & 0 & 0.0528 & 0 & -0.0528 & 0 \\ 0 & -0.263 & 0 & -0.0704 & 0 & 0.0704 & 0 & 0.263 \\ -0.263 & -0.197 & -0.0704 & 0.197 & 0.0704 & 0.0528 & 0.263 & -0.0528 \end{bmatrix} \quad (5.250)$$

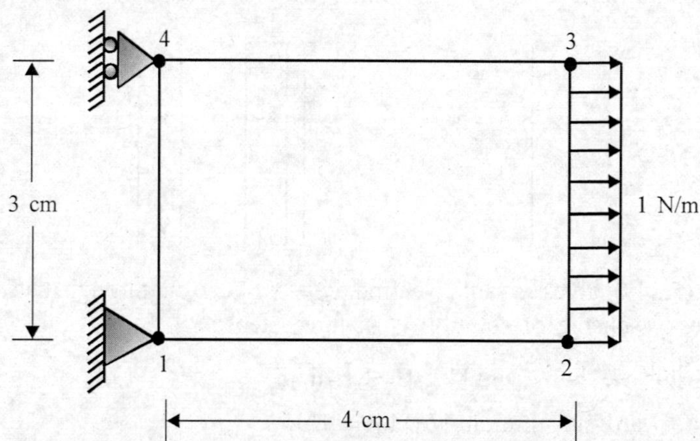

Fig. 5.35   Thin plate under uniform tension (Example 5.4).

Integration Point 2:

$$\xi = -1/\sqrt{3}, \qquad \eta = 1/\sqrt{3}$$

$$[J] = \begin{bmatrix} 2 & 0 \\ 0 & 1.5 \end{bmatrix}, \qquad |J| = 3.0$$

$$[B] = \begin{bmatrix} -0.0528 & 0 & 0.0528 & 0 & 0.197 & 0 & -0.197 & 0 \\ 0 & -0.263 & 0 & -0.0704 & 0 & 0.0704 & 0 & 0.263 \\ -0.263 & -0.0528 & -0.0704 & 0.0528 & 0.0704 & 0.197 & 0.263 & -0.197 \end{bmatrix} \quad (5.251)$$

Integration Point 3:

$$\xi = 1/\sqrt{3}, \qquad \eta = -1/\sqrt{3}$$

$$[J] = \begin{bmatrix} 2 & 0 \\ 0 & 1.5 \end{bmatrix}, \qquad |J| = 3.0$$

$$[B] = \begin{bmatrix} -0.197 & 0 & 0.197 & 0 & 0.0528 & 0 & -0.0528 & 0 \\ 0 & -0.0704 & 0 & -0.263 & 0 & 0.263 & 0 & 0.0704 \\ -0.0704 & -0.197 & -0.263 & 0.197 & 0.263 & 0.0528 & 0.0704 & -0.0528 \end{bmatrix} \quad (5.252)$$

Integration Point 4:

$$\xi = 1/\sqrt{3}, \qquad \eta = 1/\sqrt{3}$$

$$[J] = \begin{bmatrix} 2 & 0 \\ 0 & 1.5 \end{bmatrix}, \qquad |J| = 3.0$$

$$[B] = \begin{bmatrix} -0.0528 & 0 & 0.0528 & 0 & 0.197 & 0 & -0.197 & 0 \\ 0 & -0.0704 & 0 & -0.263 & 0 & 0.263 & 0 & 0.0704 \\ -0.0704 & -0.0528 & -0.263 & 0.0528 & 0.263 & 0.197 & 0.0704 & -0.197 \end{bmatrix} \quad (5.253)$$

The stress-strain matrix $[D]$ remains the same as before (see Eq. (5.123)). The element stiffness matrix is evaluated as

$$[k]^e = \sum_{i=1}^{2} \sum_{j=1}^{2} [B]^T [D][B] t |J| W_i W_j \tag{5.254}$$

After appropriate summation over all the integration points, we get the element stiffness matrix as follows:

Element stiffness matrix

$$[k]^e = 10^7 \begin{bmatrix} 0.891 \\ 0.357 & 1.17 & & & & & \text{Symmetric} \\ -0.379 & 0.0275 & 0.891 \\ -0.0275 & 0.296 & -0.357 & 1.17 \\ -0.446 & -0.357 & -0.0672 & 0.0275 & 0.891 \\ -0.357 & -0.585 & -0.0275 & -0.881 & 0.357 & 1.17 \\ -0.0672 & -0.0275 & -0.446 & 0.357 & -0.379 & 0.0275 & 0.891 \\ 0.0275 & -0.881 & 0.357 & -0.585 & -0.0275 & 0.296 & -0.357 & 1.17 \end{bmatrix} \qquad (5.255)$$

Upon incorporating the boundary conditions and using equivalent nodal forces, we obtain the final set of equations for the unknown deflections, as

$$10^7 \begin{bmatrix} 0.891 \\ -0.357 & 1.17 & & & \text{Symmetric} \\ -0.0672 & 0.0275 & 0.881 \\ -0.0275 & -0.881 & 0.357 & 1.17 \\ 0.357 & -0.585 & -0.0275 & 0.296 & 1.17 \end{bmatrix} \begin{Bmatrix} u_2 \\ v_2 \\ u_3 \\ v_3 \\ v_4 \end{Bmatrix} = \begin{Bmatrix} 1.5 \\ 0 \\ 1.5 \\ 0 \\ 0 \end{Bmatrix} \qquad (5.256)$$

Solving, we get the nodal deflection as

$$u_2 = 2 \times 10^{-7} \text{ cm}$$

$$v_2 = 0$$

$$u_3 = 2 \times 10^{-7} \text{ cm} \qquad\qquad (5.257)$$

$$v_3 = -4.5 \times 10^{-8} \text{ cm}$$

$$v_4 = -4.5 \times 10^{-8} \text{ cm}$$

Example 5.5.   *Effect of element shape on the Jacobian.* Consider the four-node quadrilateral element shown in Figure 5.36(a). The corresponding parent element in the ξ-η frame is shown in Figure 5.36(b) along with the sampling points for 2×2 Gauss quadrature rule. For each Gauss point, we will now compute the mapping point in the real x-y space and also the Jacobian. We wish to study what happens to $[J]$ as the element loses its rectangular shape.

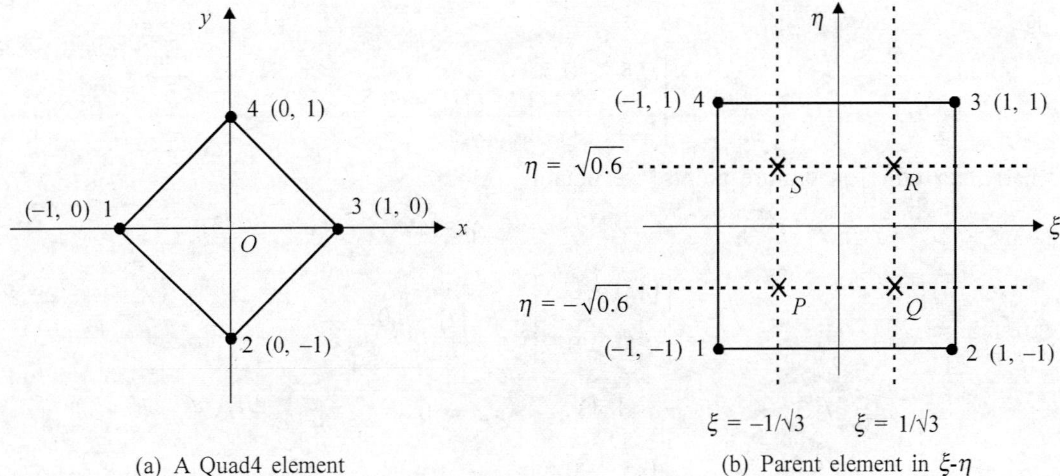

(a) A Quad4 element

(b) Parent element in ξ-η

Fig. 5.36 A regular quadrilateral element (Example 5.5).

For $\xi_1 = -1/\sqrt{3}$, $\eta_1 = -1/\sqrt{3}$, we have, from Eqs. (5.43)–(5.46), the relations

$$
\begin{aligned}
x_p &= N_1 x_1 + N_2 x_2 + N_3 x_3 + N_4 x_4 \\
&= N_1(-1) + 0 + N_3(1) + 0 \\
&= -N_1 + N_3 \\
&= \frac{1}{4}\left[-\left(1 + 1/\sqrt{3}\right)\left(1 + 1/\sqrt{3}\right) + \left(1 - 1/\sqrt{3}\right)\left(1 - 1/\sqrt{3}\right)\right] \\
&= -1/\sqrt{3}
\end{aligned}
\tag{5.258}
$$

$$
\begin{aligned}
y_p &= N_1 y_1 + N_2 y_2 + N_3 y_3 + N_4 y_4 \\
&= N_1(0) + N_2(-1) + N_3(0) + N_4(1) \\
&= N_4 - N_2 \\
&= \frac{1}{4}\left[\left(1 + 1/\sqrt{3}\right)\left(1 - 1/\sqrt{3}\right) - \left(1 - 1/\sqrt{3}\right)\left(1 + 1/\sqrt{3}\right)\right] = 0
\end{aligned}
\tag{5.259}
$$

From Eq. (5.158), we have

$$
[J] =
\begin{bmatrix}
\left(\dfrac{1+1/\sqrt{3}}{4}\right)(0+1) + \left(\dfrac{1-1/\sqrt{3}}{4}\right)(1-0) & \left(\dfrac{1-1/\sqrt{3}}{4}\right)(-1-0) + \left(\dfrac{1+1/\sqrt{3}}{4}\right)(0-1) \\[2em]
\left(\dfrac{1-1/\sqrt{3}}{4}\right)(1+0) + \left(\dfrac{1-1/\sqrt{3}}{4}\right)(1-0) & \left(\dfrac{1+1/\sqrt{3}}{4}\right)(1-0) + \left(\dfrac{1-1/\sqrt{3}}{4}\right)(1+0)
\end{bmatrix}
$$

Thus,

$$[J] = \begin{bmatrix} 0.5 & -0.5 \\ 0.5 & 0.5 \end{bmatrix}, \qquad |J| = 0.5 \tag{5.260}$$

Similarly, for other quadrature points, we obtain

$$\xi_2 = -1/\sqrt{3}, \qquad \eta_2 = 1/\sqrt{3}, \qquad x_Q = 0, \qquad y_Q = 1/\sqrt{3}$$

$$[J] = \begin{bmatrix} 0.5 & -0.5 \\ 0.5 & 0.5 \end{bmatrix}, \qquad |J| = 0.5 \tag{5.261}$$

$$\xi_3 = 1/\sqrt{3}, \qquad \eta_3 = -1/\sqrt{3}, \qquad x_R = 0, \qquad y_R = -1/\sqrt{3}$$

$$[J] = \begin{bmatrix} 0.5 & -0.5 \\ 0.5 & 0.5 \end{bmatrix}, \qquad |J| = 0.5 \tag{5.262}$$

$$\xi_4 = 1/\sqrt{3}, \qquad \eta_4 = 1/\sqrt{3}, \qquad x_S = 1/\sqrt{3}, \qquad y_S = 0$$

$$[J] = \begin{bmatrix} 0.5 & -0.5 \\ 0.5 & 0.5 \end{bmatrix}, \qquad |J| = 0.5 \tag{5.263}$$

$$\sum_1^4 |J| = 0.5 + 0.5 + 0.5 + 0.5 = 2 \tag{5.264}$$

Thus, sum of the determinants of Jacobian equals the area of the element.

We observe that $|J|$ is positive and uniform at all quadrature points. Let us now consider a distorted four-node quadrilateral as shown in Figure 5.37. The corresponding computations are as follows:

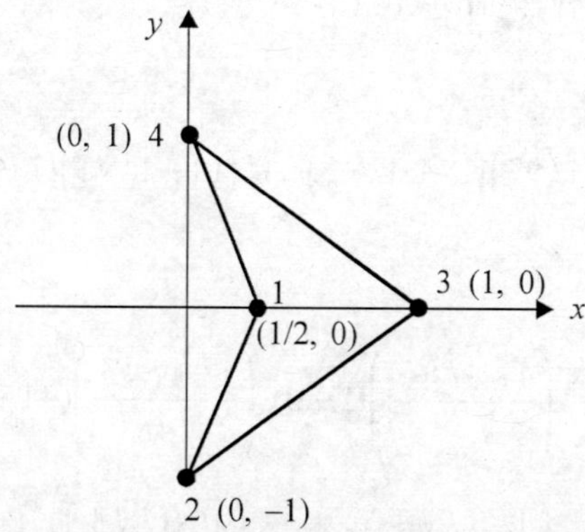

Fig. 5.37   A distorted quadrilateral element (Example 5.5).

$$\xi_1 = -1/\sqrt{3}, \qquad \eta_1 = -1/\sqrt{3}, \qquad x_P = 0.35566, \qquad y_P = 0$$

$$[J] = \begin{bmatrix} -0.0915 & -0.5 \\ -0.0915 & 0.5 \end{bmatrix}, \qquad |J| = -0.0915 \tag{5.265}$$

We observe that (x_P, y_P) falls outside the element and $|J|$ is negative.

$$\xi_2 = -1/\sqrt{3}, \qquad \eta_2 = 1/\sqrt{3}, \qquad x_Q = 0.25, \qquad y_Q = 1/\sqrt{3}$$

$$[J] = \begin{bmatrix} 0.3415 & -0.5 \\ -0.0915 & 0.5 \end{bmatrix}, \qquad |J| = 0.125 \tag{5.266}$$

$$\xi_3 = 1/\sqrt{3}, \qquad \eta_3 = -1/\sqrt{3}, \qquad x_R = 0.25, \qquad y_R = -1/\sqrt{3}$$

$$[J] = \begin{bmatrix} -0.0915 & -0.5 \\ 0.3415 & 0.5 \end{bmatrix}, \qquad |J| = 0.125 \tag{5.267}$$

$$\xi_4 = 1/\sqrt{3}, \qquad \eta_4 = 1/\sqrt{3}, \qquad x_S = 0.64434, \qquad y_S = 0$$

$$[J] = \begin{bmatrix} 0.3415 & -0.5 \\ 0.3415 & 0.5 \end{bmatrix}, \qquad |J| = 0.3415 \tag{5.268}$$

$$\sum_1^4 |J| = -0.0915 + 0.125 + 0.125 + 0.3415 = 0.5, \text{ area of the element} \tag{5.269}$$

We observe that, as the element is distorted from its rectangular shape, $|J|$ may vary from point to point and may become even negative. Though the isoparametric formulation in natural coordinates permits general quadrilateral elements and even elements with curved edges, etc., all such elements lose accuracy when distorted from a rectangular shape. Included angles at the vertices, variation of $|J|$ from one Gauss point to another, ratio of length of the largest to the smallest side (aspect ratio), etc. can all be used as indices of distortion. Such measures of quality of shape of an element are incorporated in all the automatic mesh generation algorithms available in most commercial FE software packages.

Example 5.6. *Use of plane elements for bending.* In this example, we will consider the simple problem of a cantilever beam ($EI = 0.1667 \times 10^{10}$ N-cm^2) subjected to shear load at the tip. We attempt to use the 2-d plane elements to model the bending deformation. Consider the cantilever beam shown in Figure 5.38. Under the tip load, the expected tip deflection and slope are readily obtained as

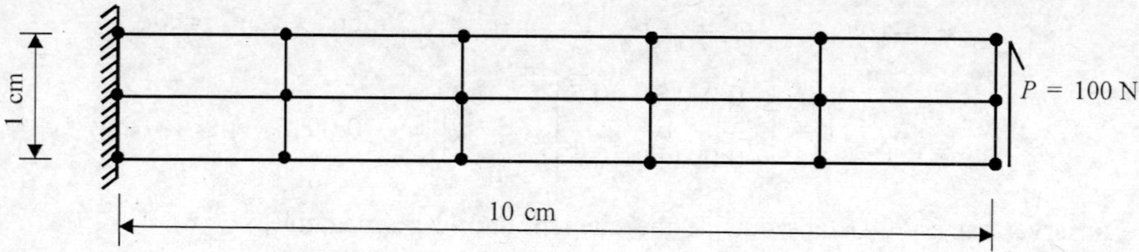

Fig. 5.38   A cantilever loaded at the tip (Example 5.6) (5 × 2 mesh of Quad4 element is shown).

$$\delta_{\text{tip}} = \frac{PL^3}{3EI} = 0.02 \text{ cm} \tag{5.270}$$

$$\theta_{\text{tip}} = \frac{PL^2}{2EI} = 0.003 \tag{5.271}$$

We will first use the plane four-node quadrilateral element to model this problem. The load at the tip is distributed between the nodes using the consistent load vector, viz., $\int [N]^T q \, dv$. The plane element discussed here does not permit rotational d.o.f. at the nodes. Thus we compute the slope at the tip as $(u_{\text{top}} - u_{\text{bottom}})/h$, where u_{top} and u_{bottom} correspond to the top and bottom nodes and h is the height of the beam. The results obtained with different meshes are listed in Table 5.3.

Table 5.3   Bending of a Cantilever (Example 5.6)

Mesh of Quad4 elements	Tip deflection (cm)	Tip slope
1 element	0.00051	0.000076
5 × 1 mesh	0.0076	0.00114
(5 elements along length; 1 element in height)		
5 × 2 mesh	0.0078	0.00117
10 × 1 mesh	0.0136	0.002
20 × 1 mesh	0.0168	0.0025
40 × 1 mesh	0.018	0.0027

We observe that the plane Quad4 element is too stiff (deflection is underestimated) for modelling such bending deformation and is not to be used for such problems. We find the tip deflection to be 10% in error even with 40 elements! From the shape functions, we observe that on any vertical line the variation of u-displacement is linear, which is the case in Euler–Bernoulli beam bending. Thus it is somewhat surprising as to why the Quad4 element's performance is so poor. The interested reader is advised to compute the nonzero shear strain in this element under a typical pure bending moment (where the shear strain is expected to be zero). The spurious shear strain energy (parasitic shear) results in poor performance of the element.

If we model the problem with the eight-noded serendipity element, the results obtained are listed in Table 5.4.

Table 5.4   Bending of a Cantilever (Example 5.6)

Mesh of Quad8 elements	Tip deflection (cm)	Tip slope
1 element	0.0188	0.00287
2 × 1 mesh	0.0194	0.0029

In fact, the Quad9 element has been found to be even better than Quad8 element in terms of sensitivity to shape distortion, ability to model bending deformation, and so on.

Example 5.7.   *Stress concentration around a circular hole.* In this example, we wish to illustrate the use of typical 2-d elements for modelling the problem of stress concentration in a thin plate with a circular hole. We also wish to discuss exploiting symmetry in typical FE analysis. Consider a plate subjected to in-plane uni-axial tension as shown in Figure 5.39. For

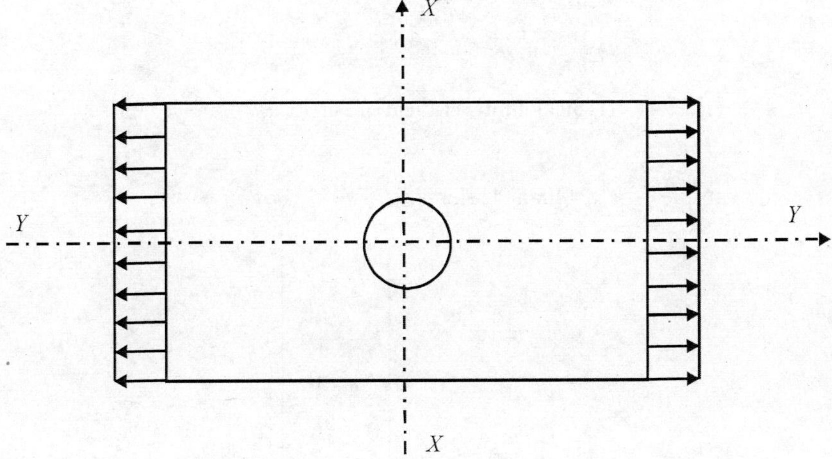

Fig. 5.39   Thin plate with a circular hole under uniform tension (Example 5.7).

a homogeneous, isotropic material, perfect symmetry exists about the *X*- and *Y*-axes shown in the figure. Thus only one quadrant as shown in Figure 5.40 needs to be modelled for finite

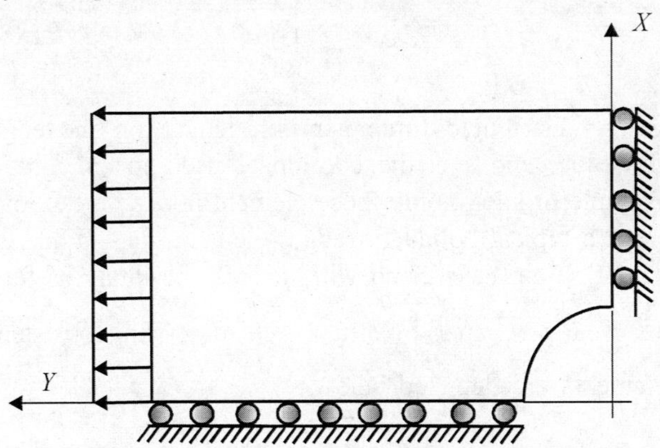

Fig. 5.40   Quadrant of plate (Example 5.7).

element analysis. For all nodes that lie on the symmetry planes, we need to stipulate symmetry boundary conditions. By virtue of symmetry, the points on the X-axis can only move along the X-axis. Similarly, for points on the Y-axis. A typical mesh for the quadrant of the structure is shown in Figure 5.41. The tensile loading on the edge is modelled as consistent load vector for

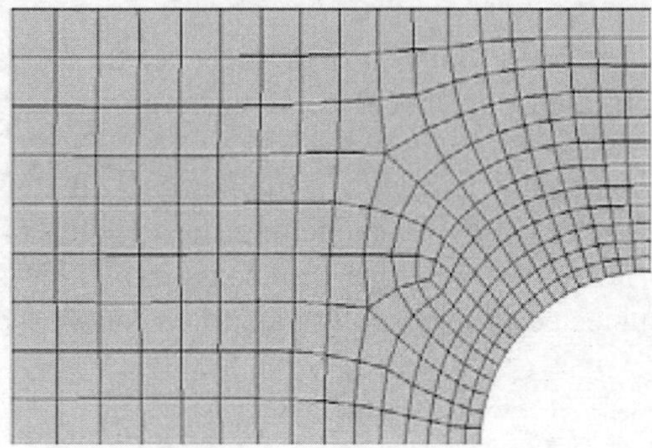

Fig. 5.41 Typical finite element mesh (Example 5.7).

each element. For example, for a Quad 4 element,

$$\{f\}^e = \int [N]^T q \, dv = \oint_s \begin{bmatrix} N_1 & 0 \\ 0 & N_1 \\ N_2 & 0 \\ 0 & N_2 \\ N_3 & 0 \\ 0 & N_3 \\ N_4 & 0 \\ 0 & N_4 \end{bmatrix} \begin{Bmatrix} q_x \\ q_y \end{Bmatrix} ds$$

where q_x, q_y represent the distributed force per unit length on the edge; shape functions N_i are evaluated on the edge, and s is the coordinate used for integration along the edge (e.g., ξ or η). In this example problem, only those element nodes on the loaded edge will have non-zero nodal forces. For a typical boundary (edge) element, $q_x = 0$ and $\xi = -1$ on the edge. Thus, for a 4-node element, $N_2 = 0 = N_3$ and with $q_x = 0$, the integrals for nodal force vector simply involve $\int N_1(\xi = -1)q_y \, d_y$ and $\int N_4(\xi = -1)q_y d_y$. The consistent nodal forces for Quad 4 and Quad 9 elements are depicted below.

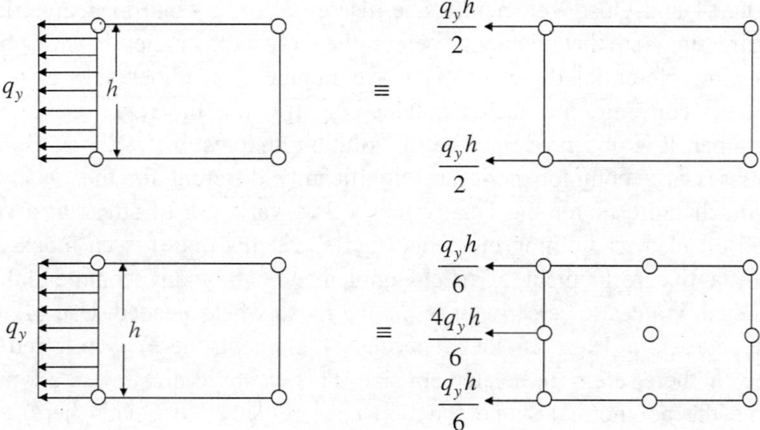

From the classical solution, it is well known that the stress concentration factor is 3.0 for this problem. The FE estimate of stress concentration along the vertical ($y = 0$) edge is shown in Figure 5.42 for a 10×15 mesh (10 elements along radial and 15 along circumferential

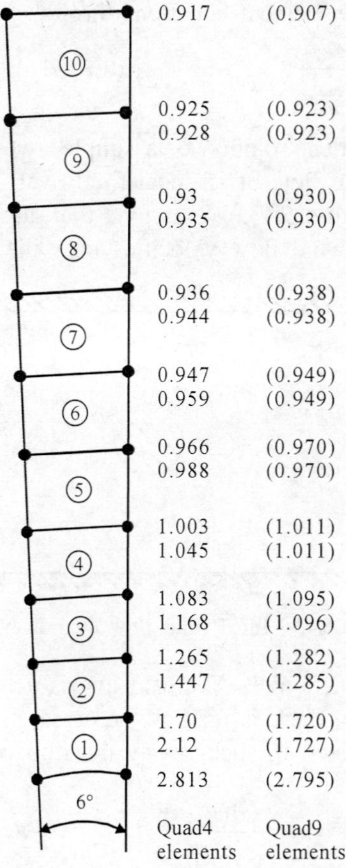

Fig. 5.42   Stress concentration factor obtained using FEM (Example 5.7).

direction) of Quad4 and Quad9 elements. The discretisation is done in geometric progression in the radial direction (so that elements nearer the hole are smaller in size) but arithmetic progression in circumferential direction. The two numbers on either side of a node are the estimates of stress concentration factor obtained by treating the node as belonging to one element or the other. It is observed that the FE solution underestimates the stress concentration factor. The stresses at a common node are significantly different for the Quad4 mesh while they are very much uniform for the Quad9 mesh. The variation of stress at a common node when computed of all the adjoining elements (that share this node) is an index of the quality of the mesh. Adaptive refinement of mesh, implemented in many commercial FE software packages, uses such indices to selectively refine the mesh where needed. Use of more elements of the same type (e.g. a 15×15 mesh of Quad4 elements here) is referred to as *h-type refinement.* The "*h*" here refers to the element size. This example illustrates "p-type" refinement where we refine the polynomial shape function used (Quad4 to Quad9 here).

We have so far discussed two-dimensional finite element analysis for structural mechanics. As discussed earlier, many of the concepts discussed are applicable to other fields also. We will now analyse the finite element solution of 2-d fluid flow problems. In contrast to the structural mechanics problems, we wish to start from the basic governing differential equation and develop the finite element equations through the weak form statement.

5.8   2-d Fluid Flow

We now discuss the finite element solution of a simple, two-dimensional fluid flow problem, viz., the steady-state irrotational flow of an ideal fluid over a cylinder, confined between two parallel plates as shown in Figure 5.43. We assume that, at the inlet, the velocity is uniform, say V_0. We wish to determine the flow velocities near the cylinder. We will formulate the

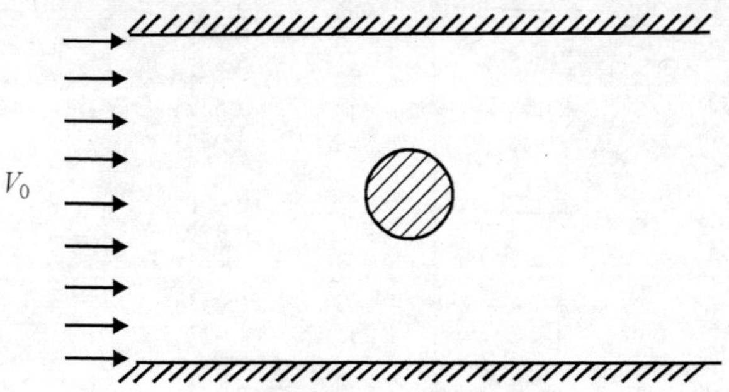

Fig. 5.43   Ideal fluid flow over a cylinder.

problem in terms of stream function Ψ. A stream line is a line that is tangent to the velocity vector. By definition, therefore, there is no flow across a stream line. The stream function is constant for a stream line defined in such a way that the velocity components are given by

$$u = \frac{\partial \psi}{\partial y}, \qquad v = -\frac{\partial \psi}{\partial x} \tag{5.272}$$

The governing differential equation, in terms of the stream function, is given by

$$\frac{\partial^2 \psi}{\partial x^2} + \frac{\partial^2 \psi}{\partial y^2} = 0 \qquad (5.273)$$

Considering the symmetry in the problem, only one quadrant as shown in Figure 5.44 needs to be modelled. By virtue of symmetry, AB becomes a stream line (no flow across it)

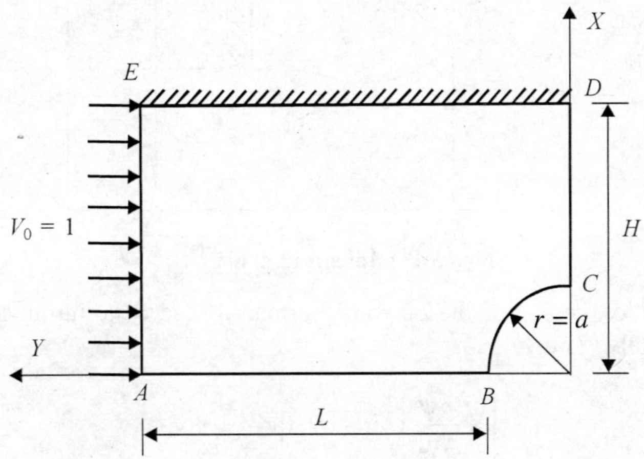

Fig. 5.44 **Domain used for finite element modelling.**

and ψ should be a constant along AB. Since the flow velocities are only derivatives of ψ, any arbitrary datum for ψ can be chosen. Let $\psi = 0$ on the line AB.

The curved boundary BC is again a stream line (and, therefore, ABC becomes a single continuous stream line) and $\psi = 0$ on this. In view of the uniform inlet velocity V_0, $\psi = V_0 x$. on the line AE. On the line DE, therefore, $\psi = V_0 H$. By virtue of symmetry, on the line CD, $\partial\psi/\partial y = 0$. We can write the weighted residual statement as

$$\iint_A W \left(\frac{\partial^2 \psi}{\partial x^2} + \frac{\partial^2 \psi}{\partial y^2} \right) dA = 0 \qquad (5.274)$$

Recall that, integrating by parts, we get

$$\int_{x_\ell}^{x_u} W \frac{d^2\psi}{dx^2}\, dx = \left(W \frac{d\psi}{dx} \right)_{x_\ell}^{x_u} - \int_{x_\ell}^{x_u} \frac{dW}{dx}\frac{d\psi}{dx}\, dx \qquad (5.275)$$

On similar lines, we write (using the Green–Gauss theorem)

$$\iint_A W \frac{\partial^2 \psi}{\partial x^2}\, dA = \int_{y_\ell}^{y_u} \left(W \frac{\partial \psi}{\partial x} \right)_{x_\ell}^{x_u} dy - \iint_A \frac{\partial W}{\partial x}\frac{\partial \psi}{\partial x}\, dA \qquad (5.276)$$

The general interpretation of x_ℓ, x_u and y_ℓ, y_u is shown in Figure 5.45. Therefore,

$$dy = \pm \ell_x\, ds \qquad (5.277)$$

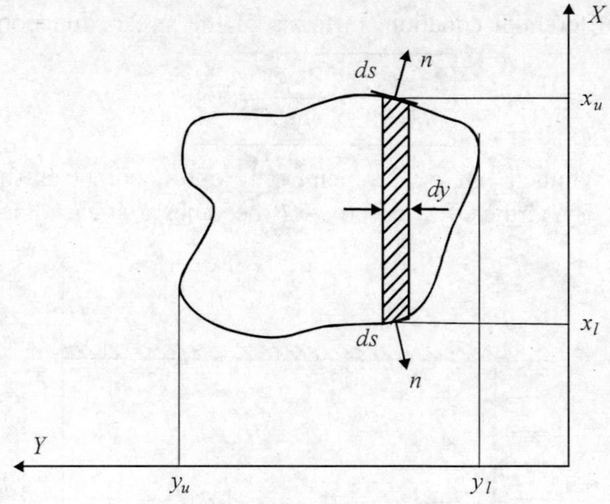

Fig. 5.45 Integration limits.

where ℓ_x is the direction cosine of the outward normal $\vec{n}$ and $\pm$ is introduced to take care of both ends appropriately. Thus,

$$\int_{y_\ell}^{y_u} \left[W\frac{\partial \psi}{\partial x} \right]_{x_\ell}^{x_u} dy = \oint_s W\frac{\partial \psi}{\partial x}\ell_x \, ds \tag{5.278}$$

Substituting in Eqs. (5.276), we get

$$\iint_A W\frac{\partial^2 \psi}{\partial x^2} dA = -\iint_A \frac{\partial W}{\partial x}\frac{\partial \psi}{\partial x} dA + \oint_s W\frac{\partial \psi}{\partial x}\ell_x \, ds \tag{5.279}$$

where s refers to the boundary curve. Similarly, we have

$$\iint_A W\frac{\partial^2 \psi}{\partial y^2} dA = -\iint_A \frac{\partial W}{\partial y}\frac{\partial \psi}{\partial y} dA + \oint_s W\frac{\partial \psi}{\partial y}\ell_y \, ds \tag{5.280}$$

Substituting Eqs. (5.279)–(5.280) in Eq. (5.274), we have

$$\iint_A W\left(\frac{\partial^2 \psi}{\partial x^2} + \frac{\partial^2 \psi}{\partial y^2} \right) dA = -\iint_A \left(\frac{\partial W}{\partial x}\frac{\partial \psi}{\partial x} + \frac{\partial W}{\partial y}\frac{\partial \psi}{\partial y} \right) dA$$

$$+ \oint_s W\left(\frac{\partial \psi}{\partial x}\ell_x + \frac{\partial \psi}{\partial y}\ell_y \right) ds \tag{5.281}$$

$$= 0$$

Therefore,

$$\iint_A \left(\frac{\partial W}{\partial x}\frac{\partial \psi}{\partial x} + \frac{\partial W}{\partial y}\frac{\partial \psi}{\partial y} \right) dA = \oint_s W\left(\frac{\partial \psi}{\partial x}\ell_x + \frac{\partial \psi}{\partial y}\ell_y \right) ds = \oint_s W\left(\frac{\partial \psi}{\partial n} \right) ds \tag{5.282}$$

where $\partial \psi/\partial n$ represents flow velocity in the negative s-direction (i.e. $V_s = -\dfrac{\partial y}{\partial n}$).

For a typical element, each node has one d.o.f., viz., the stream function value ψ. In our standard finite element notation, for a general 2-d element, we have

$$\psi = [N]\{\psi\}^e \tag{5.283}$$

where $[N]$, the shape function matrix and $\{\psi\}^e$, the nodal values of stream function take appropriate size depending on the element. For a Quad4 element, the sizes are $\{\psi\}^e_{4\times1}$, $[N]_{1\times4}$.

$$\begin{Bmatrix} \dfrac{\partial \psi}{\partial x} \\[2mm] \dfrac{\partial \psi}{\partial y} \end{Bmatrix} = [B]\{\psi\}^e \tag{5.284}$$

where $[B]$ contains the derivatives of shape functions. In the Galerkin formulation, we use the same shape functions as the weight functions also (i.e., $W = N$). Thus, Eq. (5.282) for an element can be rewritten as

$$\left(\iint_A [B]^T[B]\, dA \right)\{\psi\}^e = \oint_s [N]^T \left(\frac{\partial \psi}{\partial n} \right) ds \tag{5.285}$$

i.e.

$$\boxed{[K]^e\{\psi\}^e = \{f\}^e} \tag{5.286}$$

where

$$\boxed{[K]^e = \left(\iint_A [B]^T[B]\, dA \right), \qquad \{f\}^e = \oint_s [N]^T \left(\frac{\partial \psi}{\partial n} \right) ds} \tag{5.287}$$

For the entire finite element mesh, we have

$$\sum_1^{\text{NELEM}} \iint_A [B]^T[B]\, dA = \sum_1^{\text{NELEM}} \oint_s [N]^T \left(\frac{\partial \psi}{\partial n} \right) ds \tag{5.288}$$

i.e.

$$[K]\{\psi\} = \{F\} \tag{5.289}$$

where

$$[K] = \sum_1^{\text{NELEM}} \{K\}^e \tag{5.290}$$

$$\{F\} = \sum_1^{\text{NELEM}} \{f\}^e \tag{5.291}$$

and $\{\psi\}$ contains all the nodal d.o.f. of stream function values for the entire mesh and the summation over the elements is interpreted as per standard assembly procedure in FEM.

The term on the RHS in Eq. (5.288) is evaluated along the element boundary. For every edge of an element that is shared with another element, this term will be evaluated twice–once clockwise (for one element) and the other anti-clockwise (for the other element). Hence, when summed up for all the elements, these terms on internal boundaries of individual elements vanish, leaving out only the outer boundary of the entire domain. In the present problem, no velocity parallel to any boundary (recall that $V_s = -\partial\psi/\partial n$) has been specified and hence this term will be zero.

Following Eq. (5.164), for a typical element we write

$$\psi = [N]\{\psi\}^e \tag{5.292}$$

$$\begin{Bmatrix} \dfrac{\partial\psi}{\partial x} \\[2ex] \dfrac{\partial\psi}{\partial y} \end{Bmatrix}_{2\times1} = [B]\{\psi\}^e = [B_1]_{2\times2}[B_2]_{2\times2}[B_3]_{2\times\text{NDF}}\{\psi\}^e_{\text{NDF}\times1} \tag{5.293}$$

where, for the present case,

$$[B_1] = \begin{bmatrix} 1 & 0 \\ 0 & 1 \end{bmatrix}_{2\times2} \tag{5.294}$$

$$[B_2] = \begin{bmatrix} \dfrac{J_{22}}{|J|} & \dfrac{-J_{12}}{|J|} \\[2ex] \dfrac{-J_{21}}{|J|} & \dfrac{J_{11}}{|J|} \end{bmatrix}_{2\times2} \tag{5.295}$$

$$[B_3] = \begin{bmatrix} \dfrac{\partial N_1}{\partial\xi} & \dfrac{\partial N_2}{\partial\xi} & \cdots \\[2ex] \dfrac{\partial N_1}{\partial\eta} & \dfrac{\partial N_2}{\partial\eta} & \cdots \end{bmatrix}_{2\times\text{NDF}} \tag{5.296}$$

where NDF is the total number of d.o.f. for the element (e.g. 4 for Quad4, 9 for Quad9, etc.). Thus the solution of the fluid flow over cylinder is very much similar to the stress concentration problem discussed earlier (i.e., we can consider $[D]$ to be identity matrix), and we use the same mesh for both the problems.

The governing equations are finally obtained as

$$[K]\{\psi\} = \{0\} \tag{5.297}$$

where, for all nodes falling on the line AB or BC (Figure 5.44) $\psi = 0$; for nodes on the line DE, $\psi = V_0H$; for each node on the line AE, $\psi = V_0x$, where x represents the x-coordinate of that node. For these prescribed boundary conditions, we solve Eq. (5.297) for the remaining unknown ψ values. Once the ψ values are known, the flow velocities at any point in the element can be calculated from $[B]\{\psi\}^e$ for that element. Results obtained with a 10×15 mesh (10 elements along radial and 15 along circumferential direction) of Quad4 and Quad9 elements are shown in Figure 5.46. It can be easily verified that the finite element solution tallies very well with the exact solution.

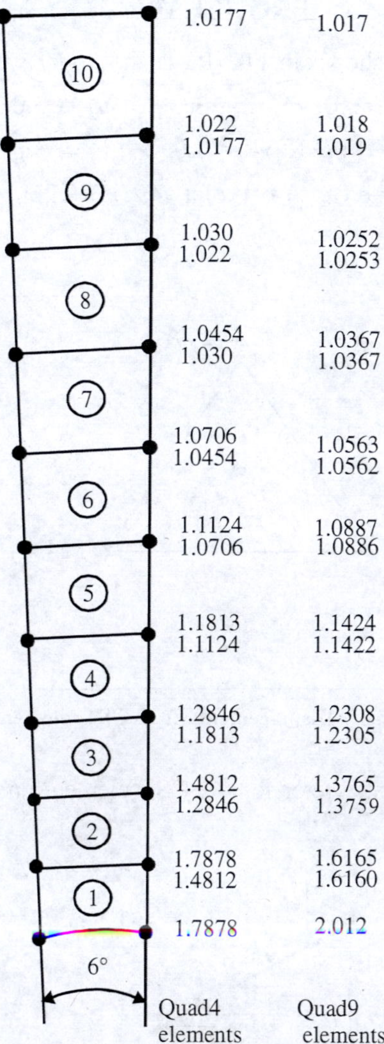

Fig. 5.46   Flow velocity obtained using FEM.

SUMMARY

We have discussed various 2-d finite elements in this chapter. Our basic procedure for finite element formulation has been the same as that used for 1-d elements. However, to allow for general curved edge elements, we introduced the natural coordinates and presented the use of numerical integration to evaluate the element matrices in a typical FE program. We also talked about incorporation of prescribed boundary conditions and Gauss elimination based algorithms for solution of equations. We studied typical structural mechanics and ideal fluid flow problems. Extension to three-dimensional problems is possible, i.e., triangular elements become tetrahedral, etc., and numerical integration will have to be carried out in 3-d. So far, however, we have discussed only steady state problems. In Chapter 6, we will discuss the finite element solution of an important class of problems, viz., 1–d and 2–d transient dynamic problems.

PROBLEMS

5.1 For the triangular element shown in Fig. P5.1, the nodal values of displacements are:

$$u_1 = 2.0, \qquad u_2 = 3.0, \qquad u_3 = 5.0$$
$$v_1 = 1.0, \qquad v_2 = 2.0, \qquad v_3 = 3.0$$

Obtain the displacements (i.e. u, v) of point $P(2, 2)$ within the element.

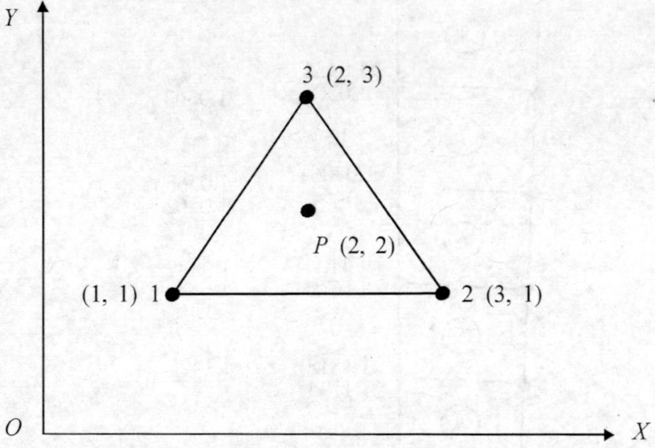

Fig. P5.1 Displacement of a CST element.

5.2 Determine the consistent load vector for the CST element under the action of the loading shown in Fig. P5.2.

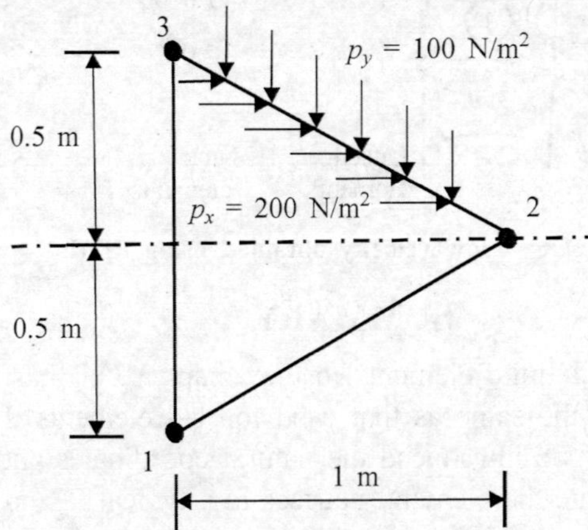

Fig. P5.2 Consistent nodal load vector for CST element (thickness = 0.1 m).

5.3 Determine the consistent load vector for the CST element under the action of gravity loading, acting in the plane of the element.

5.4 Find the consistent nodal force vector for the six-noded triangular element loaded as shown in Fig. P5.4.

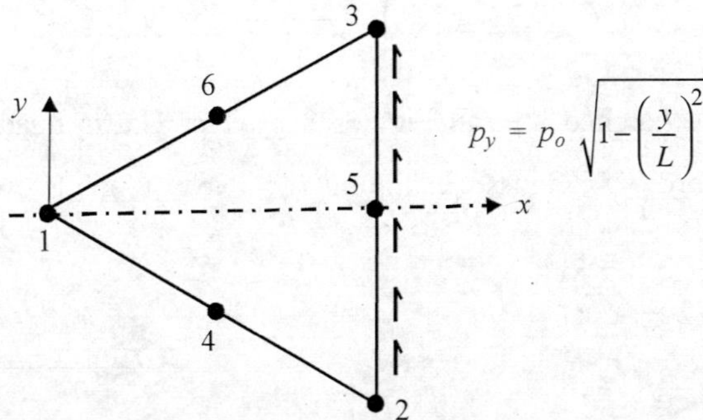

Fig. P5.4 Consistent nodal load vector for LST element (edge length = L; thickness = 1).

5.5 Obtain the consistent nodal load vector for the element shown in Fig. P5.5.

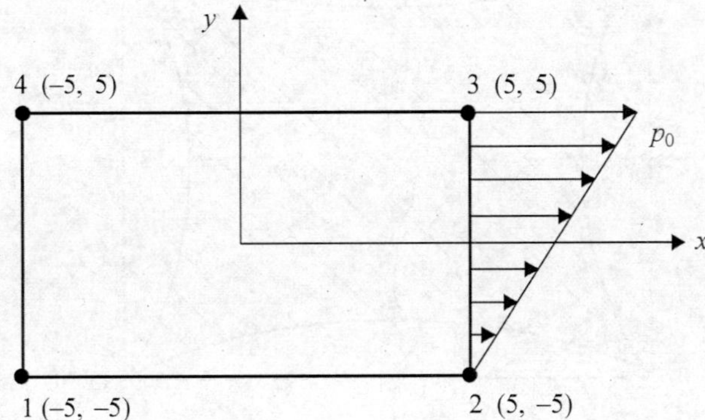

Fig. P5.5 Consistent nodal load vector for Quad4 element (thickness = 1).

5.6 For an eight-noded quadrilateral element, shown in Fig. P5.6, determine the consistent nodal load vector.

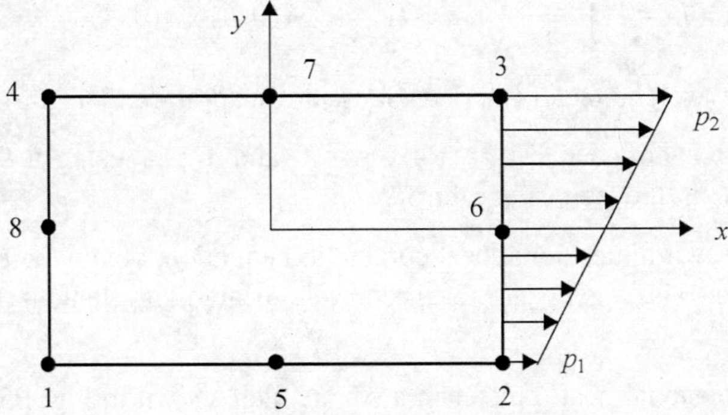

Fig. P5.6 Consistent nodal load vector for Quad8 element (edge length = L; thickness = 1).

5.7 Evaluate the following integral by a 2 × 2 Gauss quadrature rule:

$$\int_{-1}^{1} \int_{-1}^{1} \frac{2 + x}{3 + xy} \, dx \, dy$$

5.8 Determine the hatched area in Fig. P5.8 by 1-pt Gauss quadrature.

5.9 Check what order of Gauss quadrature (i.e. No. of points) would exactly integrate the following:
(a) $2 + 3x + 5x^3 + 8x^6$;
(b) $(2 - x^2)/(2 + x^2)$;
(c) $(x^2 + y^2)$;
(d) $2 + 3x + 5x^3$;
(e) $(2 - x)/(2 + x)$.

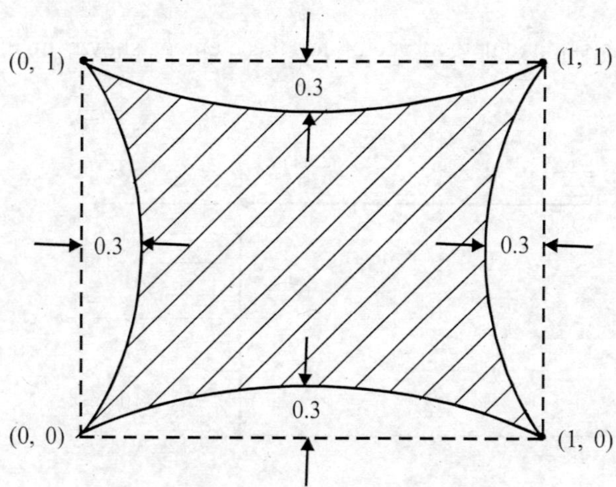

Fig. P5.8   Area of a figure.

5.10 Use a 2 × 2 Gauss quadrature rule to evaluate I.

$$I = \iint \frac{3 + x^2}{2 + y^2} \, dx \, dy, \qquad x(0,6), \ y(0,4)$$

5.11 Explain why the weight factors in 1-d Gaussian quadrature, sum up to 2.

5.12 Determine the area bounded by $y^2 = 4x$, $y = 0$ and $x = 4$ using a single-point Gauss quadrature. What is the percentage error?

5.13 Starting from the two-dimensional heat conduction equation, obtain the element nodal heat flow-temperature relationship for a three-node triangular element through Galerkin formulation.

5.14 For the two-dimensional heat transfer in a square slab shown in Fig. P5.14, each element ijk has the following equations:

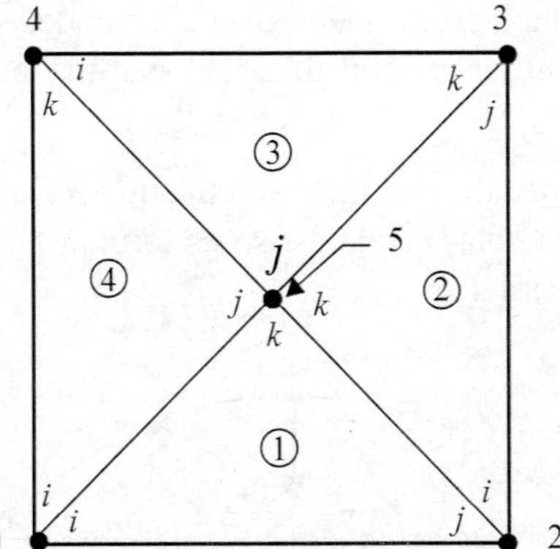

Fig. P5.14   A square slab.

$$\begin{bmatrix} 25 & -1 & 21 \\ -1 & 18 & -7 \\ 21 & -7 & 41 \end{bmatrix} \begin{Bmatrix} T_i \\ T_j \\ T_k \end{Bmatrix} = \begin{Bmatrix} 300 \\ 500 \\ 800 \end{Bmatrix}$$

All the edges are maintained at $T = 0^\circ$C. Obtain the final equation(s) after assembling element matrices and incorporating the boundary conditions.

5.15 For the three-noded triangular element shown in Fig. P5.15, calculate the temperature at point P, given the nodal temperatures as $T_1 = 100^\circ$C, $T_2 = 200^\circ$C, and $T_3 = 300^\circ$C.

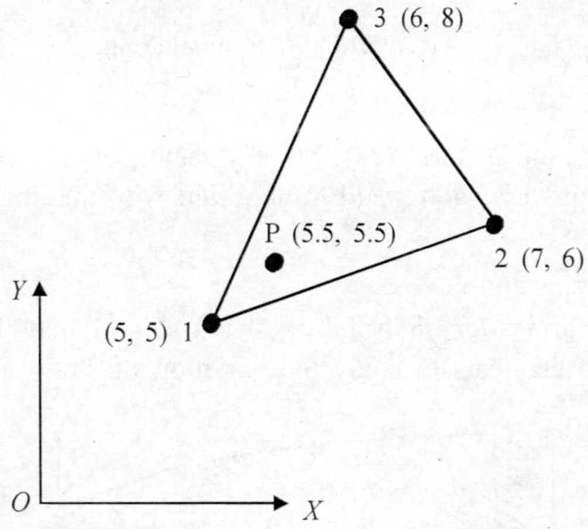

Fig. P5.15   Temperature at an interior point.

5.16 In a particular discretisation consisting of all six-node triangles for stress analysis, the temperature over an element can be taken to be constant, say, T_σ. The material properties are E and v. Determine the element nodal load vector in the absence of any other loads.

5.17 The temperature at the four corner nodes of a four-noded rectangle are T_1, T_2, T_3, and T_4. Determine the consistent load vector for a 2-d analysis, aimed to determine the thermal stresses.

5.18 Number the nodes in the following discretisation (Fig. P5.18) using eight-noded quadrilateral/six-noded triangular elements, so as to minimise the bandwidth. What is the bandwidth for such numbering?

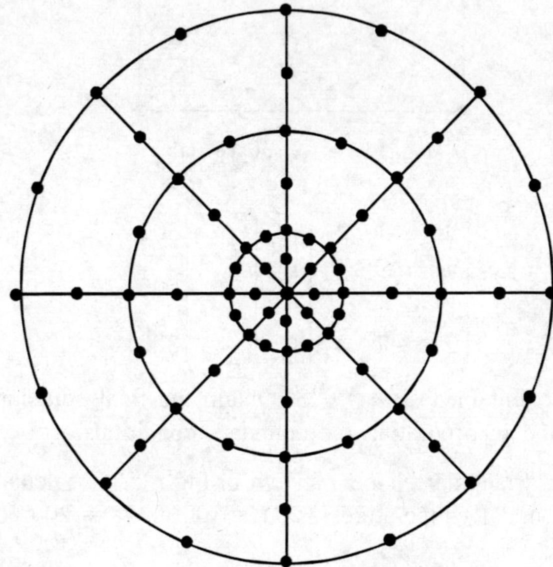

Fig. P5.18   Node numbering.

5.19 The Gauss elimination method (or Cholesky factorisation) broke down while solving a system of equations arising from FEM. Explain why this may be so.

Problems to Investigate

5.20 Consider the uniform strength beam shown in Fig. P5.20. It is aimed at using two-dimensional finite elements to study the variation of strain across the depth of such a

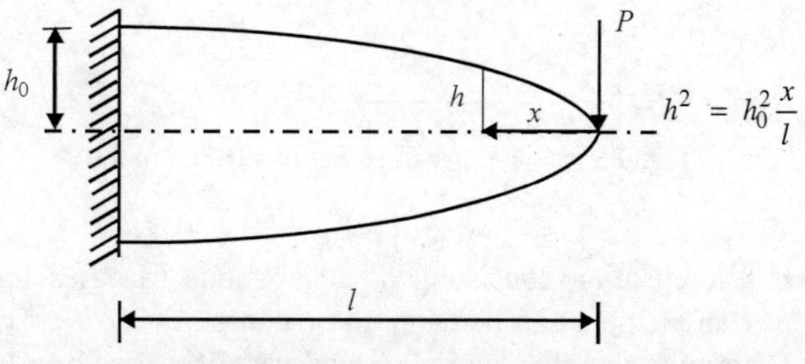

Fig. P5.20   Uniform strength beam.

beam in order to verify the accuracy of the Euler–Bernoulli assumptions. Show clearly how the curved edge may be modelled using isoparametric elements, through appropriate coordinate transformation.

5.21 Write all the shape functions for the elements shown in Fig. P5.21.

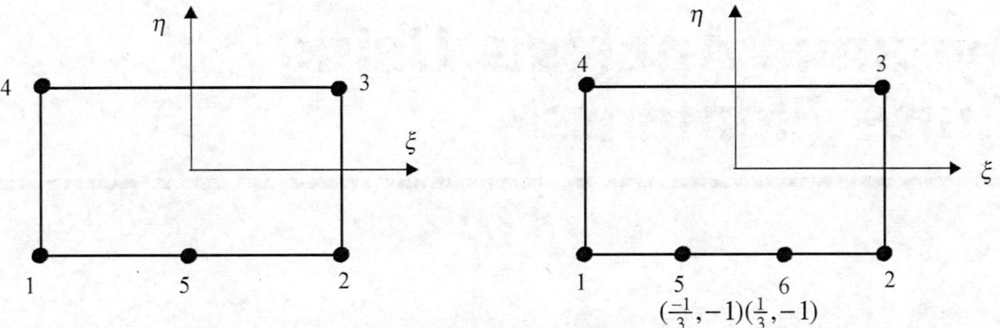

Fig. P5.21   Transition elements.

5.22 Give an approximate estimate of the total number of multiplications (n.o.m.) involved in the Gaussian elimination solution of $[k]_{n \times n}\{\delta\}_{n \times 1} = \{f\}_{n \times 1}$. Show that the n.o.m. involved in Cholesky factorising a banded, symmetric matrix $[k]_{n \times b}$ is $nb^2/2$.

5.23 What are the conditions for convergence in the displacement finite element formulation? Show that the completeness requirements are met if $\sum N_i = 1$, where N_i's are the element displacement shape functions.

5.24 For a six-noded linear strain triangular element used for 2-d stress analysis:
  (a) Show whether the equilibrium equations (ignoring body forces) are satisfied or not within the element, i.e.

$$\frac{\partial \sigma_x}{\partial x} + \frac{\partial \tau_{xy}}{\partial y} = 0, \qquad \frac{\partial \tau_{xy}}{\partial x} + \frac{\partial \sigma_y}{\partial y} = 0$$

  (b) Discuss whether the equilibrium is satisfied across inter-element boundaries.
  (c) Comment on the compatibility across element boundaries as well as within the element
  (d) State whether or not the equilibrium of nodal forces and moments is satisfied.

CHAPTER 6

Dynamic Analysis Using Finite Elements

6.1 Introduction

So far we have discussed situations where the external excitation as well as the response of the system were time invariant. However, in many practical situations, such "steady-state" conditions are reached after a period of time in which the external disturbances cause the system response to fluctuate with time ("transient" period). For example, when a certain temperature boundary condition is suddenly prescribed, thermal transients are set up in the system. The specific heat of the material is the property that resists variation of temperature with time and needs to be accounted for. In a vibrating system, the acceleration and deceleration of the structural parts are resisted by the inertia of the system and certain dissipative forces are also developed (such as viscous friction proportional to the velocity of the moving parts).

When analysing such unsteady state or transient dynamic problems, we are interested in finding the response of the system (i.e., temperature, displacement, etc.) as a function of time given the external disturbances. In this chapter we discuss the formulation and solution of such problems using the finite element method. We will first discuss the vibration problems and then transient heat transfer problems. We can develop the formulation using the governing differential equations (as in Chapter 2) or using certain energy principle (as in Chapter 3), and we will show that the formulation follows on very similar lines. The additional terms in the governing equations due to inertia, dissipation, specific heat, etc. lead to very interesting and often complex behaviour of the system. We will analyse the solution of the equations in considerable detail.

6.2 Vibration Problems

When a structure is excited by forces which vary with time, the response of the structure is also time varying, and inertia/dissipation properties of the structure affect the response. As we will be discussing in this chapter, a complete dynamic response analysis is usually much more complex than a static (time invariant) analysis. Thus it is useful to assess the need for a dynamic analysis—when the excitation forces are varying slowly with time, we term the response *quasi-static,* and a static analysis is sufficient. If the highest frequency component in the excitation is less than about one-third the lowest (fundamental) natural frequency of the structure, a static analysis is usually assumed to be sufficient.

This can be argued from the typical response of a single d.o.f. system to a harmonic excitation (Figure 6.1). The magnification factor (MF), (i.e. the factor by which the dynamic response is more than the static response), is given as

$$|\text{MF}| = \frac{1}{1 - \left(\Omega/\omega_n\right)^2} \tag{6.1}$$

when

$$\Omega = \frac{1}{3}\omega_n, \quad |\text{MF}| = 1.125 \tag{6.2}$$

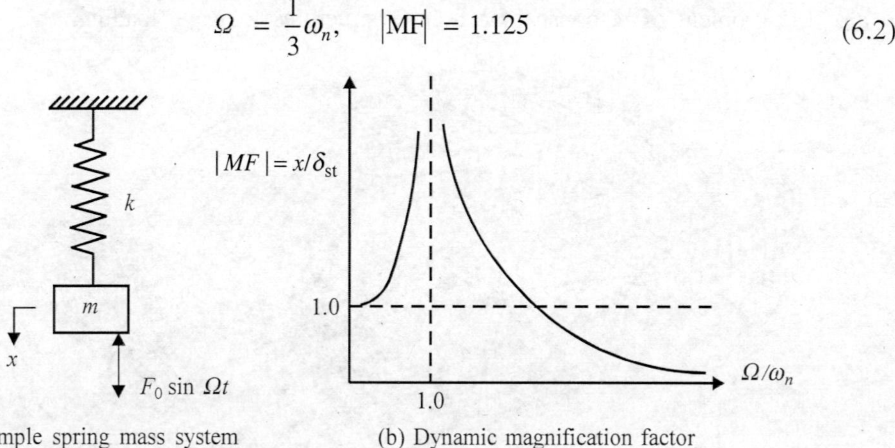

(a) Simple spring mass system   (b) Dynamic magnification factor

Fig. 6.1   Single d.o.f. system and its dynamic response.

Thus the dynamic response amplitude will only be 12.5% more than the static response for $\Omega = \omega_n/3$. The typical response of a multi-d.o.f. system (with several natural frequencies) is indicated in Figure 6.2. Static analysis is considered sufficient if the excitation frequency is less than about one-third the lowest (fundamental) natural frequency of the system.

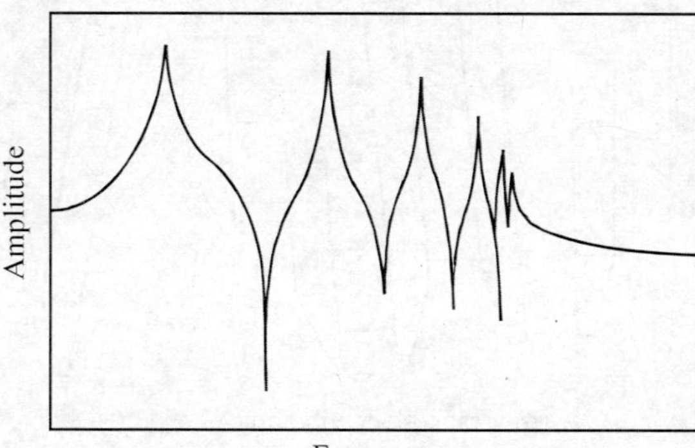

Fig. 6.2   Dynamic response of a multi d.o.f. system.

We need to distinguish two types of dynamic analysis problems, called the *wave propagation* and *structural dynamics* problems depicted in Figure 6.3. In the wave propagation problems, the excitation is usually an impact or blast force usually lasting for a fraction of a second or so. The entire structure does not instantaneously know that it has been hit. The time scales involved are comparable to the time taken for this information to traverse (i.e. the stress wave to propagate at the speed of sound in the medium) the entire structure. Crash analysis of a car shown in Figure 6.3(a) is a typical problem of this type. Impact analysis of a missile on a target structure is another typical wave propagation problem. In the structural dynamics type of problems, the entire structure simultaneously participates in the response, and the time scales involved are often several seconds. An automotive crank shaft vibration shown in Figure 6.3(b) is a typical problem of this type. A typical earthquake excitation contains frequencies up to, say 25 Hz, and is essentially a low frequency structural dynamic problem. So also the problem of response of off-shore structures to wave loading.

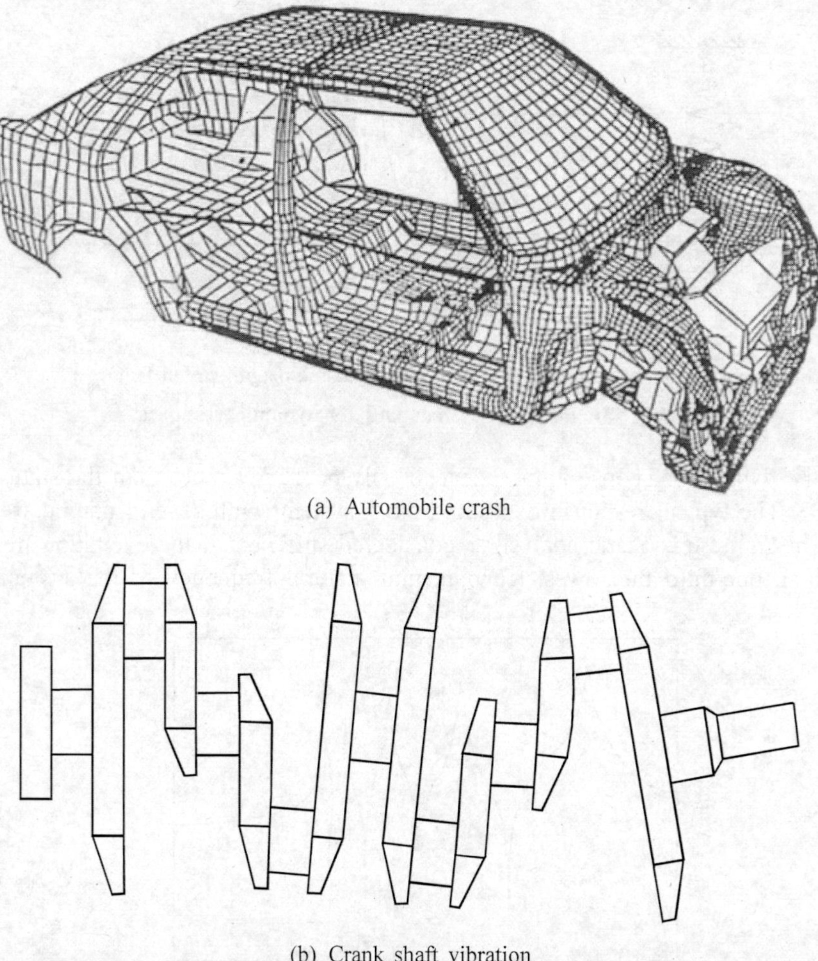

(a) Automobile crash

(b) Crank shaft vibration

Fig. 6.3   Types of dynamic problems.

Our discussion in this book will primarily centre on finite element solution of structural dynamics problems. We will analyse problems of free vibration (i.e. how the structure will vibrate when subjected to some initial conditions and left free to vibrate on its own) as well as forced vibration (i.e. response of structure under time varying forces). Free vibration problems are eigenvalue problems whose solution gives us the resonance frequencies and mode shapes of the structure. It is not only of interest for the designer to know the resonance frequencies (to be isolated from operating frequencies), but some of the solution strategies for the forced vibration problems require this information (e.g. mode superposition technique). So we will first discuss the free vibration problems and then take up determination of transient dynamic response of structures subjected to time dependent loads.

Returning to the simple example of a single d.o.f. system (Figure 6.1), we observe that the damped natural frequency is given as

$$\omega_d = \omega_n \sqrt{1 - \xi^2} \tag{6.3}$$

where ξ is the damping factor. Normally, ξ is very low, $\xi^2 \ll 1$. For example, if $\xi = 0.1$ (i.e. 10% critical damping), $\omega_d = 0.995\omega_n$. So, for many practical systems, $\omega_d \approx \omega_n$. In other words, we can ignore damping in the evaluation of natural frequencies. However, the response near a resonance is critically determined by the damping present in the system and we will include damping while estimating the dynamic response.

Finite element equations for dynamics can be derived in a manner entirely analogous to static equations discussed in earlier chapters (e.g. Chapter 2). We start with the typical partial differential equation, write down its weak form, and develop the finite element form of the weak form equation. Alternatively, we can use an energy based method (similar to Chapter 3) to derive the finite element dynamic equations. We will discuss both the methods for typical elements.

6.3 Equations of Motion Based on Weak Form

6.3.1 Axial Vibration of a Rod

The governing equation for free axial vibration of a rod (considering the dynamic equilibrium of a differential element) is given by

$$AE\frac{\partial^2 u}{\partial x^2} = \rho A \frac{\partial^2 u}{\partial t^2} \tag{6.4}$$

Using the technique of separation of variables and assuming harmonic vibration, we have

$$u(x,t) = U(x)e^{-i\omega t} \tag{6.5}$$

Substituting from Eq. (6.5) in Eq. (6.4), we obtain

$$\boxed{AE\frac{d^2 U}{dx^2} + \rho A \omega^2 U = 0} \tag{6.6}$$

The Weighted-Residual (WR) statement can be written as

$$\int_0^L W(x)\left(AE\frac{d^2U}{dx^2} + \rho A\omega^2 U \right) dx = 0 \qquad (6.7)$$

Integrating by parts, the weak form of the WR statement can be rewritten as

$$\left[W(x)AE\frac{dU}{dx} \right]_0^L - \int_0^L AE\frac{dU}{dx}\frac{dW}{dx}\, dx + \int_0^L W(x)\rho A\omega^2 U(x)\, dx = 0 \qquad (6.8)$$

We observe that the first two terms are identical to Eq. (2.97) and we have an additional term involving the mass density (inertia) effects. For a finite element mesh, the integrals are evaluated over each element and then summation is done over all elements.

We will now develop the necessary equations for a single element. For a typical bar element (Figure 6.4), we have two nodes and axial deformation d.o.f. at each node. The interpolation functions are given by Eq. (2.126).

$$U(x) = \left(1 - \frac{x}{\ell} \right)U_1 + \left(\frac{x}{\ell} \right)U_2 \qquad (6.9)$$

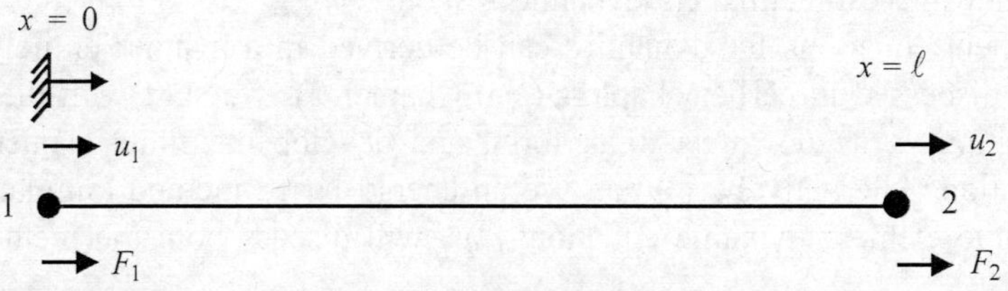

Fig. 6.4   **Typical bar element.**

Following the Galerkin formulation, the weight functions are the same as the shape (interpolation) functions. So we have

$$W_1(x) = 1 - \frac{x}{\ell}, \qquad W_2(x) = \frac{x}{\ell} \qquad (6.10)$$

Writing the weak form equations with respect to W_1 and W_2, we obtain (from Eq. (6.8)).

$$-P_0 - \int_0^\ell AE\left(\frac{U_2 - U_1}{\ell} \right)\left(\frac{-1}{\ell} \right) dx + \int_0^\ell \left(1 - \frac{x}{\ell} \right)\rho A\omega^2 \left[\left(1 - \frac{x}{\ell} \right)U_1 + \left(\frac{x}{\ell} \right)U_2 \right] dx = 0$$

$$(6.11)$$

$$P_\ell - \int_0^\ell AE\left(\frac{U_2 - U_1}{\ell} \right)\left(\frac{1}{\ell} \right) dx + \int_0^\ell \left(\frac{x}{\ell} \right)\rho A\omega^2 \left[\left(1 - \frac{x}{\ell} \right)U_1 + \left(\frac{x}{\ell} \right)U_2 \right] dx = 0 \qquad (6.12)$$

where P stands for $AE(dU/dx)$.

We can rewrite this in matrix form as

$$\frac{AE}{\ell}\begin{bmatrix} 1 & -1 \\ -1 & 1 \end{bmatrix}\begin{Bmatrix} U_1 \\ U_2 \end{Bmatrix} = \begin{Bmatrix} -P_0 \\ P_\ell \end{Bmatrix} + \rho A \omega^2 \begin{bmatrix} \int_0^\ell \left(1 - \frac{x}{\ell}\right)\left(1 - \frac{x}{\ell}\right)dx & \int_0^\ell \left(1 - \frac{x}{\ell}\right)\left(\frac{x}{\ell}\right)dx \\ \int_0^\ell \left(\frac{x}{\ell}\right)\left(1 - \frac{x}{\ell}\right)dx & \int_0^\ell \left(\frac{x}{\ell}\right)\left(\frac{x}{\ell}\right)dx \end{bmatrix}\begin{Bmatrix} U_1 \\ U_2 \end{Bmatrix}$$

$$(6.13)$$

On evaluating the integrals, we can write Eq. (6.13) as

$$\boxed{\frac{AE}{\ell}\begin{bmatrix} 1 & -1 \\ -1 & 1 \end{bmatrix}\begin{Bmatrix} U_1 \\ U_2 \end{Bmatrix} = \begin{Bmatrix} -P_0 \\ P_\ell \end{Bmatrix} + \begin{bmatrix} 2 & 1 \\ 1 & 2 \end{bmatrix}\frac{\rho A L \omega^2}{6}\begin{Bmatrix} U_1 \\ U_2 \end{Bmatrix}}$$

$$(6.14)$$

We have designated the element stiffness matrix as

$$\boxed{[k]^e = \frac{AE}{\ell}\begin{bmatrix} 1 & -1 \\ -1 & 1 \end{bmatrix}}$$

$$(6.15)$$

We now designate the element mass matrix as

$$\boxed{[m]^e = \frac{\rho A \ell}{6}\begin{bmatrix} 2 & 1 \\ 1 & 2 \end{bmatrix}}$$

$$(6.16)$$

We observe that the sum of all the coefficients in this mass matrix equals $\rho A \ell$, the mass of the element. Equation (6.14) can be rewritten as

$$[k]^e \{\delta\}^e = \begin{Bmatrix} -P_0 \\ P_\ell \end{Bmatrix} + [m]^e \omega^2 \{\delta\}^e$$

$$(6.17)$$

where $\{\delta\}^e$ contains the nodal d.o.f.

The element mass matrices can be assembled following exactly the same procedure as for the element stiffness matrices. Since free vibration problems do not involve external forces, the column of forces vanishes upon assembly and we obtain the typical free vibration equation

$$\boxed{[K]\{U\} = \omega^2 [M]\{U\}}$$

$$(6.18)$$

where $[K]$ and $[M]$ represent the global, assembled stiffness and mass matrices, respectively and $\{U\}$ contains all the nodal d.o.f. This is a typical eigenvalue problem, and we will discuss the solution of such equations in Section 6.8.

6.3.2 Transverse Vibration of a Beam

The governing equation for free transverse vibration of a beam based on the Euler–Bernoulli theory (considering the dynamic equilibrium of a differential element) is given by

$$EI\frac{\partial^4 v}{\partial x^4} + \rho A\frac{\partial^2 v}{\partial t^2} = 0 \tag{6.19}$$

Using the technique of separation of variables and assuming harmonic vibration, we have

$$V(x,t) = V(x)e^{-i\omega t} \tag{6.20}$$

Substituting in Eq. (6.19), we obtain

$$EI\frac{d^4 V}{dx^4} - \rho A\omega^2 V = 0 \tag{6.21}$$

The WR statement can be written as

$$\int_0^L W(x)\left[EI\frac{d^4 V}{dx^4} - \rho A\omega^2 V\right] dx = 0 \tag{6.22}$$

On integrating by parts, we obtain

$$\left[W(x)EI\frac{d^3 V}{dx^3}\right]_0^L - \int_0^L EI\frac{d^3 V}{dx^3}\frac{dW}{dx} \, dx - \int_0^L \rho A\omega^2 W(x)V(x) \, dx = 0 \tag{6.23}$$

Once again performing integration by parts, we get

$$\left[W(x)EI\frac{d^3 V}{dx^3}\right]_0^L - \left[\frac{dW}{dx}\,EI\frac{d^2 V}{dx^2}\right]_0^L + \int_0^L EI\frac{d^2 V}{dx^2}\frac{d^2 W}{dx^2} \, dx$$

$$- \int_0^L \rho A\omega^2 W(x)V(x) \, dx = 0 \tag{6.24}$$

For a typical Euler–Bernoulli beam element (Figure 6.5), we have

$$V(x) = N_1 V_1 + N_2\theta_1 + N_3 V_2 + N_4\theta_2 \tag{6.25}$$

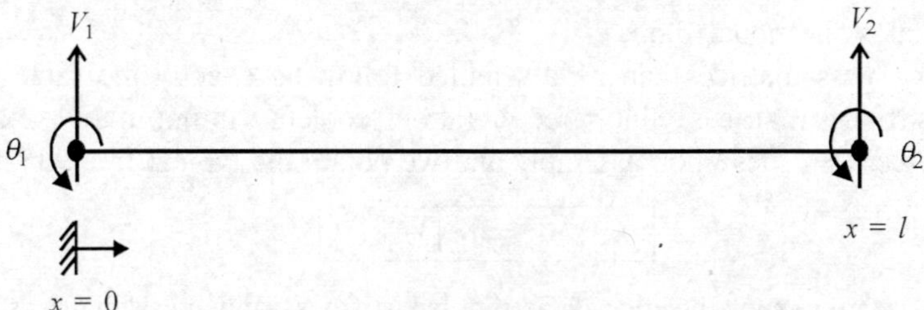

Fig. 6.5   Typical Euler–Bernoulli beam element.

where N_1, N_2, N_3 and N_4 are given in Eq. (4.117), and are reproduced here for easy reference:

$$N_1 = 1 - 3x^2/\ell^2 + 2x^3/\ell^3, \qquad N_2 = x - 2x^2/\ell + x^3/\ell^2$$

$$N_3 = 3x^2/\ell^2 - 2x^3/\ell^3, \qquad N_4 = -x^2/\ell + x^3/\ell^2 \tag{6.26}$$

Following the Galerkin procedure, we use these same shape functions as the weight functions. Substituting in the weak form [Eq. (6.24)] and evaluating the integrals with respect to each of the weighting functions one at a time, we obtain the governing equations. However, we observe that the first three terms in [Eq. (6.24)] are similar to those obtained in the static case [Eq. (2.103)]and we will therefore present here only the additional mass matrix terms. The mass matrix terms will correspond to the following integrals:

$$
\int_0^\ell \rho A W_1(x) V(x)\, dx = \int_0^\ell \left(1 - \frac{3x^2}{\ell^2} + \frac{2x^3}{\ell^3}\right)\rho A \left\{\left(1 - \frac{3x^2}{\ell^2} + \frac{2x^3}{\ell^3}\right)V_1\right.
$$

$$
+ \left(x - \frac{2x^2}{\ell} + \frac{x^3}{\ell^2}\right)\theta_1 + \left(\frac{3x^2}{\ell^2} - \frac{2x^3}{\ell^3}\right)V_2
$$

$$
\left. + \left(-\frac{x^2}{\ell} + \frac{x^3}{\ell^2}\right)\theta_2\right\} dx
$$

$$
= \frac{\rho A\ell}{420}\left[156 V_1 + 22\ell\theta_1 + 54 V_2 - 13\ell\theta_2\right] \tag{6.27}
$$

$$
\int_0^\ell \rho A W_2(x) V(x)\, dx = \int_0^\ell \left(x - \frac{2x^2}{\ell} + \frac{x^3}{\ell^2}\right)\rho A \left\{\left(1 - \frac{3x^2}{\ell^2} + \frac{2x^3}{\ell^3}\right)V_1\right.
$$

$$
+ \left(x - \frac{2x^2}{\ell} + \frac{x^3}{\ell^2}\right)\theta_1 + \left(\frac{3x^2}{\ell^2} - \frac{2x^3}{\ell^3}\right)V_2
$$

$$
\left. + \left(-\frac{x^2}{\ell} + \frac{x^3}{\ell^2}\right)\theta_2\right\} dx
$$

$$
= \frac{\rho A\ell}{420}\left(22\ell V_1 + 4\ell^2\theta_1 + 13\ell V_2 - 13\ell\theta_2\right) \tag{6.28}
$$

$$
\int_0^\ell \rho A W_3(x) V(x)\, dx = \int_0^\ell \left(\frac{3x^2}{\ell^2} - \frac{2x^3}{\ell^3}\right)\rho A \left\{\left(1 - \frac{3x^2}{\ell^2} + \frac{2x^3}{\ell^3}\right)V_1\right.
$$

$$
+ \left(x - \frac{2x^2}{\ell} + \frac{x^3}{\ell^2}\right)\theta_1 + \left(\frac{3x^2}{\ell^2} - \frac{2x^3}{\ell^3}\right)V_2
$$

$$
\left. + \left(-\frac{x^2}{\ell} + \frac{x^3}{\ell^2}\right)\theta_2\right\} dx
$$

$$
= \frac{\rho A\ell}{420}\left[54 V_1 + 13\ell\theta_1 + 156 V_2 - 22\ell\theta_2\right] \tag{6.29}
$$

$$\int_0^\ell \rho A W_4(x) V(x)\, dx = \int_0^\ell \left(\frac{-x^2}{\ell} + \frac{x^3}{\ell^2}\right) \rho A \left\{\left(1 - \frac{3x^2}{\ell^2} + \frac{2x^3}{\ell^3}\right) V_1\right.$$

$$+ \left(x - \frac{2x^2}{\ell} + \frac{x^3}{\ell^2}\right)\theta_1 + \left(\frac{3x^2}{\ell^2} - \frac{2x^3}{\ell^3}\right) V_2$$

$$\left.+ \left(-\frac{x^2}{\ell} + \frac{x^3}{\ell^2}\right)\theta_2\right\} dx$$

$$= \frac{\rho A \ell}{420}\left[-13\ell V_1 - 3\ell^2\theta_1 - 22\ell V_2 + 4\ell^2\theta_2\right] \tag{6.30}$$

Rewriting the terms in matrix form, we obtain the Euler–Bernoulli beam element mass matrix as

$$[m]^e = \frac{\rho A\ell}{420}\begin{bmatrix} 156 & & & \text{Symmetric} \\ 22\ell & 4\ell^2 & & \\ 54 & 13\ell & 156 & \\ -13\ell & -3\ell^2 & -22\ell & 4\ell^2 \end{bmatrix} \tag{6.31}$$

We observe that the coefficients of the mass matrix corresponding to translational d.o.f. $\left(\text{viz. } \dfrac{\rho A\ell}{420}\,(156 + 54 + 54 + 156)\right)$ sum up to $\rho A\ell$, the mass of the element.

If we need to write the mass matrix for a plane frame element, we can combine the bar element and beam element mass matrices in just the same way as we combined their stiffness matrices in Section 4.6.

We have illustrated the formulation of finite element equations starting from the governing differential equations. We shall now illustrate an energy based approach based on Lagrange's equations of motion.

6.4   Equations of Motion Using Lagrange's Approach

The equations of motion [e.g. Eqs. (6.4) and (6.19)] for the axial motion of a rod and the transverse motion of a beam, can be readily derived based on Newton's Second Law of motion. Many a time it is convenient to use an energy based method and Lagrange's equations* of motion are commonly used. If T represents the kinetic energy of a system and π represents

---

*Excellent discussions of the theory behind Lagrange's equations of motion are available in standard texts such as *Elements of Vibration Analysis,* 2nd ed., by L. Meirovitch, McGraw-Hill, New York, 1986 and *Lagrangian Dynamics* by D.A. Wells, Schaum Series, McGraw-Hill, New York, 1967.

potential energy, then Lagrange's equations of motion in the independent generalised coordinates q are given by

$$\boxed{\frac{d}{dt}\left(\frac{\partial T}{\partial \dot{q}}\right) - \frac{\partial T}{\partial q} + \frac{\partial \pi}{\partial q} = F\Big|_q}$$

(6.32)

where $F\big|_q$ represents the generalised force in the coordinate q and $\dot{q} = dq/dt$.

We will illustrate the use of Lagrange's approach in formulating the equations of motion through the following simple example.

Example 6.1.   Consider a two-d.o.f. spring mass system shown in Figure 6.6; $x_1(t)$ and $x_2(t)$ are the independent generalised coordinates. The kinetic energy of the system is

$$T = \frac{1}{2}m_1\dot{x}_1^2 + \frac{1}{2}m_2\dot{x}_2^2$$

(6.33)

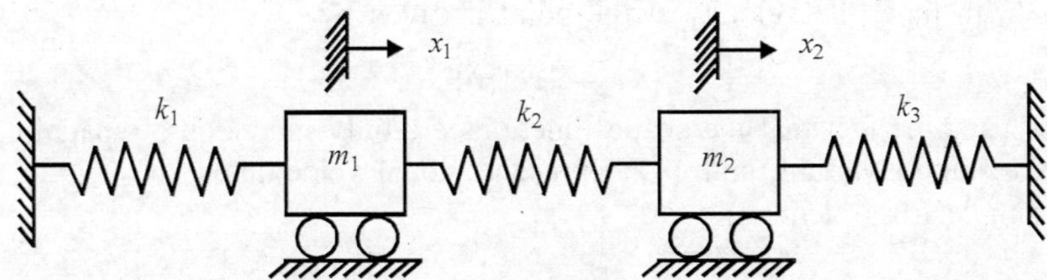

Fig. 6.6   2-d.o.f. system (Example 6.1).

The potential energy of the system is given by

$$\pi = \frac{1}{2}k_1 x_1^2 + \frac{1}{2}k_2(x_1 - x_2)^2 + \frac{1}{2}k_3 x_2^2$$

(6.34)

For further use in Lagrange's equations, we now obtain the derivatives of T and π w.r.t. x_1, x_2, $\dot{x}_1$ and $\dot{x}_2$.

$$\frac{\partial T}{\partial \dot{x}_1} = m_1\dot{x}_1, \qquad \frac{d}{dt}\left(\frac{\partial T}{\partial \dot{x}_1}\right) = m_1\ddot{x}_1$$

$$\frac{\partial T}{\partial \dot{x}_2} = m_2\dot{x}_2, \qquad \frac{d}{dt}\left(\frac{\partial T}{\partial \dot{x}_2}\right) = m_2\ddot{x}_2$$

(6.35)

$$\frac{\partial T}{\partial x_1} = 0, \qquad \frac{\partial T}{\partial x_2} = 0$$

$$\frac{\partial \pi}{\partial x_1} = k_1 x_1 + k_2(x_1 - x_2)$$

$$\frac{\partial \pi}{\partial x_2} = k_3 x_2 - k_2(x_1 - x_2)$$

Substituting from Eq. (6.35) in Eq. (6.32), we obtain the required equations of motion as

$$m_1\ddot{x}_1 + k_1 x_1 + k_2(x_1 - x_2) = F_1(t)$$
$$m_2\ddot{x}_2 + k_3 x_2 - k_2(x_1 - x_2) = F_2(t)$$

$$(6.36)$$

or, in matrix form,

$$\begin{bmatrix} (k_1 + k_2) & -k_2 \\ -k_2 & (k_2 + k_3) \end{bmatrix}\begin{Bmatrix} x_1 \\ x_2 \end{Bmatrix} + \begin{bmatrix} m_1 & 0 \\ 0 & m_2 \end{bmatrix}\begin{Bmatrix} \ddot{x}_1 \\ \ddot{x}_2 \end{Bmatrix} = \begin{Bmatrix} F_1(t) \\ F_2(t) \end{Bmatrix} \qquad (6.37)$$

6.4.1 Formulation of Finite Element Equations

Using our standard shape functions, for a typical finite element, the displacement of an interior point can be written, in terms of the nodal d.o.f., as

$$\{d\} = [N]\{\delta\}^e \qquad (6.38)$$

Differentiating with time, the velocity at the point is given by

$$\{\dot{d}\} = [N]\{\dot{\delta}\}^e \qquad (6.39)$$

where we have used the fact that the shape functions are only functions of spatial coordinates and are therefore time-invariant, and $\{\dot{\delta}\}^e$ represents nodal velocities.

For the simple truss element,

$$\{d\} = u = \left[\left(1 - \frac{x}{\ell}\right)\left(\frac{x}{\ell}\right)\right]\begin{Bmatrix} u_1 \\ u_2 \end{Bmatrix} \qquad (6.40)$$

$$\{\dot{d}\} = \dot{u} = \left[\left(1 - \frac{x}{\ell}\right)\left(\frac{x}{\ell}\right)\right]\begin{Bmatrix} \dot{u}_1 \\ \dot{u}_2 \end{Bmatrix} \qquad (6.41)$$

For the beam element,

$$\{d\} = v = \begin{bmatrix} N_1 & N_2 & N_3 & N_4 \end{bmatrix}\begin{Bmatrix} v_1 \\ \theta_1 \\ v_2 \\ \theta_2 \end{Bmatrix} \qquad (6.42)$$

$$\{\dot{d}\} = \dot{v} = \begin{bmatrix} N_1 & N_2 & N_3 & N_4 \end{bmatrix}\begin{Bmatrix} \dot{v}_1 \\ \dot{\theta}_1 \\ \dot{v}_2 \\ \dot{\theta}_2 \end{Bmatrix} \qquad (6.43)$$

For a two-dimensional element, each point can have u and v displacements and, therefore,

$$\{d\} = \begin{Bmatrix} u \\ v \end{Bmatrix} = [N]\begin{Bmatrix} u_1 \\ v_1 \\ u_2 \\ v_2 \\ \vdots \end{Bmatrix} \tag{6.44}$$

$$\{\dot{d}\} = \begin{Bmatrix} \dot{u} \\ \dot{v} \end{Bmatrix} = [N]\begin{Bmatrix} \dot{u}_1 \\ \dot{v}_1 \\ \dot{u}_2 \\ \dot{v}_2 \\ \vdots \end{Bmatrix} \tag{6.45}$$

where the size of $[N]$ and $\{\delta\}^e$ depend on the particular element (e.g. Quad4, Quad8, etc.), as discussed in Chapter 5. The kinetic energy of an elemental mass dm ($= \rho\, dV$) within the element is given by

$$dT^e = \frac{1}{2}(dm)\,(\text{velocity})^2 = \frac{1}{2}\{\dot{d}\}^T\{\dot{d}\}\rho\, dV \tag{6.46}$$

Using Eq. (6.39), we can write

$$dT^e = \frac{1}{2}\{\dot{\delta}\}^{e^T}[N]^T[N]\{\dot{\delta}\}^e\rho\, dV \tag{6.47}$$

For the whole element, the kinetic energy is obtained by integrating over the element as follows:

$$T^e = \int_v \frac{1}{2}\{\dot{\delta}\}^{e^T}[N]^T[N]\{\dot{\delta}\}^e\rho\, dV \tag{6.48}$$

Since nodal point velocities $\{\dot{\delta}\}^e$ do not vary from point to point within the element, these can be taken out of the integral, and hence

$$T^e = \frac{1}{2}\{\dot{\delta}\}^{e^T}\left(\int_v \rho[N]^T[N]\, dv\right)\{\dot{\delta}\}^e \tag{6.49}$$

$$T^e = \frac{1}{2}\{\dot{\delta}\}^{e^T}[m]^e\{\dot{\delta}\}^e \tag{6.50}$$

where $[m]^e = \left(\int_v \rho[N]^T[N]\, dv\right)$ is the consistent mass matrix for the element.

For the whole body (i.e. mesh of finite elements), we have

$$T = \sum_{1}^{\text{NOELEM}} T^e \tag{6.51}$$

where "NOELEM" is the number of elements.

The total potential energy of the system has already been obtained in Chapter 5 and is reproduced here from Eq. (5.105–5.106):

$$\Pi_p = \sum_{1}^{\text{NOELEM}} \left(\frac{1}{2} \{\delta\}^{e^T} [k]^e \{\delta\}^e - \{\delta\}^{e^T} \{f^e\} \right) \tag{6.52}$$

Interpreting the summation over all the elements as per standard assembly of finite elements, we can write

$$T = \frac{1}{2} \{\dot{\delta}\}^T [M] \{\dot{\delta}\} \tag{6.53}$$

$$\pi = \frac{1}{2} \{\delta\}^T [K] \{\delta\} - \{\delta\}^T \{F\} \tag{6.54}$$

where $\{\delta\}$ and $\{\dot{\delta}\}$ contain the displacement and velocities for all the nodes of the entire structure; $[k]$ and $[M]$ are the assembled global stiffness and mass matrices; and $\{F\}$ is the assembled global nodal force vector.

The required derivatives

$$\frac{\partial T}{\partial \{\delta\}} = 0, \qquad \frac{\partial T}{\partial \{\dot{\delta}\}} = [M] \{\dot{\delta}\}, \qquad \frac{d}{dt} \left(\frac{\partial T}{\partial \{\dot{\delta}\}} \right) = [M] \{\ddot{\delta}\} \tag{6.55}$$

can be employed for further use in Lagrange's equations,

$$\frac{\partial \pi}{\partial \{\delta\}} = [K] \{\delta\} - \{F\} \tag{6.56}$$

Substituting in Lagrange's equations of motion, we obtain

$$[M] \{\ddot{\delta}\} + [K] \{\delta\} - \{F\} = \{0\} \tag{6.57}$$

i.e.

$$\boxed{[M] \{\ddot{\delta}\} + [K] \{\delta\} = \{F\}} \tag{6.58}$$

where the global mass matrix, stiffness matrix and force vectors are obtained following the usual assembly of individual element level matrices. The stiffness matrix and force vectors have already been dealt with in the previous chapters. For dynamic problems we see that the inertia effects result in mass matrix. Since we have used the same shape functions as for the stiffness matrix, the mass matrix in Eq. (6.58) is a consistent mass matrix. The consistent mass matrices for a few elements are discussed in the following section.

6.4.2　Consistent Mass Matrices for Various Elements

Bar element

$$[m]^e = \int_v \rho [N]^T [N] \, dv = \rho A \int_0^\ell \begin{bmatrix} 1 - \dfrac{x}{\ell} \\[2mm] \dfrac{x}{\ell} \end{bmatrix} \left[\left(1 - \dfrac{x}{\ell}\right)\left(\dfrac{x}{\ell}\right) \right] dx$$

$$= \frac{\rho A \ell}{6} \begin{bmatrix} 2 & 1 \\ 1 & 2 \end{bmatrix} \tag{6.59}$$

This equation is the same as that given by Eq. (6.16) which was obtained using the weak form.

Beam element

$$[m]^e = \int_v \rho [N]^T [N] \, dv = \rho A \int_0^\ell \begin{bmatrix} N_1 \\ N_2 \\ N_3 \\ N_4 \end{bmatrix} \begin{bmatrix} N_1 & N_2 & N_3 & N_4 \end{bmatrix} dx \tag{6.60}$$

where the shape functions N_1, N_2, N_3 and N_4 are given in Eq. (6.26).

After performing all the integrations, the element mass matrix is obtained as

$$[m]^e = \frac{\rho A \ell}{420} \begin{bmatrix} 156 & & & \text{Symmetric} \\ 22\ell & 4\ell^2 & & \\ 54 & 13\ell & 156 & \\ -13\ell & -3\ell^2 & -22\ell & 4\ell^2 \end{bmatrix} \tag{6.61}$$

which is the same as the mass matrix given in Eq. (6.31) obtained using the weak form.

Two-dimensional elements

In general, it is tedious to evaluate, in closed form, the expression for the element mass matrix using

$$[m]^e = \int_v \rho [N]^T [N] \, dv \tag{6.62}$$

For example, for a Quad8 element we have $[N]_{2\times16}$. Thus, $[N]^T[N]$ will be a (16×16) matrix. Even if we use symmetry, we still need to evaluate 136 integrals! Invariably, these computations are performed numerically inside a computer program. While performing numerical integration, it must be observed that those integrals involve $[N]^T[N]$ type of terms. The terms in the integrals for a stiffness matrix [see Eq. (5.103)] are of the type $[B]^T[D][B]$, where coefficients in $[B]$ are spatial derivatives of $[N]$. Therefore, the highest degree polynomial term in the mass matrix expression is always of a degree higher than that for stiffness matrix. Thus the same rule of Gauss quadrature may not be adequate, and an appropriate rule must be carefully selected.

6.5   Consistent and Lumped Mass Matrices

The mass matrices derived above are termed *consistent mass matrices* since we used the same shape functions as for their stiffness matrices. It is, however, possible to lump the mass of the entire element at its nodes and come up with a lumped-mass equivalent representation. Such an intuitive lumping is depicted for bar and beam elements in Figure 6.7.

Fig. 6.7   **Mass lumping for bar and beam elements.**

Mathematically, the corresponding lumped mass matrices can be written as

$$[m^e]_{\text{lumped}} = \begin{bmatrix} \dfrac{\rho A\ell}{2} & 0 \\ 0 & \dfrac{\rho A\ell}{2} \end{bmatrix} \quad \text{for a bar element} \tag{6.63}$$

$$[m^e]_{\text{lumped}} = \begin{bmatrix} \dfrac{\rho A\ell}{2} & 0 & 0 & 0 \\ 0 & 0 & 0 & 0 \\ 0 & 0 & \dfrac{\rho A\ell}{2} & 0 \\ 0 & 0 & 0 & 0 \end{bmatrix} \quad \text{for a beam element} \tag{6.64}$$

From the beam element lumped mass matrix, we observe that we have assigned no lumped rotary inertia to the nodes. The mass lumping at the nodes has resulted in a diagonal mass matrix which is computationally very advantageous. The eigenvalue problem solvers for free vibration problems are typically iterative (see Section 6.8) and dynamic response calculations involve time marching schemes with small increments in time Δt (see Section 6.9), and hence are computationally intensive. Thus a diagonal mass matrix will be useful, and several researchers have proposed efficient schemes to arrive at the lumped mass matrix rather than the intuitive, *ad hoc* lumping above. We will briefly describe one such scheme, popularly used in many finite element calculations.

6.5.1   HRZ Lumping Scheme

The essential idea in this scheme is to simply use only the diagonal elements of the consistent mass matrix but to scale them in such a way that the total mass of the element is preserved. For example, for the bar element, the HRZ lumped mass matrix can be written as

$$[m^e]_{\text{HRZ}} = \frac{\rho A \ell}{2} \begin{bmatrix} 1 & 0 \\ 0 & 1 \end{bmatrix} \tag{6.65}$$

With reference to the beam element lumped mass matrix, we first write the diagonal elements of the consistent mass matrix (Eq. 6.31) as follows:

$$[m^e]_{\text{diag.}} = \frac{\rho A \ell}{420} \begin{bmatrix} 156 & 0 & 0 & 0 \\ 0 & 4\ell^2 & 0 & 0 \\ 0 & 0 & 156 & 0 \\ 0 & 0 & 0 & 4\ell^2 \end{bmatrix} \tag{6.66}$$

The total mass (as represented in this model) is obtained by summing all the diagonal elements corresponding to translational d.o.f. in one direction. Thus we get $\dfrac{\rho A \ell}{420}(156 + 156)$ $= \dfrac{312}{420}\,\rho A \ell$. Since the total mass of the element is $\rho A \ell$, the scaling factor is $\dfrac{420}{312}$. We scale all the diagonal elements with this scale factor to generate the HRZ lumped mass matrix for a beam as

$$[m^e]_{\text{HRZ}} = \rho A \ell \begin{bmatrix} 1/2 & 0 & 0 & 0 \\ 0 & \ell^2/78 & 0 & 0 \\ 0 & 0 & 1/2 & 0 \\ 0 & 0 & 0 & \ell^2/78 \end{bmatrix} \tag{6.67}$$

Simple intuitive lumping and HRZ lumping schemes can be readily applied to two-dimensional elements also. For example, for the Quad8 element, the HRZ mass distribution will yield the mass lumping at the nodes as shown in Figure 6.8.

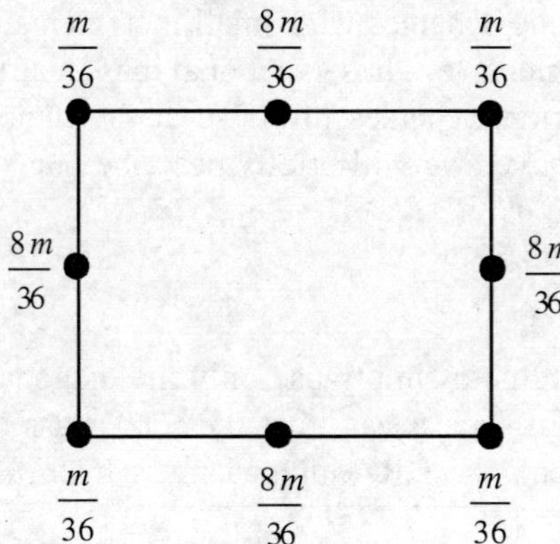

Fig. 6.8   HRZ lumped mass matrix for Quad8 element (based on 2×2 Gauss quadrature) (m = total mass of the element).

We will now solve a few simple free vibration problems. In the following example, we compare the first few natural frequencies obtained using both consistent and lumped mass matrices.

Example 6.2.   Consider a uniform cross-section bar (Figure 6.9) of length L made up of a material whose Young's modulus and density are given by E and ρ. Estimate the natural frequencies of axial vibration of the bar using both consistent and lumped mass matrices.

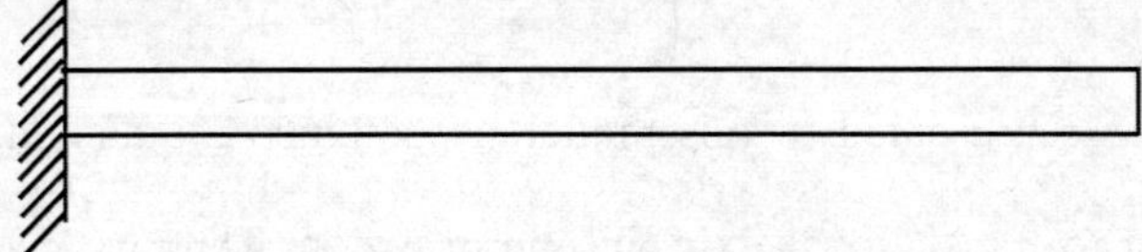

Fig. 6.9   A uniform bar (Example 6.2).

Using just one element for the entire rod (i.e. $\ell = L$) and using lumped mass matrix, we have

$$\frac{AE}{\ell}\begin{bmatrix} 1 & -1 \\ -1 & 1 \end{bmatrix}\begin{Bmatrix} u_1 \\ u_2 \end{Bmatrix} = \omega_{\text{lump}}^2 \rho A \ell \begin{bmatrix} \dfrac{1}{2} & 0 \\ 0 & \dfrac{1}{2} \end{bmatrix}\begin{Bmatrix} u_1 \\ u_2 \end{Bmatrix} \tag{6.68}$$

In view of the boundary condition at node 1 ($u_1 = 0$), we have

$$\frac{AE}{L}u_2 = \omega_{\text{lump}}^2 \frac{\rho A L}{2}u_2 \tag{6.69}$$

Hence,

$$\omega_{\text{lump}} = \sqrt{\frac{2E}{\rho L^2}} = \frac{1.414}{L}\sqrt{\frac{E}{\rho}} \qquad (6.70)$$

With one element and consistent mass matrix, we have

$$\frac{AE}{L}\begin{bmatrix} 1 & -1 \\ -1 & 1 \end{bmatrix}\begin{Bmatrix} u_1 \\ u_2 \end{Bmatrix} = \omega_{\text{cons.}}^2 \frac{\rho AL}{6}\begin{bmatrix} 2 & 1 \\ 1 & 2 \end{bmatrix}\begin{Bmatrix} u_1 \\ u_2 \end{Bmatrix} \qquad (6.71)$$

and with $u_1 = 0$,

$$\frac{AE}{L}u_2 = \omega_{\text{cons.}}^2\left(\frac{\rho AL}{6}\right)(2u_2) \qquad (6.72)$$

Therefore,

$$\omega_{\text{cons.}} = \sqrt{\frac{3E}{\rho L^2}} = \frac{1.732}{L}\sqrt{\frac{E}{\rho}} \qquad (6.73)$$

The exact solution for the fundamental frequency of a fixed-free bar is given as $1.571\sqrt{E/\rho}$. Thus we see that the consistent mass matrix overestimates, and the lumped mass matrix underestimates, the natural frequency.

If we now use a two-element mesh, i.e. $\ell = L/2$, see Figure 6.10, we get the following results with lumped and consistent mass matrices.

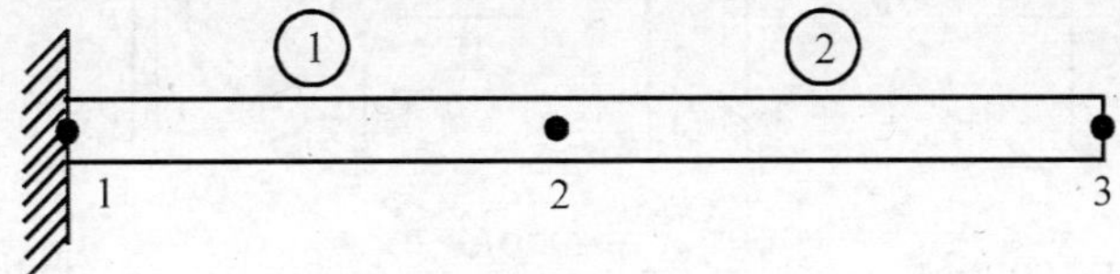

Fig. 6.10   Two-element model of a bar (Example 6.2).

Results with lumped mass matrix

The assembled equations can be readily verified to be given by

$$\frac{AE}{\left(\dfrac{L}{2}\right)}\begin{bmatrix} 1 & -1 & 0 \\ -1 & 1+1 & -1 \\ 0 & -1 & 1 \end{bmatrix}\begin{Bmatrix} u_1 \\ u_2 \\ u_3 \end{Bmatrix} = \rho A\left(\frac{L}{2}\right)\omega^2\begin{bmatrix} 1/2 & 0 & 0 \\ 0 & 1/2+1/2 & 0 \\ 0 & 0 & 1/2 \end{bmatrix}\begin{Bmatrix} u_1 \\ u_2 \\ u_3 \end{Bmatrix} \qquad (6.74)$$

With $u_1 = 0$ (boundary condition), we have

$$\frac{2AE}{L}\begin{bmatrix} 2 & -1 \\ -1 & 1 \end{bmatrix}\begin{Bmatrix} u_2 \\ u_3 \end{Bmatrix} = \frac{\rho AL}{2}\omega^2\begin{bmatrix} 1 & 0 \\ 0 & 1/2 \end{bmatrix}\begin{Bmatrix} u_2 \\ u_3 \end{Bmatrix} \qquad (6.75)$$

i.e.

$$\begin{bmatrix} 2-\lambda & -1 \\ -1 & 1-(\lambda/2) \end{bmatrix} \begin{Bmatrix} u_2 \\ u_3 \end{Bmatrix} = \begin{Bmatrix} 0 \\ 0 \end{Bmatrix} \tag{6.76}$$

where $\lambda = \dfrac{\rho L^2}{4E}\omega^2$.

For a nontrivial solution, we have

$$\begin{vmatrix} 2-\lambda & -1 \\ -1 & 1-(\lambda/2) \end{vmatrix} = 0 \text{ yielding } \lambda_1 = 0.586,\ \lambda_2 = 3.414 \tag{6.77}$$

Thus the natural frequencies are

$$\omega_1 = \frac{1.531}{L}\sqrt{\frac{E}{\rho}}, \qquad \omega_2 = \frac{3.695}{L}\sqrt{\frac{E}{\rho}} \tag{6.78}$$

Results with consistent mass matrix

Using consistent mass matrices, we obtain

$$\frac{AE}{L/2}\begin{bmatrix} 1 & -1 & 0 \\ -1 & 2 & -1 \\ 0 & -1 & 1 \end{bmatrix}\begin{Bmatrix} u_1 \\ u_2 \\ u_3 \end{Bmatrix} = \frac{\rho AL\omega^2}{12}\begin{bmatrix} 2 & 1 & 0 \\ 1 & 4 & 1 \\ 0 & 1 & 2 \end{bmatrix}\begin{Bmatrix} u_1 \\ u_2 \\ u_3 \end{Bmatrix} \tag{6.79}$$

Using $u_1 = 0$ (boundary condition) and $\lambda = \dfrac{\omega^2\rho L^2}{24E}$, for nontrivial solution, we obtain the equation

$$\begin{vmatrix} 2-4\lambda & -1-\lambda \\ -1-\lambda & 1-2\lambda \end{vmatrix} = 0 \tag{6.80}$$

On solving, we get $\lambda_1 = 0.108$ and $\lambda_2 = 1.32$. Thus the natural frequencies are

$$\omega_1 = \frac{1.61}{L}\sqrt{\frac{E}{\rho}}, \qquad \omega_2 = \frac{5.63}{L}\sqrt{\frac{E}{\rho}} \tag{6.81}$$

The exact solution can be readily verified to be

$$\omega_i = \frac{i\pi}{2L}\sqrt{\frac{E}{\rho}}, \qquad i = 1,\ 3,\ 5,\ ...,\ \infty \tag{6.82}$$

Thus,

$$\omega_1 = \frac{\pi}{2L}\sqrt{\frac{E}{\rho}}, \qquad \omega_2 = \frac{3\pi}{2L}\sqrt{\frac{E}{\rho}}, \qquad \omega_3 = \frac{5\pi}{2L}\sqrt{\frac{E}{\rho}}, \quad \dots \qquad (6.83)$$

We observe that the given rod is a continuous system and hence has infinitely many d.o.f. and therefore infinitely many natural frequencies. Our finite element model with one element has just one independent d.o.f., and that with two elements has just two independent d.o.f. Thus we can predict only one or two natural frequencies, for the one-element and two-element model, respectively. As we refine the mesh with more and more finite elements, we will actually be admitting more and more d.o.f., and therefore, predict higher natural frequencies also. The accuracy of predicted natural frequencies also improves.

Table 6.1 shows, for the axial vibrations of a rod, the predicted natural frequencies as we refine the mesh.[*]

Table 6.1 Natural Frequencies of a Fixed-free Bar

($L = 1$ m, $E = 2 \times 10^{11}$ N/m², $\rho = 7800$ kg/m³, $A = 30 \times 10^{-6}$ m²)

Mode \ No. of elements	1	2	3	4	8	16	Exact
1	1140.0	1234.0	1252.0	1258.0	1264.0	1265.0	
	1396.0	1299.0	1280.0	1274.0	1268.0	1266.0	1265.9
2		2978.0	3420.0	3582.0	3743.0	3784.0	
		4537.0	4188.0	4019.0	3853.0	3812.0	3797.8
3			4670.0	5366.0	6078.0	6266.0	
			7597.0	7301.0	6586.0	6393.0	6329.6
4				6319.0	8180.0	8688.0	
				10,560.0	9563.0	9037.0	8861.5
5					10,000.0	11,030.0	
					12,850.0	11,770.0	11,393.3

In each case, the upper row indicates the frequencies obtained with lumped mass and the lower row, the frequencies obtained with consistent mass matrices. We observe that typically these two provide the lower and upper bounds on the frequencies and the exact frequency is in between. It is conceivable that a mass matrix $[m]$ which is an average of the two can provide greater accuracy than either of them!

Example 6.3. Consider the simply supported beam, shown in Figure 6.11. Let the length $L = 1.0$ m, $E = 2 \times 10^{11}$ N/m²; area of cross-section, $A = 30$ cm²; moment of inertia $I = 100$ mm⁴; density $\rho = 7800$ kg/m³. We will obtain the first five natural frequencies using the three types of mass matrices, viz., simple lumped, HRZ lumped, and consistent mass matrices. Table 6.2 gives the results.

---

[*]See, for example, L. Meirovitch, *Elements of Vibration Analysis*, 2nd ed., McGraw-Hill, New York, 1986.

Fig. 6.11   Simply supported beam (Example 6.3).

Table 6.2   Natural Frequencies (Hz) of a Simply Supported Beam

(For each mode, the first row gives the results obtained with simple lumping,
the second row with HRZ lumping, and the third row with consistent mass matrix.)

Mode \ No. of elements	2	3	4	8	Exact
1	14.42	14.52	14.52	14.52	
	14.21	14.46	14.51	14.52	14.52
	14.58	14.53	14.52	14.52	
2		57.67	58.07	58.09	
	104.3	56.84	57.84	58.03	58.11
	64.47	58.32	58.11	58.09	
3		122.4	130.5	130.7	
	149.2	120.2	129.3	130.4	130.75
	162.1	133.1	130.9	130.7	
4			230.7	232.3	
	180.0	416.2	227.4	231.4	232.45
	295.5	257.9	233.3	232.4	
5			354.7	362.8	
		481.3	348.0	360.6	363.20
		409.9	366.4	363.3	

We observe that the simple lumped mass matrix assigns zero rotary inertia, and hence has fewer nonzero mass matrix elements. Therefore, it predicts fewer natural frequencies of the system. The HRZ lumping scheme corrects this by assigning appropriate rotational inertia at the nodes and accurately predicts as many frequencies as the full consistent mass matrix. A good mass lumping scheme, e.g. the HRZ scheme, is computationally advantageous, excellent in accuracy and is, therefore, popularly used.

The exact mode shapes for the simply supported beam are given by

$$v_i(x) = \sqrt{\frac{2}{\rho AL}}\, \sin \frac{i\pi x}{L}, \qquad i = 1, 2, 3, \ldots \tag{6.84}$$

The fifth mode shape, for example, is plotted in Figure 6.12. Within each beam element, we recall that the interpolation functions used permit cubic variation. Thus, even with eight elements, our approximation to the mode shape is inaccurate and we commit approximately 4% error in the natural frequency using the HRZ lumped mass. The anticipated mode shape vis-à-vis the individual element shape/interpolation functions must always be borne in mind while deciding the fineness of finite element mesh required for dynamics problems.

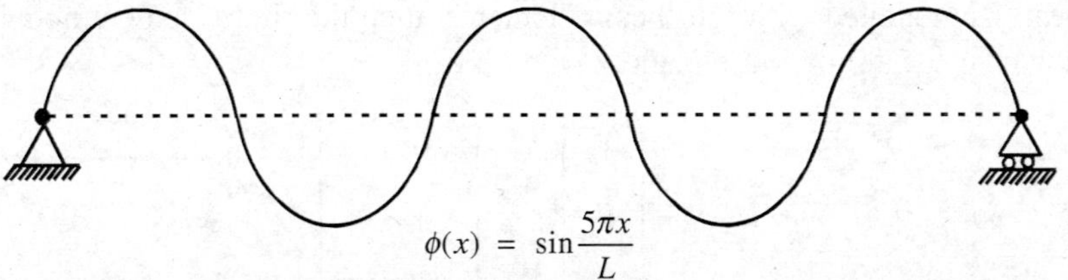

$$\phi(x) \;=\; \sin\frac{5\pi x}{L}$$

Fig. 6.12   Fifth mode shape of a simply supported beam.

6.6   Form of Finite Element Equations for Vibration Problems

A typical finite element formulation would yield the governing equations as

$$[M]\{\ddot{X}\} + [C]\{\dot{X}\} + [K]\{X\} = \{F(t)\} \tag{6.85}$$

In the case of free vibration, there is no external loading $\{F\} = 0$ and damping has negligible role. Therefore, for undamped free vibration problems,

$$[M]\{\ddot{X}\} + [K]\{X\} = 0 \tag{6.86}$$

Assuming harmonic vibration at a frequency ω_i, $\{X_i\} = \{U_i\} \sin \omega_i t$, we have

$$-\omega_i^2[M]\{U_i\} + [K]\{U_i\} = 0 \tag{6.87}$$

or

$$\boxed{[K]_{n\times n}\{U_i\}_{n\times 1} = \omega_i^2[M]_{n\times n}\{U_i\}_{n\times 1},} \qquad i = 1, 2, \cdots, n \tag{6.88}$$

This is the basic equation, in matrix form, governing the undamped free vibration of the structure. Given the stiffness and mass matrices $[K]$ and $[M]$, we have to find the natural frequencies ω_i and mode shapes $\{U_i\}$, $i = 1, 2, \cdots, n$. These represent certain characteristic states of free vibration of the system, each such state characterised by its eigenvalue ω_i and eigenvector $\{U_i\}$.

Equation (6.88) is a set of n equations in $(n + 1)$ unknowns, viz., ω_i and n elements of $\{U_i\}$. So a unique solution cannot be obtained. We choose to determine ω_i, and the ratios of elements of $\{U_i\}$. Thus the mode shapes are defined only within a multiple of themselves, i.e. if we have an eigenvector $\{U\}$, then $\alpha\{U\}$ ($\alpha \neq 0$) is also an eigenvector.

It is to be understood that there exist n pairs of natural frequencies ω and mode shape $\{U\}$. If the structure is excited at the frequency ω_i, it would vibrate such that the various points will have "relative" amplitudes as given in the mode shape eigenvector $\{U_i\}$. For example, the first mode of a simply supported beam would be as shown in Figure 6.13.

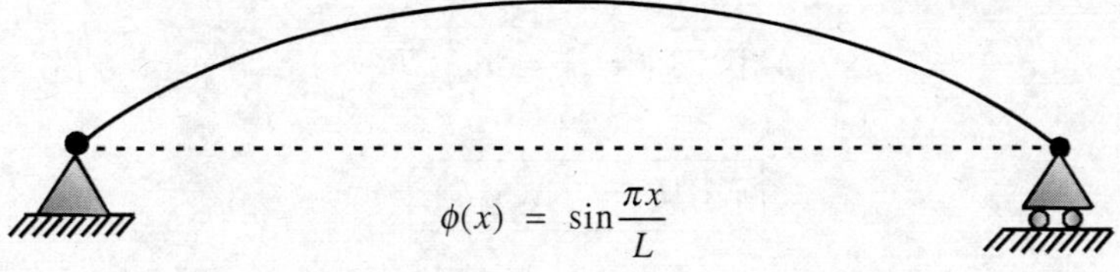

$$\phi(x) \;=\; \sin\frac{\pi x}{L}$$

Fig. 6.13   First mode shape of a simply supported beam.

If the beam is modelled with four beam elements, then the finite element nodal points will vibrate as shown in the following equation:

$$\{U_1\} = \begin{Bmatrix} v_1 \\ \theta_1 \\ v_2 \\ \theta_2 \\ v_3 \\ \theta_3 \\ v_4 \\ \theta_4 \\ v_5 \\ \theta_5 \end{Bmatrix} = \begin{Bmatrix} 0 \\ 1 \\ 0.707 \\ 0.707 \\ 1 \\ 0 \\ 0.707 \\ -0.707 \\ 0 \\ -1 \end{Bmatrix} \qquad (6.89)$$

The "mode shape" only indicates the relative amplitudes of vibration (viz., overall shape of vibration) and may readily be "scaled" to any amplitude. Thus we could write $\{U_1\}$ assuming $v_3 = 10$ units, and all the d.o.f. will be also scaled by the same factor. Thus there is no unique amplitude of free vibration. There are certain commonly accepted norms for scaling, which we will discuss later on.

Another way of interpreting the natural frequency and mode shape is that, if the structure were given an initial displacement to all its d.o.f. according to the relative amplitudes given in the mode shape, and left free to vibrate on its own, it will vibrate at the natural frequency ω, always maintaining these relative amplitudes.

Since the finite element mesh of the structure has "n" d.o.f., our model would yield n pairs of natural frequency ω_i and mode shape $\{U_i\}$ ($i = 1, 2, \ldots, n$). It is to be appreciated that the structure, being a continuous system, has infinitely many d.o.f., and hence infinitely many natural frequencies and mode shapes. Thus it is expected that our finite element model will become more and more accurate as we refine the mesh. The general form of the governing equations for the undamped free vibration is given by

$$\boxed{[K]_{n \times n} \{U_i\}_{n \times 1} = \omega_i^2 [M]_{n \times n} \{U_i\}_{n \times 1}} \qquad (6.90)$$

We may rewrite Eq. (6.90) as

$$[M]^{-1}[K]\{U_i\} = \omega_i^2 \{U_i\} \qquad (6.91)$$

i.e.

$$\boxed{[A]\{U_i\} = \lambda_i \{U_i\}} \qquad (6.92)$$

where $[A] = [M]^{-1}[K]$ and $\lambda_i = \omega_i^2$. We can also rewrite Eq. (6.90) as

$$\frac{1}{\omega_i^2}\{U_i\} = [K]^{-1}[M]\{U_i\} \tag{6.93}$$

i.e.

$$\boxed{\lambda_i\{U_i\} = [A]\{U_i\}} \tag{6.94}$$

where $[A] = [K]^{-1}[M]$ and $\lambda_i = 1/\omega_i^2$.

The form of representation

$$\boxed{[A]\{U\} = \lambda\{U\}} \tag{6.95}$$

is known as the standard form of eigenvalue problem, whereas the form $[K]\{U\} = \lambda [M]\{U\}$ is known as the nonstandard form. With the understanding that these two forms can be converted from one to the other easily, we will use both the forms in our subsequent discussion. However, it must be appreciated that, in typical finite element computations, $[M]^{-1}$ may not be possible to compute (e.g. because of diagonal zeros in a lumped mass matrix), or $[K]^{-1}$ may not be desirable (e.g. heavy computational overhead for large size matrices and also bandedness will be lost). Thus we will normally avoid any explicit computation of an inverse.

An important aspect to be observed here is the size of the problem. A typical finite element mesh for complex real-life problems may involve several thousand d.o.f. and thus $[K]$ and $[M]$ matrices are of this size. Hence we can potentially determine several thousand natural frequencies of the system. However, in most practical systems, the excitation frequencies are in a limited range (mostly less than a few kHz) and only those modes of the system that are near this range participate significantly in the response. Thus we are interested in determining a few (may be 10, 20 or 50) natural frequencies that lie in the frequency range of interest to us. Another aspect specific to finite element models of typical structural eigenvalue problems is the symmetry of $[K]$ and $[M]$ matrices. While several general purpose algorithms are available for solution of matrix eigenvalue problems, we will discuss here a few that have been popularly used in typical finite element analysis.

It is helpful to understand some of the basic properties of the eigenvalues and eigenvectors before we proceed to discuss the algorithms for their actual determination. In fact, these properties are extensively used in the actual determination of the eigenpairs.

6.7　Some Properties of Eigenpairs

Basis vectors

The eigenvectors constitute the least number of linearly independent mode shapes of the system. Any deflected shape of the structure can thus be represented as a linear combination of these eigenvectors $\{U_i\}$, i.e.,

$$\{X\} = c_1\{U_1\} + c_2\{U_2\} + c_3\{U_3\} + \cdots + c_n\{U_n\} \tag{6.96}$$

Orthogonality of eigenvectors (for symmetric matrices)

Consider two distinct eigenpairs $(\omega_i, \{U_i\})$ and $(\omega_j, \{U_j\})$:

$$[K]\{U_i\} = \omega_i^2 [M]\{U_i\} \tag{6.97}$$

$$[K]\{U_j\} = \omega_j^2 [M]\{U_j\} \tag{6.98}$$

Premultiplying Eq. (6.97) by $\{U_j\}^T$, we obtain

$$\{U_j\}^T [K]\{U_i\} = \omega_i^2 \{U_j\}^T [M]\{U_i\} \tag{6.99}$$

Taking the transpose of Eq. (6.98) and postmultiplying by $\{U_i\}$, we get

$$\{U_j\}^T [K]^T \{U_i\} = \omega_j^2 \{U_j\}^T [M]^T \{U_i\} \tag{6.100}$$

If $[K]$ and $[M]$ are symmetric, which is normally the case for a finite element mesh, we can subtract Eq. (6.100) from Eq. (6.99) and obtain

$$0 = (\omega_i^2 - \omega_j^2)\{U_j\}^T [M]\{U_i\} \tag{6.101}$$

Since they are distinct eigenpairs, $\omega_i \neq \omega_j$. Therefore,

$$\{U_i\}^T [M]\{U_j\} = 0 \qquad (i \neq j) \tag{6.102}$$

i.e. the eigenvectors are mutually orthogonal with respect to the mass matrix $[M]$.

Eigenvalue shifting

$$[K]\{U\} = \omega^2 [M]\{U\} \tag{6.103}$$

Using a scalar μ, we can write

$$([K] - \mu[M])\{U\} = (\omega^2 - \mu)[M]\{U\} \tag{6.104}$$

i.e.

$$[\bar{K}]\{U\} = \lambda[M]\{U\} \tag{6.105}$$

Thus we see that by effecting a shift in the eigenvalues of the system, we are not changing the eigenvectors. This property is used in typical computer procedures, to avoid zero eigenvalues (i.e. rigid body modes) as also to improve the convergence of two close eigenvalues.

Sturm sequence property

We have

$$[K]\{U\} = \lambda [M]\{U\} \tag{6.106}$$

Let us use a shift μ yielding

$$[[K] - \mu[M]]\{U\} = (\lambda - \mu)[M]\{U\} \tag{6.107}$$

If we now Cholesky factorise such that

$$[K] - \mu[M] = [L][D][L]^T \tag{6.108}$$

where $[L]$ is a lower triangular matrix and $[D]$ is a diagonal matrix, then an important property of $[D]$ is that the number of negative elements in $[D]$ equals the number of eigenvalues smaller than μ. This property is typically used to check our eigenvalue solution. If we have obtained, say m eigenvalues, we can use a trial value (say, μ) slightly higher than the highest eigenvalue obtained and perform a Sturm sequence check. A Sturm check fail is indicated if the number of eigenvalues below μ turns out to be more than m. This means that our eigenvalue solver missed out some eigenvalues, indicating that we need to redo the analysis.

6.8   Solution of Eigenvalue Problems

There are essentially three groups of methods of solution of eigenvalue problems:

1. Determinant based methods
2. Transformation based methods
3. Vector iteration based methods

The **determinant based methods** are primarily based on the following:

$$[A]\{U\} = \lambda\{U\} \tag{6.109}$$

$$[[A] - \lambda[I]]\{U\} = \{0\} \tag{6.110}$$

Hence,

$$\|[A] - \lambda[I]\| = 0 \text{ for a nontrivial } \{U\} \tag{6.111}$$

Thus, in principle, we can take a trial value of λ and compute the determinant $\|[A] - \lambda[I]\|$. With several trial values, we can generate a plot such as the one shown in Figure 6.14. By suitably monitoring the sign changes in the value of the determinant, we can iterate towards the eigenvalues λ.

Such a scheme is, however, never implemented in practice because of the heavy computational cost. Evaluation of each determinant of size $(n \times n)$ requires of the order of n^3

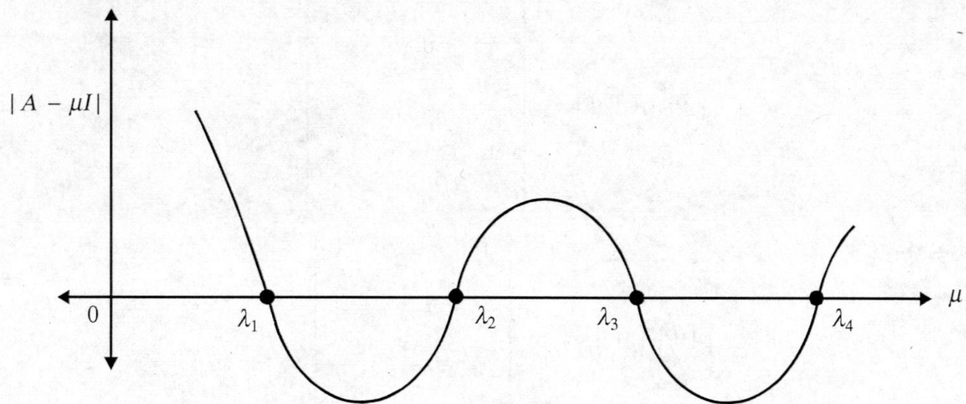

Fig. 6.14   Determinant-based method.

floating point operations, and several iterations may be required to determine all the eigenvalues of interest. We will therefore not discuss these methods in this book.

The second group of methods, viz., the **transformation based methods,** attempt to transform the eigenvalue problem as follows: Given

$$[A]\{U\} = \lambda\{U\} \tag{6.112}$$

Transform $[A]$ into a diagonal matrix or tridiagonal matrix, using a series of matrix transformations of the type $[\bar{A}] = [T]^T[A][T]$, where $[T]$, the transformation matrix, is usually an orthogonal matrix, i.e. $[T]^T = [T]^{-1}$. If we are able to transform $[A]$ completely into a diagonal matrix, then the elements on the diagonal themselves are the required eigenvalues. Among the well known methods of this type are the Givens method, Householders method, Jacobi method, and the Lanczos method. We will discuss the Jacobi method in this text, which is commonly implemented in many commercial finite element software packages.

The final group of methods, viz., **vector iteration based methods**, involve assuming a trial eigenvector and performing repeated matrix manipulations to converge to the desired eigenvector. We will discuss the basis of vector iteration methods, followed by the forward and inverse iteration methods, and simultaneous and subspace iteration methods. Vector iteration methods are normally available in many commercial finite element software packages.

6.8.1   Transformation Based Methods

The central idea behind these methods is based on the orthogonality property of the eigenvectors. We recall that, for a symmetric eigenvalue problem given by

$$[K]\{U\} = \omega^2[M]\{U\} \tag{6.113}$$

We have

$$\{U_i\}^T[M]\{U_j\} = 0 \quad \text{for } i \neq j \tag{6.114}$$

If we scale the eigenvectors such that $\{U_i\}^T[M]\{U_i\} = 1$ and we arrange together individual eigenvectors as columns of a matrix, we can write

$$[\Phi] = [\{U_1\} \; \cdots \; \{U_n\}] \tag{6.115}$$

$$[\Phi]^T[K][\Phi] = \begin{bmatrix} \lambda_1 & & & 0 \\ & \lambda_2 & & \\ & & \ddots & \\ 0 & & & \lambda_n \end{bmatrix} \tag{6.116}$$

$$[\Phi]^T[M][\Phi] = \begin{bmatrix} 1 & & & 0 \\ & 1 & & \\ & & \ddots & \\ 0 & & & 1 \end{bmatrix} \tag{6.117}$$

where λ_i are the eigenvalues.

Thus we can try to construct the eigenvector set $[\Phi]$ through a series of matrix manipulations through which we seek to transform $[K]$ and $[M]$ into diagonal matrices. We shall explain the basic procedure with respect to the eigenvalue problem in its standard form, i.e.

$$[A]\{U\} = \lambda\{U\} \tag{6.118}$$

Let us denote the transformation matrices used by $[T]$. The iterative procedure can be outlined as follows:

$$\text{Denote } [A_1] = [A] \tag{6.119}$$

$$\text{Compute } [A_2] = [T_1]^T[A_1][T_1] \tag{6.120}$$

where $[T_1]$ has been so chosen as to make any one off-diagonal element (say, a_{12}) zero. Compute

$$[A_3] = [T_2]^T[A_2][T_2] = ([T_1][T_2])^T[A]([T_1][T_2]) \tag{6.121}$$

where $[T_2]$ is formed in such a way as to make another off-diagonal element (say, a_{13}) zero. It is to be appreciated at this stage that in this process the $(1, 2)$-element may have once again become nonzero! Thus we need to perform several iterations. Compute

$$[\Phi] = [T_1][T_2] \tag{6.122}$$

Repeat this process till we reach the $(n - 1, n)$ off-diagonal element and make it zero. At this stage we have typically completed 'one sweep' of all the off-diagonal elements. Now we repeat the entire cycle.

After several such sweeps, we would have "effectively diagonalised" the matrix $[A]$. We talk of effective diagonalisation because, all computations being numerical, we will never be able to achieve exact diagonalisation. We can define a threshold value (e.g. 1% of the smallest diagonal element) and, if an off-diagonal element is less than this "threshold", we can assume it to be zero! Our estimate of the eigenvector set at any stage is clearly given by the product of the transformation matrices

$$[\Phi] = [T_1][T_2] \cdots \tag{6.123}$$

and the diagonal elements themselves are our approximation of the eigenvalues.

Jacobi method:

If an off-diagonal element a_{ij} is to be reduced to zero, the corresponding matrix $[T]$ is chosen as

$$[T] = \begin{bmatrix}
1 & & 0 & 0 & & & & & 0 \\
 & 1 & & & & & & & \\
 & & \ddots & & & & & & \\
 & & & 1 & & & & & \\
 & & & & \cos\theta & 0 & -\sin\theta & & \\
 & & & & 0 & 1 & \vdots & & \\
 & & & & \vdots & & \vdots & & \\
 & & & & \sin\theta & & \cos\theta & & \\
 & & & & & & & 1 & \\
 & & & & & & & & \ddots \\
0 & & & & & & & & 1
\end{bmatrix}
\begin{array}{l} \\ \\ \\ \\ \leftarrow \; i\text{th row} \\ \\ \\ \leftarrow \; j\text{th row} \\ \\ \\ \end{array} \tag{6.124}$$

$$\uparrow \qquad \uparrow$$
$$i\text{th col.} \qquad j\text{th col.}$$

where

$$\tan 2\theta = \frac{2a_{ij}}{a_{ii} - a_{jj}} \quad \text{for} \quad a_{ii} \neq a_{jj},$$

(6.125)

$$\theta = \frac{\pi}{4} \quad \text{if} \quad a_{ii} = a_{jj}$$

In a typical computer implementation, we actually do not have to evaluate $[T]^T[A][T]$ through complete matrix multiplication. It is just the linear combination of the corresponding two rows and columns, while the rest of the matrix remains the same.

When we have both $[K]$ and $[M]$ as in the typical FEM, we cannot directly apply the Jacobi procedure outlined above, even if $[M]$ is a diagonal lumped mass matrix. If $[T]$ has been chosen to make a k_{ij} vanish, then $[T]^T[M][T]$ will remain diagonal only if $m_{ii} = m_{jj}$. We can, however, extend the basic Jacobi scheme to the nonstandard eigenvalue problem.

$$[K]\{U\} = \omega^2[M]\{U\}$$

(6.126)

Let us say that we wish to make k_{ij} and m_{ij} vanish. Let us write the transformation matrix as

$$[T] = \begin{bmatrix} 1 & & & 0 \\ & 1 & \alpha & \\ & \beta & 1 & \\ 0 & 0 & 0 & 1 \end{bmatrix} \begin{array}{l} \leftarrow ith\ row \\ \leftarrow jth\ row \end{array}$$

(6.127)

$$\begin{array}{cc} \uparrow & \uparrow \\ ith & jth \\ \text{col.} & \text{col.} \end{array}$$

and let us write

$$[\bar{K}] = [T]^T[K][T], \qquad [\bar{M}] = [T]^T[M][T]$$

(6.128)

It is easily verified that

$$\bar{k}_{ij} = k_{ij} + \alpha k_{ii} + \beta k_{jj} + \alpha\beta k_{ij}$$

(6.129)

$$\bar{m}_{ij} = m_{ij} + \alpha m_{ii} + \beta m_{jj} + \alpha\beta m_{ij}$$

(6.130)

α and β can be obtained in such a way that $\bar{k}_{ij} = 0$, $\bar{m}_{ij} = 0$. The solution for α and β can be obtained as follows:[*] Let

$$\mu_1 = k_{ii}m_{ij} - m_{ii}k_{ij}$$

$$\mu_2 = k_{jj}m_{ij} - m_{jj}k_{ij}$$

$$\mu_3 = k_{ii}m_{jj} - m_{ii}k_{jj}$$

---

[*]K.J. Bathe, *Finite Element Procedures*, Prentice-Hall of India, New Delhi, 2001.

$$\mu_4 = \frac{\mu_3}{2} + \text{sign}\,(\mu_3) \sqrt{\left(\frac{\mu_3}{2}\right)^2 + \mu_1\mu_2} \tag{6.131}$$

then

$$\alpha = \frac{\mu_2}{\mu_4}, \qquad \beta = \frac{-\mu_1}{\mu_4} \tag{6.132}$$

In the special case that $[M]$ is a diagonal lumped mass matrix, we have

$$\mu_1 = -m_{ii}k_{ij}, \qquad \mu_2 = -m_{jj}k_{ij} \tag{6.133}$$

We will illustrate the generalised Jacobi method using the following example.

Example 6.4.   Consider the three-element model of the fixed-free bar (shown in Figure 6.15) undergoing axial vibrations. Let $L = 1$ m, $A = 30 \times 10^{-6}$ m^2, $E = 2 \times 10^{11}$ N/m^2, $\rho = 7800$

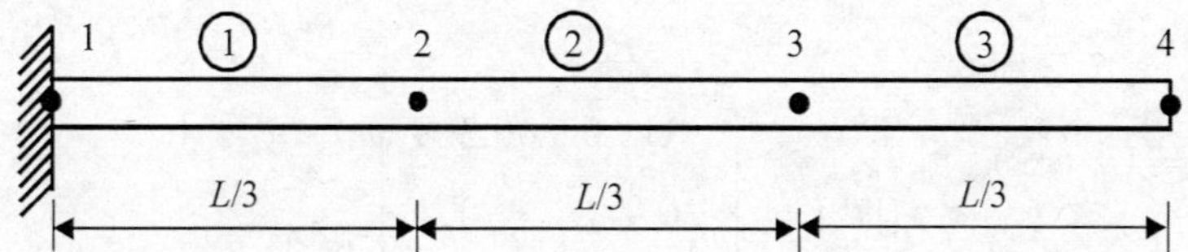

Fig. 6.15   Three-element model of a fixed-free bar (Example 6.4).

kg/m^3. Using consistent mass matrix, the assembled stiffness and mass matrices are readily obtained as

$$[K_1] = \begin{bmatrix} 0.360 \times 10^8 & -0.180 \times 10^8 & 0 \\ -0.180 \times 10^8 & 0.360 \times 10^8 & -0.180 \times 10^8 \\ 0 & -0.180 \times 10^8 & 0.180 \times 10^8 \end{bmatrix},$$

$$[M_1] = \begin{bmatrix} 0.052 & 0.013 & 0 \\ 0.013 & 0.052 & 0.013 \\ 0 & 0.013 & 0.026 \end{bmatrix} \tag{6.134}$$

where the fixed-end boundary condition is already incorporated (i.e. the corresponding row and column have not been assembled).

We will obtain the eigenpairs of this system using the generalised Jacobi method. One sweep will consist of zeroing the elements (1, 2), (1, 3) and (2, 3) in $[K]$ and $[M]$ matrices. First sweep $i = 1$, $j = 2$. From Eqs. (6.131)–(6.132),

$$\mu_1 = 140{,}4000.0, \qquad \mu_2 = 140{,}4000.0, \qquad \mu_3 = 0, \qquad \mu_4 = 140{,}4000.0$$

$$[T_1] = \begin{bmatrix} 1 & 1 & 0 \\ -1 & 1 & 0 \\ 0 & 0 & 1 \end{bmatrix} \tag{6.135}$$

$$[K_2] = [T_1]^T [K_1][T_1] = 10^8 \begin{bmatrix} 1.08 & 0 & 0.180 \\ 0 & 0.360 & -0.18 \\ 0.18 & -0.18 & 0.18 \end{bmatrix} \tag{6.136}$$

$$[M_2] = [T_1]^T [M_1][T_1] = \begin{bmatrix} 0.078 & 0 & -0.013 \\ 0 & 0.13 & 0.013 \\ -0.013 & 0.013 & 0.026 \end{bmatrix} \tag{6.137}$$

First sweep $i = 1$, $j = 3$.

$$[T_2] = \begin{bmatrix} 1 & 0 & -0.309 \\ 0 & 1 & 0 \\ 1.24 & 0 & 1 \end{bmatrix} \tag{6.138}$$

$$[K_3] = [T_2]^T [K_2][T_2] = 10^8 \begin{bmatrix} 1.80 & -0.222 & 0 \\ -0.222 & 0.36 & -0.18 \\ 0 & -0.18 & 0.172 \end{bmatrix} \tag{6.139}$$

$$[M_3] = [T_2]^T [M_2][T_2] = \begin{bmatrix} 0.0856 & 0.0161 & 0 \\ 0.0161 & 0.130 & 0.013 \\ 0 & 0.013 & 0.0415 \end{bmatrix} \tag{6.140}$$

First sweep $i = 2$, $j = 3$.

$$[T_3] = \begin{bmatrix} 1 & 0 & 0 \\ 0 & 1 & -0.47 \\ 0 & 1.36 & 1 \end{bmatrix} \tag{6.141}$$

$$[K_4] = [T_3]^T [K_3][T_3] = 10^8 \begin{bmatrix} 1.80 & -0.222 & 0.105 \\ -0.222 & 0.188 & 0 \\ 0.105 & 0 & 0.421 \end{bmatrix} \tag{6.142}$$

$$[M_4] = [T_3]^T [M_3][T_3] = \begin{bmatrix} 0.0856 & 0.0161 & -0.00756 \\ 0.0161 & 0.242 & 0 \\ -0.00756 & 0 & 0.058 \end{bmatrix} \qquad (6.143)$$

We observe that the off-diagonal elements, once zeroed, do not remain zero. Also, these are not negligible yet. We therefore need to perform another sweep from (1, 2) to (2, 3) element. We now present the results after the third such sweep has been completed. So,

$$[T_9] = \begin{bmatrix} 1 & 0 & 0 \\ 0 & 1 & -0.132 \times 10^{-6} \\ 0 & 0.373 \times 10^{-6} & 1 \end{bmatrix} \qquad (6.144)$$

$$[K_{10}] = [T_9]^T [K_9][T_9] = 10^8 \begin{bmatrix} 1.96 & -0.216\times10^{-10} & 0.284\times10^{-17} \\ -0.216\times10^{-10} & 0.161 & 0.222\times10^{-22} \\ 0.416\times10^{-16} & 0.421\times10^{-16} & 0.424 \end{bmatrix} \qquad (6.145)$$

$$[M_{10}] = [T_9]^T [M_9][T_9] = \begin{bmatrix} 0.0859 & 0.156\times10^{-11} & -0.204\times10^{-18} \\ 0.156\times10^{-11} & 0.248 & 0 \\ -0.195\times10^{-18} & 0.694\times10^{-17} & 0.0612 \end{bmatrix} \qquad (6.146)$$

We conclude that the off-diagonal elements in $[K_{10}]$ and $[M_{10}]$ are negligibly small. Our estimates of the eigenvalues of the bar are, therefore,

$$f_1 = \frac{1}{2\pi} \sqrt{\frac{k_{11}}{m_{11}}} = \frac{1}{2\pi} \sqrt{\frac{1.96\times10^8}{0.0859}} = 7597 \text{ Hz} \qquad (6.147)$$

$$f_2 = \frac{1}{2\pi} \sqrt{\frac{k_{22}}{m_{22}}} = \frac{1}{2\pi} \sqrt{\frac{0.161\times10^8}{0.248}} = 1280.43 \text{ Hz} \qquad (6.148)$$

$$f_3 = \frac{1}{2\pi} \sqrt{\frac{k_{33}}{m_{33}}} = \frac{1}{2\pi} \sqrt{\frac{0.424\times10^8}{0.0612}} = 4187.64 \text{ Hz} \qquad (6.149)$$

which are the same as those given against consistent mass in Table 6.1 for three-element case, though the order is interchanged. Our estimate of the eigenvectors is given by

$$[\bar{\Phi}] = [T_1][T_2] \cdots [T_9] \qquad (6.150)$$

$$[\bar{\Phi}] = \begin{bmatrix} 0.697 & 0.745 & -0.886 \\ -1.21 & 1.29 & 0 \\ 1.39 & 1.49 & 0.886 \end{bmatrix} \tag{6.151}$$

However, these approximate mode shapes have not been scaled such that $[\bar{\Phi}]^T [M] [\bar{\Phi}] = [I]$. We can extract the $[M]$-orthonormalised mode shapes $[\Phi]$ from $[\bar{\Phi}]$, following the procedure discussed later in the section. We then obtain

$$[\Phi] = \begin{bmatrix} 2.38 & 1.50 & -3.58 \\ -4.13 & 2.58 & -0.005 \\ 4.74 & 3.01 & 3.59 \end{bmatrix} \tag{6.152}$$

We observe that the original matrices were banded (semi-bandwidth = 2). During Jacobi transformations, bandedness is not guaranteed. So, in a practical implementation of the scheme on a several thousand d.o.f. finite element mesh (though usually bandwidth may be much small—a couple of hundred at most), we need to store the entire ($n \times n$) stiffness and mass matrices. It can be readily shown that even a diagonal lumped mass matrix will not remain diagonal after Jacobi transformation. Moreover, numerical round-off errors slowly creep into the computations. Thus the most common implementation of the generalised Jacobi scheme has been to solve the reduced eigenvalue problem of the subspace iteration technique discussed towards the end of this section.

6.8.2   Vector Iteration Methods

Basis of vector iteration method

Consider the standard eigenvalue problem $[A]\{U\} = \lambda\{U\}$. Let its eigenpairs be ($\lambda_1\{U_1\}$), ($\lambda_2\{U_2\}$), $\cdots$ ($\lambda_n\{U_n\}$). Since any deflected shape can be represented as a linear combination of these eigenvectors, a trial vector $\{X^1\}$ can be written as

$$\{X^1\} = c_1\{U_1\} + c_2\{U_2\} + c_3\{U_3\} + \cdots + c_n\{U_n\} \tag{6.153}$$

Compute

$$\{X^2\} = [A]\{X^1\} = c_1[A]\{U_1\} + c_2[A]\{U_2\} + \cdots + c_n[A]\{U_n\}$$

$$= c_1\lambda_1\{U_1\} + c_2\lambda_2\{U_2\} + \cdots + c_n\lambda_n\{U_n\} \tag{6.154}$$

After performing premultiplications with $[A]$ m times, we obtain

$$\{X^{m+1}\} = c_1\lambda_1^m\{U_1\} + c_2\lambda_2^m\{U_2\} + \cdots + c_n\lambda_n^m\{U_n\}$$

$$= c_1\lambda_1^m\left[\{U_1\} + c_2\left(\frac{\lambda_2}{\lambda_1}\right)^m\{U_2\} + \cdots + c_n\left(\frac{\lambda_n}{\lambda_1}\right)^m\{U_n\}\right] \tag{6.155}$$

If $\lambda_1 > \lambda_2 > \lambda_3, \cdots > \lambda_m$, i.e. $(\lambda_2/\lambda_1 < 1, \ldots, \lambda_n/\lambda_1 < 1)$, then after several iterations,

$$\left(\frac{\lambda_i}{\lambda_1}\right)^m \to 0 \quad (i = 2, 3, 4, \ldots, n), \text{ or}$$

$$\boxed{\{X^{m+1}\} \approx c_1\lambda_1^m\{U_1\}} \tag{6.156}$$

For example, if an eigenvalue is one-fifth the highest eigenvalue, after 10 iterations, we have $(1/5)^{10} = 0.0000001024$, which is negligibly small. If the individual eigenvalues are well separated, the trial vector $\{X^{m+1}\}$ will essentially contain components of the mode shape $\{U_1\}$ while all other terms would vanish. We say that we have achieved "convergence" to the eigenvector $\{U_1\}$. Since mode shapes are only relative amplitudes, the scale factor $c_1\lambda^m$ is of little consequence. If we now compute $\left(\{X\}^T[A]\{X\}/\{X\}^T\{X\}\right)$, we get an approximate estimate of λ_1. It should be observed that, irrespective of our choice for trial vector $\{X^1\}$, convergence is always to the largest eigenvalue and the corresponding eigenvector.[*]

The rate of convergence (i.e. how quickly we can estimate the mode shape components) depends on how well separated the eigenvalues are. If an eigenvalue is close to the highest eigenvalue, say, the ratio is 0.9, then, even after several iterations, its contribution in $\{X\}$ will not vanish. To improve the rate of convergence, we need to separate the eigenvalues and, for this purpose, we can successfully exploit the eigenvalue shift property. For example, let $\lambda_1 = 100$ and $\lambda_2 = 90$, yielding $\lambda_2/\lambda_1 = 0.9$. If we use a shift of $\mu = 85$, i.e. we operate on the matrix $[\overline{A}] = [A] - \mu[I]$, then we will effectively have $\overline{\lambda}_1 = 100 - 85 = 15$ and $\overline{\lambda}_2 = 90 - 85 = 5$ yielding $\overline{\lambda}_2/\overline{\lambda}_1 = 0.33$. After 10 iterations, $(\lambda_2/\lambda_1)^{10} = (90/100)^{10} = 0.3486$ whereas $(\overline{\lambda}_2/\overline{\lambda}_1)^{10} = (5/15)^{10} = 0.000015$!

Forward and inverse iteration

In typical finite element computations, we have the nonstandard eigenvalue problem

$$[K]\{U\} = \omega^2[M]\{U\} \tag{6.157}$$

When we extend the basic vector iteration scheme to this nonstandard form, we have two options $[K]\{U\} = \omega^2[M]\{U\}$ can be written as

$1/\omega^2\{U\} = [K]^{-1}[M]\{U\}$	$[M]^{-1}[K]\{U\} = \omega^2\{U\}$
$\therefore \lambda\{U\} = [A]\{U\}$	$[A]\{U\} = \lambda\{U\}$
Assume $\{U\}$ and premultiply by $[A]$ to converge to the largest λ, i.e. the lowest ω^2	Assume $\{U\}$ and premultiply by $[A]$ to converge to the largest λ, i.e. the largest ω^2
INVERSE ITERATION	FORWARD ITERATION

Since we are usually interested in the lowest eigenvalue (i.e. the first or the lowest critical speed of a rotor, etc.), the inverse iteration is commonly employed.

In each iteration an estimate of the eigenvalue can be obtained by computing a quotient called Rayleigh Quotient (RQ), defined as

$$RQ = \frac{\{X\}^T[K]\{X\}}{\{X\}^T[M]\{X\}} \tag{6.158}$$

---

[*]This does not hold good in the degenerate case when our trial vector $\{X\}$ is perfectly orthogonal to $\{U_1\}$.

This is based on the fact that

$$[K]\{U\} = \omega^2[M]\{U\} \tag{6.159}$$

Therefore,

$$\{U\}^T[K]\{U\} = \omega^2\{U\}^T[M]\{U\} \tag{6.160}$$

Thus,

$$\omega^2 \approx \frac{\{U\}^T[K]\{U\}}{\{U\}^T[M]\{U\}} \tag{6.161}$$

This is, in fact, a measure of the potential energy/kinetic energy. We have already noted that $\{U\}$ is a set of relative amplitudes. To fix the magnitude of the vector $\{U\}$, we usually normalise $\{U\}^T[M]\{U\} = 1$. So, in each iteration simply compute the RQ, and continue the iterative process till the eigenvalue converges within the preset tolerance limit.

Algorithm for inverse iteration scheme

Step 1: Formulate the global $[K]$ and $[M]$ for the structure, i.e. form $[K]\{U\} = \omega^2[M]\{U\}$.

Step 2: Assume a trial vector $\{X^1\}$.

Step 3: Compute $\{R\} = [M]\{X^1\}$.

Step 4: Solve $[K]\{\overline{X}\} = \{R\}$.

Thus, this iteration scheme is similar to premultiplying the trial vector $\{X^1\}$ by $[K]^{-1}[M]$. Instead of explicit computation of $[K]^{-1}$, we have effective algorithms for solving $[K]\{\overline{X}\} = \{R\}$, as discussed in Chapter 5 (see section 5.7).

Step 5: Obtain $\{X^2\}$ from $\{\overline{X}\}$ such that

$$\{X^2\}^T[M]\{X^2\} = 1$$

i.e.

$$\{X^2\} = \frac{1}{\sqrt{\{\overline{X}\}^T[M]\{\overline{X}\}}}\{\overline{X}\} \tag{6.162}$$

Step 6: Compute $\lambda = \{X^2\}^T[K]\{X^2\}$.

Repeat steps (3)–(6) till λ converges to within a preset tolerance.

The inverse iteration scheme forms one of the most powerful tools for the determination of the first eigenpair of a system. Since several modes usually participate in a vibration response, we are usually interested in determining a few (and not just the lowest) eigenpairs. It is reiterated that our trial vector always converges to the largest eigenpair. Thus, if we need three eigenpairs and we take three distinct trial vectors, they will all converge to the largest eigenpair only. Recall that if it so happens that the trial vector is exactly orthogonal to the actual eigenvector, then, it can never converge to this eigenvector. This leads us to a scheme of obtaining higher eigenpairs from this inverse iteration process.

Gram–Schmidt deflation

We have

$$[K]\{U\} = \omega^2[M]\{U\} \tag{6.163}$$

Let $\lambda_1\{U_1\}$, the lowest eigenpair, be already known by, for example, inverse iteration. Then from any trial vector we choose $\{X_2\}$ to get $\{U_2\}$, subtract $\beta_1\{U_1\}$, where

$$\beta_1 = \{U_1\}^T[M]\{X_2\} \tag{6.164}$$

$$\{\overline{X}_2\} = \{X_2\} - \beta_1\{U_1\} \tag{6.165}$$

Then, after proper iteration, $\{\overline{X}_2\}$ will converge to $\{U_2\}$. Let us see how this happens. Let

$$\{X_2\} = c_1\{U_1\} + c_2\{U_2\} + c_3\{U_3\} + \cdots + c_n\{U_n\} \tag{6.166}$$

Hence,

$$\{U_1\}^T[M]\{X_2\} = c_1\{U_1\}^T[M]\{U_1\} + c_2\{U_1\}^T[M]\{U_2\}$$
$$+ \cdots + c_n\{U_1\}^T[M]\{U_n\} \tag{6.167}$$

Since the eigenvectors are $[M]$ orthogonal,

$$\{U_1\}^T[M]\{U_j\} = 0 \quad \text{for } j = 2, 3, \cdots, n \tag{6.168}$$

$$\{U_1\}^T[M]\{U_1\} = 1 \tag{6.169}$$

Therefore,

$$\beta_1 = \{U_1\}^T[M]\{X_2\} = c_1 \tag{6.170}$$

Thus,

$$\{\overline{X}_2\} = \{X_2\} - \beta_1\{U_1\}$$
$$= c_1\{U_1\} + c_2\{U_2\} + c_3\{U_3\} + \cdots + c_n\{U_n\} - c_1\{U_1\}$$
$$= c_2\{U_2\} + c_3\{U_3\} + \cdots + c_n\{U_n\} \tag{6.171}$$

We have thus eliminated $\{U_1\}$ component from the trial vector $\{\overline{X}_2\}$. Therefore, $\{\overline{X}_2\}$. Therefore will actually be orthogonal to, and will not converge to, $\{U_1\}$. In its implementation in inverse iteration scheme, we need to perform this Gram–Schmidt deflation in every iteration. Our trial vector will then converge to the eigenpair λ_2 and $\{U_2\}$. After determining λ_2 and $\{U_2\}$ to determine λ_3 and $\{U_3\}$, we will use a deflated trial vector given by

$$\{\overline{X}\} = \{X\} - \beta_1\{U_1\} - \beta_2\{U_2\} \tag{6.172}$$

where β_2 is similarly determined.

Algorithm for inverse iteration with Gram–Schmidt deflation

Step 1: Formulate the assembled global $[K]$ and $[M]$ for the entire structure.

Step 2: Assume a trial vector $\{X_1\}$.

If iterating for the lowest eigenpair, go to step 4. Else, go to step 3.

Step 3: Perform deflation, i.e.

i.e.

$$\{\bar{X}_1\} = \{X_1\} - \sum_{i=1}^{m} \beta_i \{U_i\} \tag{6.173}$$

where $\beta_i = \{U_i\}^T [M]\{X_1\}$ and m denotes the No. of eigenpairs already found.

Step 4: Compute $\{R\} = [M]\{\bar{X}_1\}$.

Step 5: Solve $[K]\{\bar{\bar{X}}_1\} = \{R\}$ (similar to solution of static problem).

Step 6: Obtain $\{X_2\}$ from $\{\bar{\bar{X}}_1\}$ such that

$$\{X_2\}^T [M]\{X_2\} = 1 \tag{6.174}$$

$$\{X_2\} = \sqrt{\frac{1}{\{\bar{\bar{X}}\}^T [M]\{\bar{\bar{X}}\}}} \tag{6.175}$$

Step 7: Compute the $(m + 1)^{\text{th}}$ eigenvalue, i.e. $\lambda_{m+1} = \{X_2\}^T [K]\{X_2\}$.

Repeat steps (3)–(7) till a satisfactory convergence is achieved.

Simultaneous iteration method

In the Gram–Schmidt deflation technique, each trial vector was made orthogonal to all the eigenvectors obtained earlier. Thus we can ensure that convergence was obtained to the eigenpairs in a sequential manner, starting from the first till the mth desired vector. In the present method, we start with a set of trial vectors and iterate on them simultaneously.

This iteration scheme is similar to the basic vector iteration method, but with multiple trial vectors. The basic idea is to ensure that the individual trial vectors are mutually orthogonal in each iteration so that they will converge to distinct eigenvectors. We have

$$[K]\{U\} = \omega^2 [M]\{U\} \tag{6.176}$$

The basic steps in this method are now outlined.

Step 1: Choose the trial vector set $\{\bar{X}_1\}_{n\times m}$. For any arbitrary trial vector set, individual vectors may not be orthogonal to one another. Let us assume that by some procedure, we are able to extract an orthonormal set of vectors $\{X_1\}_{n\times m}$ from $\{\bar{X}_1\}_{n\times m}$, i.e.,

$$\{X_1\}^T [M]\{X_1\} = [I] \tag{6.177}$$

Step 2: Compute $\{R\}_{n\times m} = [M]_{n\times n}\{X_1\}_{n\times m}$.

Step 3: Solve for $\{\bar{X}_2\}$ from $[K]_{n\times n}\{\bar{X}_2\}_{n\times m} = \{R\}_{n\times m}$.

Step 4: Compute $[\lambda]_{m\times m} = \{X_1\}^T [K]\{X_1\}$. $[\lambda]$ is now an $(m \times m)$ matrix, the diagonal elements of which are our current approximation to the eigenvalues. The iterations are continued till $[\lambda]$ converges fully. When converged, $[\lambda]$ will be of the form

$$\begin{bmatrix} \lambda_1 & & & \approx 0 \\ & \lambda_2 & & \\ & & \ddots & \\ \approx 0 & & & \lambda_n \end{bmatrix} \tag{6.178}$$

and the corresponding $\{X\}$ will have the eigenvectors.

We see that these basic steps are identical to those of inverse iteration except that we operate on multiple trial vectors. The significant deviation is in the procedure to get orthonormalised vectors $\{X\}$ from original $\{\overline{X}\}$, which is now described[*].

To get $\{X\}$ from $\{\overline{X}\}$, we use the following condition:

$$\{X\}^T_{m\times n} [M]_{n\times n} \{X\}_{n\times m} = [I]_{m\times m} \tag{6.179}$$

Let

$$\{X\}_{n\times m} = \{\overline{X}\}_{n\times m} [S]_{m\times m} \tag{6.180}$$

Therefore,

$$\{X\}^T [M]\{X\} = [S]^T \{\overline{X}\}^T [M]\{\overline{X}\}[S] = [S]^T [D][S] = [I] \tag{6.181}$$

where

$$[D] = \{\overline{X}\}^T [M]\{\overline{X}\} \tag{6.182}$$

Cholesky factorise (see Section 5.7)

$$[D] = [L][L]^T \tag{6.183}$$

Then, from Eq. (6.181),

$$[S] = [L]^{-T} \tag{6.184}$$

Thus, our method of finding the orthonormal vector set $\{X\}$ from $\{\overline{X}\}$ involves use of the inverse of the lower triangular matrix $[L]$ obtained from Cholesky factorisation of $\{\overline{X}\}^T [M]\{\overline{X}\}$. We therefore write

$$\{X\} = \{\overline{X}\}[L]^{-T} \tag{6.185}$$

Postmultiplying with $[L]^T$, we get

$$\{X\}[L]^T = \{\overline{X}\} \tag{6.186}$$

Avoiding the explicit computation of the inverse $[L]^{-1}$, we prefer to obtain $\{X\}$ as a solution of Eq. (6.186) by backward substitution (see Section 5.7).

---

*The interested reader should refer V. Ramamurti, *Computer Aided Mechanical Design and Analysis*, McGraw-Hill, New York, 1998.

Algorithm for simultaneous iteration scheme

Step 1: Formulate the assembled global stiffness and mass matrices $[K]$ and $[M]$.

Step 2: Choose a trial vector set $\{\bar{X}_1\}_{n\times m}$.

Step 3: Compute $[D]_{m\times m} = \{\bar{X}_1\}^T_{m\times n}[M]_{n\times n}\{\bar{X}_1\}_{n\times m}$.

Step 4: Cholesky factorise $[D]_{m\times m} = [L]_{m\times m}[L]^T_{m\times m}$.

Step 5: Solve for $\{X_1\}_{n\times m}$ from $\{X_1\}_{n\times m}[L]^T_{n\times m} = \{\bar{X}_1\}_{n\times m}$ (by back substitution).

Step 6: Compute $[R]_{n\times m} = [M]_{n\times n}\{X_1\}_{n\times m}$.

Step 7: Solve for $\{\bar{X}_2\}_{n\times m}$ from $[K]_{n\times n}\{\bar{X}_2\}_{n\times m} = \{R\}_{n\times m}$.

Step 8: Compute the Rayleigh quotient matrix $[\lambda] = \{X_1\}^T_{m\times m}[K]_{n\times n}\{X_1\}_{n\times m}$.

Repeat steps 3–8 till satisfactory convergence is obtained in $[\lambda]$, i.e. the off-diagonal elements must be negligibly small, and the change in the diagonal elements from one iteration to the next must be negligibly small.

We will now illustrate the simultaneous vector iteration procedure using the following example.

Example 6.5. *Axial vibrations of a fixed-free bar.* Consider the 3-element model of a uniform rod shown in Figure 6.15. Let $L = 1$ m, $A = 30 \times 10^{-6}$ m^2, $E = 2 \times 10^{11}$ N/m^2, $\rho = 7800$ kg/m^3. Using consistent mass matrix, the assembled stiffness and mass matrices are readily obtained as

$$[K] = 10^8 \begin{bmatrix} 0.360 & -0.180 & 0 \\ -0.180 & 0.360 & -0.180 \\ 0 & -0.180 & 0.180 \end{bmatrix} \tag{6.187}$$

$$[M] = \begin{bmatrix} 0.052 & 0.013 & 0 \\ 0.013 & 0.052 & 0.013 \\ 0 & 0.013 & 0.026 \end{bmatrix} \tag{6.188}$$

which are the same as those in Example 6.4. Let us obtain the three eigenpairs of this system using the simultaneous vector iteration method.

Iteration 1 Let

$$\{\bar{X}_1\}_{3\times 3} = \begin{bmatrix} 1 & 0 & 0 \\ 0 & 1 & 0 \\ 0 & 0 & 1 \end{bmatrix} \tag{6.189}$$

$$[D] = \{\overline{X}_1\}^T[M]\{\overline{X}_1\} = \begin{bmatrix} 0.052 & 0.013 & 0 \\ 0.013 & 0.052 & 0.013 \\ 0 & 0.013 & 0.026 \end{bmatrix} \tag{6.190}$$

The orthonormalised trial vectors are given by

$$\{X_1\} = \begin{bmatrix} 4.39 & -1.13 & 0.444 \\ 0 & 4.53 & -1.78 \\ 0 & 0 & 6.66 \end{bmatrix} \tag{6.191}$$

$$[R] = [M]\{X_1\} = \begin{bmatrix} 0.228 & 0 & 0 \\ 0.057 & 0.221 & 0 \\ 0 & 0.0589 & 0.15 \end{bmatrix} \tag{6.192}$$

Solving for $\{\overline{X}_2\}$ from $[K]\{\overline{X}_2\} = [R]$, we have

$$\{\overline{X}_2\} = 10^{-7} \begin{bmatrix} 0.158 & 0.155 & 0.0834 \\ 0.190 & 0.311 & 0.167 \\ 0.190 & 0.343 & 0.250 \end{bmatrix} \tag{6.193}$$

Extracting the orthonormal set of vectors $\{X_2\}$, from $\{\overline{X}_2\}$, we proceed to the second iteration. After five iterations, our approximation to mode shapes is obtained as

$$\{\phi\} = \begin{bmatrix} 1.50 & -3.59 & 2.37 \\ 2.59 & 0.008 & -4.12 \\ 2.99 & 3.57 & 4.76 \end{bmatrix} \tag{6.194}$$

The natural frequencies are obtained from the diagonal elements of the Rayleigh quotient matrix as

$$f_1 = 1280 \text{ Hz}, \quad f_2 = 4188 \text{ Hz}, \quad f_3 = 7597 \text{ Hz} \tag{6.195}$$

The mode shapes corresponding to each frequency (i.e. respective column of $[\phi]$) match with those obtained from generalised Jacobi scheme in Example 6.4.

Subspace iteration technique

This technique can essentially be viewed as being similar to the simultaneous iteration technique, varying only in the way the orthonormalisation of trial vectors is done, i.e. how do we get $\{X\}$ from $\{\overline{X}\}$? In this scheme we evaluate

$$[\overline{K}]_{m\times m} = \{\overline{X}\}^T_{m\times n}[K]_{n\times n}\{\overline{X}\}_{n\times m} \tag{6.196}$$

$$[\bar{M}]_{m\times m} = \{\bar{X}\}^T_{m\times n}[M]_{n\times n}\{\bar{X}\}_{n\times m} \tag{6.197}$$

Let us consider the eigenvalue problem

$$[\bar{K}]_{m\times m}\{\phi\}_{m\times m} = [\bar{\lambda}]_{m\times m}[\bar{M}]_{m\times m}\{\phi\}_{m\times m} \tag{6.198}$$

This is said to constitute the subspace (m-dimensional) of the vector basis of the parent problem (n-dimensional). Since the typical finite element mesh d.o.f. (n) is very large and we are usually interested in a few eigenpairs only, $m \ll n$. The typical values may be $n = 10,000$ and $m = 20$! The size of the subspace eigenvalue problem is thus much smaller than the parent problem.

Now, we solve for **all** the eigenpairs $\{\bar{\lambda},\phi\}$, using, for example, the generalised Jacobi transformation scheme. Obtain $\{X\}$ as

$$\{X\}_{n\times m} = \{\bar{X}\}_{n\times m}\{\phi\}_{m\times m} \tag{6.199}$$

This is because

$$\{X\}^T[M]\{X\} = \{\phi\}^T\{\bar{X}\}^T[M]\{\bar{X}\}\{\phi\} = \{\phi\}^T[\bar{M}]\{\phi\} = [I] \tag{6.200}$$

since the eigenvectors $\{\phi\}$ are $[\bar{M}]$ orthogonal.

Thus, instead of using $\{X\} = \{\bar{X}\}[L]^{-T}$, as in simultaneous vector iteration, we use here $\{X\} = \{\bar{X}\}\{\phi\}$. $[\phi]$ has to be obtained as a solution of the subspace eigenvalue problem.

While a rigorous discussion on the convergence characteristics of various algorithms is beyond the scope of the book, a useful guideline is that it is common practice to take $2m$ or $(m + 10)$, whichever is less, number of trial vectors, where m is the desired number of eigenpairs.

6.9   Transient Vibration Analysis

6.9.1   Modelling of Damping

We have so far discussed free vibration problems, where we studied the formulation of element mass matrices as also several solution algorithms. We observed that damping had little effect, in most cases, on the natural frequencies, and hence we studied undamped free vibration problems. The results of such an analysis gave us insights into the behaviour of the system when left free to vibrate on its own. In most practical situations, however, time dependent excitation forces are always present and it is of interest to find the actual dynamic response of the structure to such an excitation.

We also observed that a particular mode of vibration participates significantly in the dynamic response if that natural frequency closely matches any frequency component in the excitation. For such a situation, the dynamic response is limited essentially by the damping present in the system. In a typical structure, there exist many sources of dissipation such as viscous friction (for example, a shock absorber in an automobile), Coulombic friction, internal material damping (hysterisis), air drag, etc. In most cases, these various mechanisms cannot be reliably modelled and, quite often, an experimental estimate of overall damping present in the system is our only guideline. Thus, an equivalent viscous damping model is commonly used— viscous friction force, being linearly proportional to the relative velocity, offers significant modelling advantage. Analogous to the stiffness and mass matrices, we can use a damping matrix and write the generalised governing equations as

$$[M]\{\ddot{X}\} + [C]\{\dot{X}\} + [K]\{X\} = \{F(t)\} \tag{6.201}$$

While we have discussed, in detail, the derivation of stiffness and mass matrices, it is not entirely obvious at this stage how the individual elements of the damping matrix $[C]_{n\times n}$ can be obtained. We will address this issue a little later. Given the loading and the system parameters, we are interested in determining the response $\{X(t)\}$. There are generally two categories of methods of solving Eq. (6.201), viz., direct solution using a numerical integration scheme (called *direct integration methods*) and, secondly, transform these equations into a reduced set of uncoupled equations using the mode shapes prior to the numerical integration (called *mode superposition method*). We will discuss both these methods in this chapter.

We have already noted that any deflected shape of the structure can be represented as a linear combination of its eigenvectors, i.e.

$$\{X\} = c_1\{U_1\} + c_2\{U_2\} + \cdots + c_n\{U_n\} \tag{6.202}$$

or, since it is normally observed that a few modes (in the range of excitation frequencies) predominantly contribute to the structural response,

$$
\begin{Bmatrix} X_1(t) \\ X_2(t) \\ \vdots \\ X_n(t) \end{Bmatrix}
=
\left[
\begin{Bmatrix} u_1 \\ u_2 \\ \vdots \\ u_n \end{Bmatrix}^1
\begin{Bmatrix} u_1 \\ u_2 \\ \vdots \\ u_n \end{Bmatrix}^2
\cdots
\begin{Bmatrix} u_1 \\ u_2 \\ \vdots \\ u_n \end{Bmatrix}^m
\right]_{n\times m}
\begin{Bmatrix} p_1(t) \\ p_2(t) \\ \vdots \\ p_m(t) \end{Bmatrix}_{m\times 1}
\tag{6.203}
$$

or, simply,

$$\{X(t)\} = [U]\{p(t)\} \tag{6.204}$$

The p_i are the mode participation factors indicating how much each mode contributes to the response and these factors vary with time. $\{X(t)\}$ contains all the nodal d.o.f. values for the entire finite element mesh. Substituting from Eq. (6.204) in Eq. (6.201), we obtain

$$[M]\{U\}\{\ddot{p}\} + [C]\{U\}\{\dot{p}\} + [K]\{U\}\{p\} = \{F(t)\} \tag{6.205}$$

Premultiplying by $\{U\}^T$, we get

$$\{U\}^T[M]\{U\}\{\ddot{p}\} + \{U\}^T[C]\{U\}\{\dot{p}\} + \{U\}^T[K]\{U\}\{p\} = \{U\}^T\{F(t)\} \tag{6.206}$$

From the $[M]$ orthonormality property of eigenvectors,

$$\{U\}^T[M]\{U\} = [I] \tag{6.207}$$

From the Rayleigh quotient,

$$\{U\}^T[K]\{U\} = \begin{bmatrix} \ddots & & 0 \\ & \omega^2 & \\ 0 & & \ddots \end{bmatrix} \tag{6.208}$$

Thus, $\{U\}^T[M]\{U\}$ and $\{U\}^T[K]\{U\}$ are both diagonal, and Eq. (6.206) can be rewritten as

$$\{\ddot{p}\} + \{U\}^T[C]\{U\}\{\dot{p}\} + \begin{bmatrix} \searchfill & & 0 \\ & \omega^2 & \\ 0 & & \searrow \end{bmatrix}\{p\} = \{U\}^T\{F(t)\} \tag{6.209}$$

If $[C]$ is such that $\{U\}^T[C]\{U\}$ is also diagonal, we will have effectively *decoupled* the equations of motion. The independent equations so obtained are typical second order, ordinary differential equations (representative of spring-mass-damper system) for which the solution is well known. For a single d.o.f. spring-mass-damper system, we have

$$m\ddot{x} + c\dot{x} + kx = f(t) \tag{6.210}$$

Equation (6.210) can be rewritten as

$$\ddot{x} + 2\xi\omega\,\dot{x} + \omega^2 x = \frac{f(t)}{m} \tag{6.211}$$

where ξ is the damping factor given by $\xi = c/2\sqrt{(km)}$.

The m equations in modal coordinates contained in Eq. (6.209) can be written, in a manner analogous to Eq. (6.211), as

$$\ddot{p}_i + 2\xi_i\omega_i\dot{p}_i + \omega_i^2 p_i = \bar{f}(t), \qquad i = 1, 2, ..., m \tag{6.212}$$

The natural frequencies ω and the mode shapes $\{\phi\}$ required to obtain the right-hand side transformed forces have already been determined during the solution of the undamped free vibration problem. The modal damping factors can be directly determined experimentally and plugged into Eq. (6.212), and these equations can be used to determine the dynamic response. Such a technique is known as the "mode superposition technique". Thus we see that we need not form an explicit damping matrix $[C]$. However, if we do desire to determine the damping matrix $[C]$ we can, in principle, proceed in the following way[*]:

$$\{U\}^T[C]\{U\} = [c] = \begin{bmatrix} \searchfill & & 0 \\ & 2\xi\omega & \\ 0 & & \searrow \end{bmatrix} \tag{6.213}$$

Let us assume that $\{U\}$ is a square matrix containing all the n mode shapes of the system. Thus,

$$[C] = [U]^{-T}[c][U]^{-1} \tag{6.214}$$

If the entire set of mode shapes is known, we can then determine the desired damping matrix $[C]$ from the experimentally measured modal damping factors. However, in practice, this is not directly implemented but we use the relation

$$[U]^T[M][U] = [I] \tag{6.215}$$

---

[*]The interested reader should refer to *Dynamics of Structures,* R.W. Clough and J. Penzien, McGraw-Hill, New York, 1982.

Postmultiplying both sides of Eq. (6.215) with $[U]^{-1}$, we get

$$[U]^{-1} = [U]^T[M] \tag{6.216}$$

Substituting in Eq. (6.214), (since $[M]^T = [M]$), we get

$$[C] = [M][U][c][U]^T[M] \tag{6.217}$$

If m modes are likely to participate in the response and we have experimental estimates of modal damping factors for these m modes, then we can write Eq. (6.217) as

$$[C]_{n\times n} = [M]_{n\times n}[U]_{n\times m}[c]_{m\times m}[U]^T_{m\times n}[M]_{n\times n} \tag{6.218}$$

Such an explicit formulation of the system damping matrix is needed when we solve for the dynamic response $\{X(t)\}$ from Eq. (6.201) using direct integration methods, as will be discussed in Section 6.9.3. A system damping matrix $[C]$ satisfying the orthogonality with respect to the mode shapes [as in Eq. (6.213)] is referred to as Rayleigh (Proportional) Damping. We shall use only this form of damping in the rest of our discussion.

Having addressed the issue of modelling system damping, we will now proceed to the discussion of solution techniques for transient response determination. We will first discuss the Mode Superposition technique and subsequently the direct integration methods.

6.9.2 The Mode Superposition Scheme

Using the orthogonality property of the mode shapes and assuming Rayleigh (proportional) damping, we can transform the original equations of motion given in Eq. (6.201) into a set of uncoupled single degree of freedom oscillator equations in modal coordinates as follows:

$$\begin{aligned}
\ddot{p}_1 + 2\xi_1\omega_1\dot{p}_1 + \omega_1^2 p_1 &= \bar{f}_1(t) \\
\ddot{p}_2 + 2\xi_2\omega_2\dot{p}_2 + \omega_2^2 p_2 &= \bar{f}_2(t) \\
&\vdots \\
\ddot{p}_m + 2\xi_m\omega_m\dot{p}_m + \omega_m^2 p_m &= \bar{f}_m(t)
\end{aligned} \tag{6.219}$$

where $\bar{f}_1(t)$, $\bar{f}_2(t)$, etc. are obtained from $\{U\}^T\{F(t)\}$. It must be observed here that we have used the mode shapes obtained for the undamped system, assuming that the effect of damping on the mode shapes is negligible.

These are a set of independent second order differential equations which can be easily solved. For example, consider any one of the equations in Eq. (6.219) above:

$$\ddot{p}_i + 2\xi_i\omega_i\dot{p}_i + \omega_i^2 p_i = \bar{f}_i(t) \tag{6.220}$$

This is a second order, ordinary differential equation representing the equilibrium of a single d.o.f. oscillator. Its solution can be calculated using the well-known Duhamel integral

$$\dot{p}_i(t) = \frac{1}{\omega_i}\int_0^t \bar{f}_i(t) \sin\omega_i(t-\tau)\,d\tau + \alpha_i \sin\omega_i t + \beta_i \cos\omega_i t \tag{6.221}$$

where α_i, β_i are determined from the initial conditions on p_i and $\dot{p}_i$.

The initial conditions are, however, known in terms of physical coordinates $\{X_0\}$ and $\{\dot{X}_0\}$. We need to estimate $\{p_0\}$ and $\{\dot{p}_0\}$ from the known $\{X_0\}$ and $\{\dot{X}_0\}$. We have

$$\{X\} = [U]\{p\} \tag{6.222}$$

Therefore,

$$[U]^T[M]\{X\} = [U]^T[M][U]\{p\} \tag{6.223}$$

Since $[U]^T[M][U] = [I]$, we have

$$\{p_0\} = [U]^T[M]\{X_0\} \tag{6.224}$$

$$\{\dot{p}_0\} = [U]^T[M]\{\dot{X}_0\} \tag{6.225}$$

The integral in Eq. (6.221) may have to be evaluated numerically. Also, any other convenient strategy can be adopted for the solution of Eq. (6.220).

Algorithm for mode superposition

Step 1: Find all the m eigenpairs ω_i, $\{U_i\}$, $i = 1, 2, \ldots, m$.

Step 2: Evaluate $\{U\}^T_{m \times n}\{F\}_{n \times 1} = \{\bar{f}\}_{m \times 1}$.

Step 3: Transform the initial conditions $\{X_0\}$ and $\{\dot{X}_0\}$ into initial conditions in modal coordinates $\{p_0\}$ and $\{\dot{p}_0\}$.

Step 4: Use the initial displacement and velocity conditions and the force $\{\bar{f}(t)\}$ to evaluate $\{p(t)\}_{m \times 1}$, $\{\dot{p}(t)\}_{m \times 1}$.

Step 5: Use the transformations

$$\{X(t)\}_{n \times 1} = \{U\}_{n \times m}\{p\}_{m \times 1}$$

$$\{\dot{X}(t)\}_{n \times 1} = \{U\}_{n \times m}\{\dot{p}\}_{m \times 1}$$

to obtain actual displacement and velocity at all points of the structure.

Step 6: From $X(t)$, obtain strains and then stresses as a function of time. Thus the complete structural response is available.

Most of the computational effort usually goes into formulating the global $[K]$ and $[M]$ and computing the eigenpairs. The number of modes (m) participating in the response is usually very small, e.g. $m = 10$ or 20 while n, the d.o.f. of the finite element mesh, may be several thousands. Thus mode superposition itself consumes very little extra computational effort and hence is a popular scheme for evaluation of structural response.

We will now illustrate the mode superposition technique through the following simple examples.

Example 6.6. Consider the undamped, 2-d.o.f. system shown in Figure 6.16. Find the response of the system when the first mass alone is given an initial displacement of unity and released from rest.

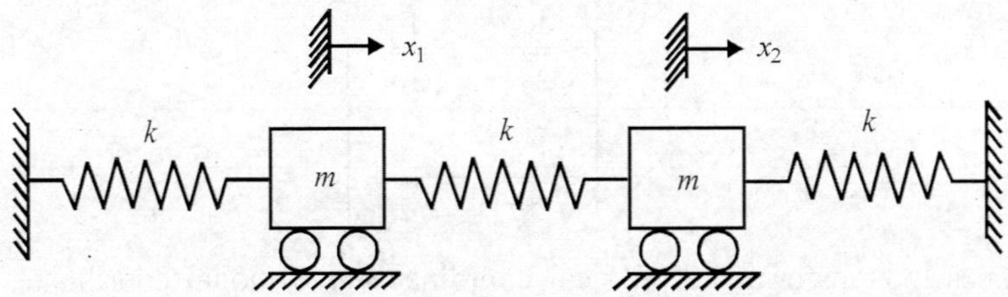

Fig. 6.16 Two-d.o.f. system (Example 6.6).

The mathematical representation of the system for free, harmonic vibration is given by

$$\begin{bmatrix} 2k & -k \\ -k & 2k \end{bmatrix}\begin{Bmatrix} x_1 \\ x_2 \end{Bmatrix} = \omega^2 \begin{bmatrix} m & 0 \\ 0 & m \end{bmatrix}\begin{Bmatrix} x_1 \\ x_2 \end{Bmatrix} \tag{6.226}$$

For a nontrivial solution,

$$\begin{vmatrix} 2k-m\omega^2 & -k \\ -k & 2k-m\omega^2 \end{vmatrix} = 0 \tag{6.227}$$

yielding

$$\omega_1 = \sqrt{\frac{k}{m}}, \qquad \omega_2 = \sqrt{\frac{3k}{m}} \tag{6.228}$$

The corresponding mode shapes are $\begin{Bmatrix} 1 \\ 1 \end{Bmatrix}$ and $\begin{Bmatrix} 1 \\ -1 \end{Bmatrix}$.

If we organise the two mode shapes as columns of a matrix, $[U]$, then we have

$$[U] = \begin{bmatrix} 1 & 1 \\ 1 & -1 \end{bmatrix} \tag{6.229}$$

$$[U]^T[M][U] = \begin{bmatrix} 1 & 1 \\ 1 & -1 \end{bmatrix}\begin{bmatrix} m & 0 \\ 0 & m \end{bmatrix}\begin{bmatrix} 1 & 1 \\ 1 & -1 \end{bmatrix} = \begin{bmatrix} 2m & 0 \\ 0 & 2m \end{bmatrix} \tag{6.230}$$

Orthonormalising the mode shapes with respect to the mass matrix such that $[U]^T[M][U] = [I]$, the identity matrix, we have the orthonormalised mode shapes as

$$[U] = \begin{bmatrix} \dfrac{1}{\sqrt{2m}} & \dfrac{1}{\sqrt{2m}} \\[3mm] \dfrac{1}{\sqrt{2m}} & -\dfrac{1}{\sqrt{2m}} \end{bmatrix} \tag{6.231}$$

Using these shapes to transform the physical coordinates into modal coordinates, we write

$$\{x\} = \begin{Bmatrix} x_1(t) \\ x_2(t) \end{Bmatrix} = \begin{bmatrix} \dfrac{1}{\sqrt{2m}} & \dfrac{1}{\sqrt{2m}} \\[3mm] \dfrac{1}{\sqrt{2m}} & -\dfrac{1}{\sqrt{2m}} \end{bmatrix} \begin{Bmatrix} p_1(t) \\ p_2(t) \end{Bmatrix} \tag{6.232}$$

and the modal equations are given by

$$\ddot{p}_1 + \frac{k}{m}\,p_1 = \overline{f}_1(t), \qquad \ddot{p}_2 + \frac{3k}{m}\,p_2 = \overline{f}_2(t) \tag{6.233}$$

Since the external forces are zero, $\overline{f}_1(t) = 0 = \overline{f}_2(t)$. Let us convert the initial conditions in real physical coordinates x_1 and x_2 into those on modal coordinates p_1 and p_2. The initial conditions in modal coordinates are given by

$$\begin{Bmatrix} p_{10} \\ p_{20} \end{Bmatrix} = \begin{bmatrix} \dfrac{1}{\sqrt{2m}} & \dfrac{1}{\sqrt{2m}} \\[3mm] \dfrac{1}{\sqrt{2m}} & -\dfrac{1}{\sqrt{2m}} \end{bmatrix} \begin{bmatrix} m & 0 \\ 0 & m \end{bmatrix} \begin{Bmatrix} 1 \\ 0 \end{Bmatrix} = \begin{Bmatrix} \dfrac{\sqrt{m}}{\sqrt{2}} \\[3mm] \dfrac{\sqrt{m}}{\sqrt{2}} \end{Bmatrix} \tag{6.234}$$

Since $\dot{x}_1(0) = 0 = \dot{x}_2(0)$, $\dot{p}_{10} = 0 = \dot{p}_{20}$. The solution for $p_1(t)$ can be written as

$$p_1(t) = A \cos \omega_1 t + B \sin \omega_1 t \tag{6.235}$$

from the initial conditions, $p_1(0) = \sqrt{m}/\sqrt{2}$ and $\dot{p}_1(0) = 0$, we have

$$A = \sqrt{m}/\sqrt{2}, \qquad B = 0 \tag{6.236}$$

Therefore,

$$p_1(t) = \sqrt{\frac{m}{2}} \, \cos \sqrt{\frac{k}{m}}\,t \tag{6.237}$$

Similarly,

$$p_2(t) = \sqrt{\frac{m}{2}} \, \cos \sqrt{\frac{3k}{m}}\,t \tag{6.238}$$

Thus,

$$\{x(t)\} = \begin{Bmatrix} x_1(t) \\ x_2(t) \end{Bmatrix} = [U]\{p\}$$

$$= \begin{bmatrix} \dfrac{1}{\sqrt{2m}} & \dfrac{1}{\sqrt{2m}} \\[3mm] \dfrac{1}{\sqrt{2m}} & -\dfrac{1}{\sqrt{2m}} \end{bmatrix} \begin{Bmatrix} \sqrt{\dfrac{m}{2}} \; \cos \sqrt{\dfrac{k}{m}}t \\[4mm] \sqrt{\dfrac{m}{2}} \; \cos \sqrt{\dfrac{3k}{m}}t \end{Bmatrix}$$

$$= \begin{Bmatrix} \dfrac{1}{2} \cos \sqrt{\dfrac{k}{m}}t + \dfrac{1}{2} \cos \sqrt{\dfrac{3k}{m}}t \\[4mm] \dfrac{1}{2} \cos \sqrt{\dfrac{k}{m}}t - \dfrac{1}{2} \cos \sqrt{\dfrac{3k}{m}}t \end{Bmatrix} \qquad (6.239)$$

6.9.3   Direct Integration Methods

The governing equation for the transient dynamic response of a structure is given by

$$[M]\{\ddot{X}\} + [C]\{\dot{X}\} + [K]\{X\} = \{F(t)\} \qquad (6.240)$$

Direct integration methods, as the name suggests, attempt to directly integrate equation (6.240) in time without any transformation of coordinates (e.g. to modal coordinates as done in the mode superposition method). The key ideas in such a method of direct numerical integration are as follows:

1. The above equation of motion is assumed to hold good at particular, discrete instants of time t_n, i.e. we write

$$[M]\{\ddot{X}_n\} + [C]\{\dot{X}_n\} + [K]\{X_n\} = \{F(t_n)\} \qquad (6.241)$$

2. The dynamic response is obtained by solution of the above equation.
3. The solution obtained in the above step holds good till we write the equation of motion again at time $t_{n+1} = t_n + \Delta t$.
4. The time increment Δt is chosen to be sufficiently small to ensure accuracy and stability of the solution.
5. Some solution methods require that Δt be less than a specific limit, i.e.

$$\Delta t \leq \Delta t_{\text{crit}} \qquad (6.242)$$

without which the resulting solution could actually grow unbounded. These are referred to as *conditionally stable methods*. It must be appreciated that this is purely a numerical phenomenon. The actual dynamic response of the structure may be perfectly normal.

Unconditionally stable methods are usually preferred. In such methods, Δt needs to be chosen purely from accuracy point of view, which may permit much larger Δt to be used.

6. Since we use a single Δt, it must be observed that **all** the individual equations contained in Eq. (6.241) are integrated with the same Δt. Thus it is easy to see that a uniform Δt may result in very poor integration of some of the high frequency (i.e. low T_p) terms. On the other hand, Δt required to represent the highest frequency term accurately may be infeasibly small. Moreover, as evident from Tables 6.1 and 6.2, for a given finite element mesh, the highest frequencies may not be accurate after all. Also, depending on the frequency content of the excitation, only a few modes are predominant in the response. Thus the correct choice of Δt, keeping in mind all these factors becomes crucial.

7. Since we attempt to directly integrate the equation of motion, we do not require that the eigenvalues and eigenvectors be first determined.

Let us now study a simple method of solving the equation

$$[M]\{\ddot{X}_n\} + [C]\{\dot{X}_n\} + [K]\{X_n\} = F(t_n)\} \tag{6.243}$$

Central difference method

In this method, the time derivatives of $\{X\}$, viz., $\{\dot{X}\}$ and $\{\ddot{X}\}$, are approximated as

$$\{\dot{X}_n\} = \frac{1}{2(\Delta t)}\{\{X_{n+1}\} - \{X_{n-1}\}\} \tag{6.244}$$

$$\{\ddot{X}_n\} = \frac{1}{(\Delta t)^2}\{\{X_{n+1}\} - 2\{X_n\} + \{X_{n-1}\}\} \tag{6.245}$$

where we have used the standard Newton's central difference formulae. Substituting in Eq. (6.243), we get

$$\frac{1}{(\Delta t)^2}[M]\{\{X_{n+1}\} - 2\{X_n\} + \{X_{n-1}\}\} + \frac{1}{2(\Delta t)^2}[C]\{\{X_{n+1}\} - \{X_{n-1}\}\}$$

$$+ [K]\{X_n\} = \{F(t_n)\} \tag{6.246}$$

Rearranging the terms, we obtain

$$\left(\frac{1}{(\Delta t)^2}[M] + \frac{1}{2(\Delta t)^2}[C]\right)\{X_{n+1}\} = \{F(t_n)\} - \left([K] - \frac{2}{(\Delta t)^2}[M]\right)\{X_n\}$$

$$- \left(\frac{1}{(\Delta t)^2}[M] - \frac{1}{2(\Delta t)^2}[C]\right)\{X_{n-1}\} \tag{6.247}$$

For an undamped system,

$$\frac{1}{(\Delta t)^2}[M]\{X_{n+1}\} = \{F(t_n)\} - \left([K] - \frac{2}{(\Delta t)^2}[M]\right)\{X_n\} - \frac{1}{(\Delta t)^2}[M]\{X_{n-1}\} \quad (6.248)$$

For a lumped mass matrix with no nonzero diagonal terms, the solution for $\{X_{n+1}\}$ can be directly written as

$$\{X_{n+1}\} = [M]^{-1}\left(\{F(t_n)\} - [K]\{X_n\}\right)(\Delta t)^2 + 2\{X_n\} - \{X_{n-1}\} \quad (6.249)$$

where

$$[M]^{-1} = \begin{bmatrix} \diagdown & & 0 \\ & \dfrac{1}{m_{ii}} & \\ 0 & & \diagdown \end{bmatrix}$$

For diagonal $[M]$ and $[C]$, the solution of Eq. (6.247) for $\{X_{n+1}\}$ is straightforward. Otherwise, the solution of Eq. (6.247) at each time step may be computationally very time consuming.

Once we have $\{X_{n+1}\}$, we can proceed in exactly the same manner to find $\{X_{n+2}\}$, etc. using the recurrence relation given in Eq. (6.247). If required, velocities and accelerations can be readily obtained using Eqs. (6.244)–(6.245).

We observe from Eq. (6.247) that the solution for $\{X_{n+1}\}$ requires knowledge of $\{X_n\}$ as well as $\{X_{n-1}\}$. When we start off the solution process at time $t = 0$, we know the initial conditions. Thus to get the response at $t = \Delta t$, we need $\{X_0\}$ as also $\{X_{-\Delta t}\}$. Obviously, $\{X_{-\Delta t}\}$ has no physical meaning but can be computed from the known initial conditions $\{X_0\}$ and $\{\dot{X}_0\}$ as

$$\{\dot{X}_0\} = \frac{1}{2\Delta t}\left(\{X_{\Delta t}\} - \{X_{-\Delta t}\}\right) \quad (6.250)$$

$$\{\ddot{X}_0\} = \frac{1}{(\Delta t)^2}\left(\{X_{\Delta t}\} - 2\{X_0\} + \{X_{-\Delta t}\}\right) \quad (6.251)$$

Eliminating $\{X_{\Delta t}\}$ from these two equations, we get

$$(\Delta t)^2\{\ddot{X}_0\} - 2(\Delta t)\{\dot{X}_0\} = -2\{X_0\} + 2\{X_{-\Delta t}\} \quad (6.252)$$

Thus,

$$\{X_{-\Delta t}\} = \{X_0\} - (\Delta t)\{\dot{X}_0\} + \frac{(\Delta t)^2}{2}\{\ddot{X}_0\} \quad (6.253)$$

It must be observed that the initial conditions are specified only on position and velocity and $\{\ddot{X}_0\}$ is not an initial condition but must be obtained from solving the equation of motion at time $t = 0$, i.e.

$$[M]\{\ddot{X}_0\} + [C]\{\dot{X}_0\} + [K]\{X_0\} = \{F(0)\} \tag{6.254}$$

Thus,

$$\{\ddot{X}_0\} = [M]^{-1}\big(\{F(0)\} - [C]\{\dot{X}_0\} - [K]\{X_0\}\big) \tag{6.255}$$

We can now summarise our solution procedure.

Algorithm for central difference method

Step 1: Form assembled $[K]$, $[M]$ and $[C]$.

Step 2: From the prescribed $\{X_0\}$ and $\{\dot{X}_0\}$, find $\{\ddot{X}_0\}$ using Eq. (6.255).

Step 3: Using Eq. (6.253), find $\{X_{-\Delta t}\}$.

Step 4: Solve for $\{X_{\Delta t}\}$, $\{X_{2\Delta t}\}$, $\{X_{3\Delta t}\}$, etc. by repeated use of Eq. (6.247).

The *major drawback of central difference method is that it is only conditionally stable,* i.e. $\Delta t < \dfrac{2}{\omega_{\max}}$, where $\omega_{\max}$ is the maximum frequency of $[K]\{U\} = \omega^2[M]\{U\}$. If Δt chosen is larger than this, the solution obtained will be erroneous.

Since the central difference method discussed here finds $\{X_{n+1}\}$ purely based on historical information, i.e., $\{X_n\}$, $\{X_{n-1}\}$, it is known as an *explicit method. Implicit methods,* on the other hand, require the knowledge of time derivatives of $\{X_{n+1}\}$ for the determination of $\{X_{n+1}\}$. We will now discuss some popular implicit methods. Most of those methods are *unconditionally stable* and place no restriction on Δt except as required for accuracy.

Average acceleration method (trapezoidal rule)

Let $\{\ddot{X}_n\}$ and $\{\ddot{X}_{n+1}\}$ be the accelerations at time t_n and t_{n+1}, respectively. In the interval (t_n, t_{n+1}), the acceleration is assumed to be constant at $1/2(\{\ddot{X}_n\} + \{\ddot{X}_{n+1}\})$, as indicated in Figure 6.17. Therefore, we can write[*]

$$\{\dot{X}_{n+1}\} = \{\dot{X}_n\} + \left(\frac{1}{2}\big(\{\ddot{X}_n\} + \{\ddot{X}_{n+1}\}\big)\right)\Delta t \tag{6.256}$$

$$\{X_{n+1}\} = \{X_n\} + \{\dot{X}_n\}\Delta t + \frac{1}{2}\left(\frac{1}{2}\big(\{\ddot{X}_n\} + \{\ddot{X}_{n+1}\}\big)\right)(\Delta t)^2 \tag{6.257}$$

---

[*]Using the standard relations $v = u + at$ and $s = ut + \frac{1}{2}at^2$.

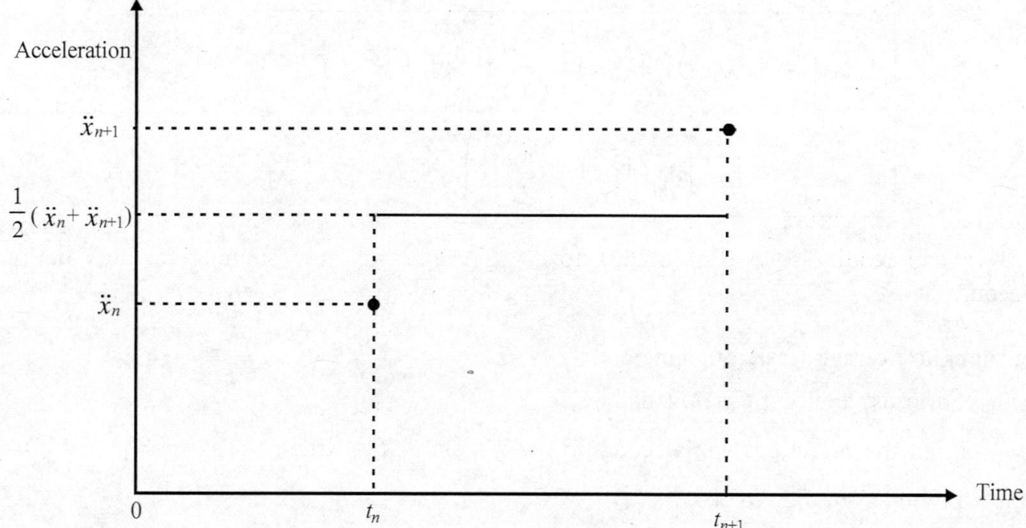

Fig. 6.17 Average acceleration scheme.

Substituting Eq. (6.256) in Eq. (6.257), we obtain

$$\{X_{n+1}\} = \{X_n\} + \frac{1}{2}\big(\{\dot{X}_n\} + \{\dot{X}_{n+1}\}\big)\Delta t \tag{6.258}$$

Rearranging the terms in Eq. (6.258), we get

$$\{\dot{X}_{n+1}\} = \frac{2}{\Delta t}\big[\{X_{n+1}\} - \{X_n\}\big] - \{\dot{X}_n\} \tag{6.259}$$

From Eq. (6.257),

$$\{\ddot{X}_{n+1}\} = \frac{4}{(\Delta t)^2}\big[\{X_{n+1}\} - \{X_n\}\big] - \frac{4}{\Delta t}\{\dot{X}_n\} - \{\ddot{X}_n\} \tag{6.260}$$

Writing the equation of motion [Eq. (6.243)] at time t_{n+1}, we have

$$[M]\{\ddot{X}_{n+1}\} + [C]\{\dot{X}_{n+1}\} + [K]\{X_{n+1}\} = \{F(t_{n+1})\} \tag{6.261}$$

Substituting from Eq. (6.259) and Eq. (6.260) in Eq. (6.261), we get

$$\boxed{[\bar{K}]\{X_{n+1}\} = \{\bar{F}_{n+1}\}} \tag{6.262}$$

where

$$\boxed{[\bar{K}] = \frac{4}{(\Delta t)^2}[M] + \frac{2}{\Delta t}[C] + [K]} \tag{6.263}$$

$$\boxed{\begin{aligned}\{\bar{F}_{n+1}\} &= \{F(t_{n+1})\} + [M]\left(\frac{4}{(\Delta t)^2}\{X_n\} + \frac{4}{\Delta t}\{\dot{X}_n\} + \{\ddot{X}_n\}\right) \\ &\quad + [C]\left(\frac{2}{\Delta t}\{X_n\} + \{\dot{X}_n\}\right)\end{aligned}} \qquad (6.264)$$

We can readily solve Eq. (6.262) for $\{X_{n+1}\}$, and we now summarise the solution procedure.

Algorithm for average acceleration method

Step 1: Form assembled $[K]$, $[M]$ and $[C]$.

Step 2: From the prescribed initial conditions $\{X_0\}$ and $\{\dot{X}_0\}$, find $\{\ddot{X}_0\}$ using Eq. (6.255).

Step 3: Form $[\bar{K}]$ of Eq. (6.263).

Step 4: Form $\{\bar{F}\}$ and solve for $\{X\}$ from Eq. (6.262) repeatedly for $t = \Delta t$, $2\Delta t$, $3\Delta t$, $\cdots$

We observe that, for linear problems, $[\bar{K}]$ does not change with time. If the problem involves material nonlinearity, $[K]$, and hence $[\bar{K}]$ would depend on $\{X\}$. If we are using Cholesky factorisation scheme for solution of $[\bar{K}]\{X\} = [\bar{F}]$, for linear problems we need to factorise $[\bar{K}]$ only once. Repeated solution for $[\bar{X}]$ is therefore fairly straightforward. We also observe that $[\bar{K}]$ has off-diagonal elements whether or not $[M]$ is diagonal. Therefore, there is no particular advantage in using diagonal, lumped mass matrix. Thus the more accurate consistent mass matrices are commonly used with implicit integration methods. On the other hand, the central difference method is computationally attractive if we use a diagonal lumped mass matrix. This method does not require, unlike the central difference method, any special starting procedure. It is unconditionally stable and, therefore, Δt can be chosen such that acceptable accuracy is achieved.

Linear acceleration method

The average acceleration method assumed acceleration to be uniform in the interval (t_n, t_{n+1}). Instead, we may assume that the acceleration varies linearly from $\{\ddot{X}_n\}$ to $\{\ddot{X}_{n+1}\}$ as indicated in Figure 6.18.

Thus we have, for any time $t_n + \tau$, the relations

$$\{\ddot{X}_{t_n+\tau}\} = \{\ddot{X}_n\} + \frac{\tau}{\Delta t}(\{\ddot{X}_{n+1}\} - \{\ddot{X}_n\}) \qquad (6.265)$$

Integrating Eq. (6.265) with respect to time, we get

$$\{\dot{X}_{t_n+\tau}\} = \{\dot{X}_n\} + \{\ddot{X}_n\}\tau + \frac{\tau^2}{2(\Delta t)}(\{\ddot{X}_{n+1}\} - \{\ddot{X}_n\}) \qquad (6.266)$$

$$\{X_{t_n+\tau}\} = \{X_n\} + \{\dot{X}_n\}\tau + \frac{\tau^2}{2}\{\ddot{X}_n\} + \frac{\tau^3}{6(\Delta t)}(\{\ddot{X}_{n+1}\} - \{\ddot{X}_n\}) \qquad (6.267)$$

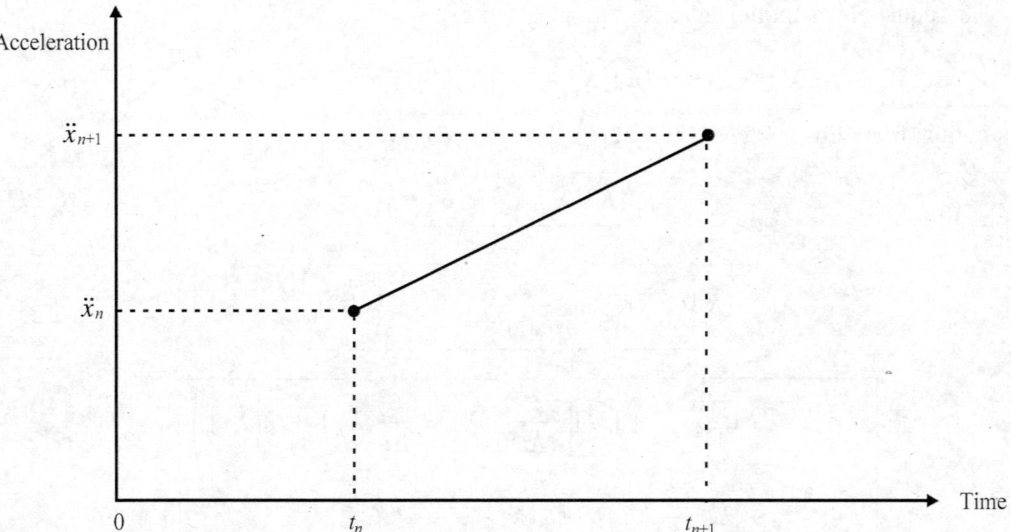

Fig. 6.18 Linear acceleration scheme.

At time t_{n+1} we, therefore, have

$$\{\dot{X}_{n+1}\} = \{\dot{X}_n\} + \{\ddot{X}_n\}\Delta t + \frac{\Delta t}{2}(\{\ddot{X}_{n+1}\} - \{\ddot{X}_n\}) \tag{6.268}$$

$$\{X_{n+1}\} = \{X_n\} + \{\dot{X}_n\}\Delta t + \frac{(\Delta t)^2}{2}\{\ddot{X}_n\} + \frac{(\Delta t)^2}{6}(\{\ddot{X}_{n+1}\} - \{\ddot{X}_n\}) \tag{6.269}$$

Equation (6.269) can be rewritten as

$$\{\ddot{X}_{n+1}\} = \{\ddot{X}_n\} + \frac{6}{(\Delta t)^2}\left[\{X_{n+1}\} - \{X_n\} - \{\dot{X}_n\}\Delta t - \frac{(\Delta t)^2}{2}\{\ddot{X}_n\}\right]$$

$$= \frac{6}{(\Delta t)^2}\{X_{n+1}\} - \frac{6}{(\Delta t)^2}\{X_n\} - \frac{6}{\Delta t}\{\dot{X}_n\} - 2\{\ddot{X}_n\} \tag{6.270}$$

Using Eq. (6.270), Eq. (6.268) can be rewritten as

$$\{\dot{X}_{n+1}\} = \{\dot{X}_n\} + \{\ddot{X}_n\}\Delta t - \frac{\Delta t}{2}\{\ddot{X}_n\}$$

$$+ \frac{\Delta t}{2}\left[\frac{6}{(\Delta t)^2}\{X_{n+1}\} - \frac{6}{(\Delta t)^2}\{X_n\} - \frac{6}{\Delta t}\{\dot{X}_n\} - 2\{\ddot{X}_n\}\right]$$

$$= \frac{3}{\Delta t}\{X_{n+1}\} - \frac{3}{\Delta t}\{X_n\} - 2\{\dot{X}_n\} - \frac{\Delta t}{2}\{\ddot{X}_n\} \tag{6.271}$$

The equation of motion at time t_{n+1} is given as

$$[M]\{\ddot{X}_{n+1}\} + [C]\{\dot{X}_{n+1}\} + [K]\{X_{n+1}\} = \{F(t_{n+1})\} \tag{6.272}$$

Substituting from Eqs. (6.270)–(6.271) in Eq. (6.272), we get

$$\boxed{[\bar{K}]\{X_{n+1}\} = \{\bar{F}\}} \tag{6.273}$$

where

$$\boxed{[\bar{K}] = [K] + \frac{6}{(\Delta t)^2}[M] + \frac{3}{\Delta t}[C]} \tag{6.274}$$

$$\boxed{\begin{aligned} \{\bar{F}\} = \{F(t_{n+1})\} \;&+ [M]\left(\frac{6}{(\Delta t)^2}\{X_n\} + \frac{6}{\Delta t}\{\dot{X}_n\} + 2\{\ddot{X}_n\}\right) \\ &+ [C]\left(\frac{3}{\Delta t}\{X_n\} + 2\{\dot{X}_n\} + \frac{\Delta t}{2}\{\ddot{X}_n\}\right) \end{aligned}} \tag{6.275}$$

A commonly used variant of the linear acceleration method is the Wilson-θ method, where the acceleration is assumed to be linear, not just within the interval $(t, t + \Delta t)$ but over an extended time interval $(t, t + \theta\Delta t)$, where $\theta \geq 1$, as shown in Figure 6.19. It can be shown that if $\theta \geq 1.37$, unconditional stability can be ensured. Thus, normally, $\theta = 1.4$ is used.

We proceed on exactly similar lines as for the simple linear acceleration scheme, except that we consider the equations of motion at time $t_n + \theta\,\Delta t$, rather than at $t_n + \Delta t$. Thus we use an extrapolated (often linearly) force vector at time $t_n + \theta\Delta t$. The final equations are obtained as

$$\boxed{[\bar{K}]\{X_{t_n+\theta\Delta t}\} = \{\bar{F}\}} \tag{6.276}$$

where

$$\boxed{[\bar{K}] = [K] + \frac{6}{(\theta\Delta t)^2}[M] + \frac{3}{\theta\Delta t}[C]} \tag{6.277}$$

$$\boxed{\begin{aligned} \{\bar{F}\} = \{F(t_n)\} &+ \theta\left(\{F_{t_n+\Delta t}\} - \{F_{t_n}\}\right) \\ &+ [M]\left(\frac{6}{(\theta\Delta t)^2}\{X_n\} + \frac{6}{\theta\Delta t}\{\dot{X}_n\} + 2\{\ddot{X}_n\}\right) \\ &+ [C]\left(\frac{3}{\theta\Delta t}\{X_n\} + 2\{\dot{X}_n\} + \frac{\theta\Delta t}{2}\{\ddot{X}_n\}\right) \end{aligned}} \tag{6.278}$$

The solution steps in the Wilson-θ method are identical to steps 1–4 of average acceleration method except that we use Eqs. (6.277)–(6.278) for $[\bar{K}]$ and $[\bar{F}]$.

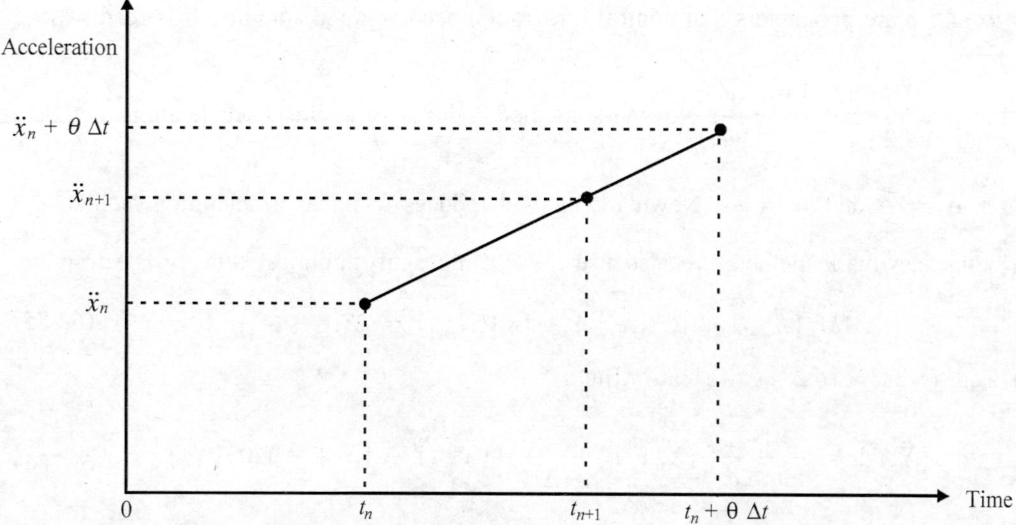

Fig. 6.19   Wilson-θ linear acceleration scheme.

Newmark family of methods

The Newmark method can be viewed as a generalisation of the average acceleration method and linear acceleration method. Let us now reproduce the basic equations used in these two methods.

Average acceleration method
From Eqs. (6.256)–(6.257),

$$\{\dot{X}_{n+1}\} = \{\dot{X}_n\} + (\Delta t)\left[\frac{1}{2}\{\ddot{X}_n\} + \frac{1}{2}\{\ddot{X}_{n+1}\}\right] \tag{6.279}$$

$$\{X_{n+1}\} = \{X_n\} + (\Delta t)\{\dot{X}_n\} + \frac{(\Delta t)^2}{2}\left[\frac{1}{2}\{\ddot{X}_n\} + \frac{1}{2}\{\ddot{X}_{n+1}\}\right] \tag{6.280}$$

Linear acceleration method
From Eqs. (6.268) and (6.269),

$$\{\dot{X}_{n+1}\} = \{\dot{X}_n\} + (\Delta t)\left[\left(1 - \frac{1}{2}\right)\{\ddot{X}_n\} + \frac{1}{2}\{\ddot{X}_{n+1}\}\right] \tag{6.281}$$

$$\{X_{n+1}\} = \{X_n\} + (\Delta t)\{\dot{X}_n\} + \frac{(\Delta t)^2}{2}\left[\left(1 - \frac{1}{3}\right)\{\ddot{X}_n\} + \frac{1}{3}\{\ddot{X}_{n+1}\}\right] \tag{6.282}$$

In the Newmark method, we write

$$\{\dot{X}_{n+1}\} = \{\dot{X}_n\} + (\Delta t)\left[(1 - \delta)\{\ddot{X}_n\} + (\delta)\{\ddot{X}_{n+1}\}\right] \tag{6.283}$$

$$\{X_{n+1}\} = \{X_n\} + (\Delta t)\{\dot{X}_n\} + \frac{(\Delta t)^2}{2}\left[(1 - 2\alpha)\{\ddot{X}_n\} + (2\alpha)\{\ddot{X}_{n+1}\}\right] \tag{6.284}$$

where α and δ are parameters that control integration accuracy and stability. It is readily seen that

for $\delta = \dfrac{1}{2}$ and $\alpha = \dfrac{1}{4}$, Newmark method reduces to average acceleration method

for $\delta = \dfrac{1}{2}$ and $\alpha = \dfrac{1}{6}$, Newmark method reduces to linear acceleration method

In the Newmark method, we also use the equations of motion at time $t_n + \Delta t$, i.e.

$$[M]\{\ddot{X}_{n+1}\} + [C]\{\dot{X}_{n+1}\} + [K]\{X_{n+1}\} = \{F(t_n + \Delta t)\} \tag{6.285}$$

From Eqs. (6.283)–(6.284), we can write

$$\{\ddot{X}_{n+1}\} = -\frac{1-2\alpha}{2\alpha}\{\ddot{X}_n\} + \frac{1}{\alpha(\Delta t)^2}\Big[\{X_{n+1}\} - \{X_n\} - (\Delta t)\{\dot{X}_n\}\Big] \tag{6.286}$$

$$\{\dot{X}_{n+1}\} = \{\dot{X}_n\} + (\Delta t)\Big[(1-\delta)\{\ddot{X}_n\} + (\delta)\Big\{-\frac{1-2\alpha}{2\alpha}\{\ddot{X}_n\}$$

$$+ \frac{1}{\alpha(\Delta t)^2}\Big[\{X_{n+1}\} - \{X_n\} - (\Delta t)\{\dot{X}_n\}\Big]\Big\}\Big] \tag{6.287}$$

Substituting these in Eq. (6.285), we can write

$$\boxed{[\bar{K}]\{X_{n+1}\} = \{\bar{F}\}} \tag{6.288}$$

where

$$\boxed{[\bar{K}] = [K] + \frac{1}{\alpha(\Delta t)^2}[M] + \frac{\delta}{\alpha(\Delta t)}[C]} \tag{6.289}$$

$$\boxed{\begin{aligned}[\bar{F}] = \{F_{t_n+\Delta t}\} &+ [M]\left(\frac{1}{\alpha(\Delta t)^2}\{X_n\} + \frac{1}{\alpha(\Delta t)}\{\dot{X}_n\} + \left(\frac{1}{2\alpha - 1}\right)\{\ddot{X}_n\}\right) \\ &+ [C]\left(\frac{\delta}{\alpha(\Delta t)}\{X_n\} + \left(\frac{\delta}{\alpha - 1}\right)\{\dot{X}_n\} + \frac{(\Delta t)}{2(\delta/\alpha - 1)}\{\ddot{X}_n\}\right)\end{aligned}} \tag{6.290}$$

The solution steps are identical to those given earlier for average acceleration method except for the fact that we use Eqs. (6.289)–(6.290) for $[\bar{K}]$ and $\{\bar{F}\}$.

Newmark originally proposed the unconditionally stable scheme with $\delta = \frac{1}{2}$ and $\alpha = \frac{1}{4}$ (i.e. the constant average acceleration scheme). For general, α, δ, it can be shown that the method is unconditionally stable only if $2\alpha \geq \delta \geq 1/2$.

We will now illustrate the use of direct integration methods using the example of thermal transients, viz., unsteady heat transfer in a fin.

6.10   Thermal Transients—Unsteady Heat Transfer in a Pin-Fin

In Section 4.7, we discussed the solution of steady-state heat conduction problems with convection from lateral surfaces. We studied the particular example of a pin-fin attached to a wall. We shall now consider the situation when the temperature of the wall to which the fin is attached is suddenly raised to and maintained at T_w, while the rest of the fin is initially at ambient temperature. We expect that the temperature of the fin will also gradually increase with time and we wish to determine the temperature in the fin as a function of time. Obviously, the temperature in the fin stabilises over a period of time and we reach the steady-state solution obtained in Section 4.7. The differential equation governing this unsteady heat transfer problem is given by

$$k\frac{\partial^2 T}{\partial x^2} + q = \frac{P}{A_c}h(T - T_\infty) + \rho c\frac{\partial T}{\partial t} \tag{6.291}$$

where k = coefficient of thermal conductivity of the material, T = temperature, q = internal heat source per unit volume, P = perimeter, A_c = cross-sectional area, h = convective heat transfer coefficient, T_∞ = ambient temperature, ρ = density, and c = specific heat.

The weighted-residual statement can be written as

$$\int_0^L W\left(k\frac{\partial^2 T}{\partial x^2} + q - \left(\frac{P}{A_c}\right)h(T - T_\infty) - \rho c\frac{\partial T}{\partial t}\right)dx = 0 \tag{6.292}$$

Let us assume that there is no internal heat source (i.e. $q = 0$). The weak form of the differential equation can be obtained by performing integration by parts as

$$\left[Wk\frac{\partial T}{\partial x}\right]_0^L - \int_0^L k\frac{\partial W}{\partial x}\frac{\partial T}{\partial x}\,dx - \int_0^L W\left(\frac{P}{A_c}\right)h(T - T_\infty)\,dx - \int_0^L W\rho c\frac{\partial T}{\partial t}\,dx = 0 \tag{6.293}$$

i.e.

$$\int_0^L k\frac{dW}{dx}\frac{\partial T}{\partial x}\,dx + \int_0^L W\left(\frac{P}{A_c}\right)hT\,dx + \int_0^L W\rho c\frac{\partial T}{\partial t}\,dx$$

$$= \int_0^L W\frac{P}{A_c}hT_\infty\,dx + \left[Wk\frac{\partial T}{\partial x}\right]_0^L \tag{6.294}$$

The weak form for a typical mesh of "n" finite elements can be written as

$$\sum_1^n\left[\int_0^\ell k\frac{dW}{dx}\frac{\partial T}{\partial x}\,dx + \int_0^\ell W\frac{P}{A_c}hT\,dx + \int_0^\ell W\rho c\frac{\partial T}{\partial t}\,dx\right]$$

$$= \sum_1^n\left[\int_0^\ell W\left(\frac{PhT_\infty}{A_c}\right)dx + \left[Wk\frac{\partial T}{\partial x}\right]_0^\ell\right] \tag{6.295}$$

We observe that all the terms are the same as used earlier [see Eq. (4.159)] except for the additional term involving the specific heat. For the standard linear element, we have

$$T = \left(1 - \frac{x}{\ell}\right)T_i + \left(\frac{x}{\ell}\right)T_j \tag{6.296}$$

Differentiating with respect to time, we get

$$\dot{T} = \left(1 - \frac{x}{\ell}\right)\dot{T}_i + \left(\frac{x}{\ell}\right)\dot{T}_j \tag{6.297}$$

The weight functions being the same as the shape functions, we have

$$W_1 = \left(1 - \frac{x}{\ell}\right) \tag{6.298}$$

$$W_2 = \left(\frac{x}{\ell}\right) \tag{6.299}$$

We can now perform the necessary integration for the additional specific heat term as follows:

$$\int_0^\ell W_1 \rho c \frac{\partial T}{\partial t}\, dx = \int_0^\ell \rho c\left(1 - \frac{x}{\ell}\right)\left[\left(1 - \frac{x}{\ell}\right)\dot{T}_i + \left(\frac{x}{\ell}\right)\dot{T}_j\right] dx \tag{6.300}$$

$$\int_0^\ell W_2 \rho c \frac{\partial T}{\partial t}\, dx = \int_0^\ell \rho c\left(\frac{x}{\ell}\right)\left[\left(1 - \frac{x}{\ell}\right)\dot{T}_i + \left(\frac{x}{\ell}\right)\dot{T}_j\right] dx \tag{6.301}$$

Carrying out the integrations and rewriting the results in matrix form, we get the element level specific heat matrix as

$$[c]^e = \frac{\rho c \ell}{6}\begin{bmatrix} 2 & 1 \\ 1 & 2 \end{bmatrix}\begin{Bmatrix} \dot{T}_i \\ \dot{T}_j \end{Bmatrix} \tag{6.302}$$

The matrix $[c]^e$ is observed to be analogous to the element consistent mass matrix. Here too, it is possible to use a lumped matrix similar to the mass lumping we discussed earlier.

Now the final element level equations are given by

$$\left(\frac{k}{\ell}\begin{bmatrix} 1 & -1 \\ -1 & 1 \end{bmatrix} + \frac{Ph\ell}{6A_c}\begin{bmatrix} 2 & 1 \\ 1 & 2 \end{bmatrix}\right)\begin{Bmatrix} T_i \\ T_j \end{Bmatrix} + \frac{\rho c \ell}{6}\begin{bmatrix} 2 & 1 \\ 1 & 2 \end{bmatrix}\begin{Bmatrix} \dot{T}_i \\ \dot{T}_j \end{Bmatrix} = \left(\frac{Ph}{A_c}T_\infty\right)\begin{Bmatrix} \ell/2 \\ \ell/2 \end{Bmatrix} + \begin{Bmatrix} -Q_0 \\ Q_\ell \end{Bmatrix} \tag{6.303}$$

Upon assembly, for a mesh of finite elements, we get

$$[K]\{T\} + [C]\{\dot{T}\} = \{F(t)\} \tag{6.304}$$

For example, using the trapezoidal rule, the final equations can be readily obtained as

$$\left([K] + \frac{2}{\Delta t}[C]\right)\{T_{n+1}\} = \{F_{n+1}\} + \{F_n\} - \left([K] - \frac{2}{\Delta t}[C]\right)\{T_n\} \qquad (6.305)$$

We will illustrate the solution of this problem using the following example.

Example 6.7. *Temperature distribution in a pin-fin.* Consider a 1 mm diameter, 50 mm long aluminium pin-fin as shown in Figure 6.20 used to enhance the heat transfer from a surface wall maintained at 300°C. Use k = 200 W/m/°C, ρ = 2700 kg/m^3, c = 900 J/kg°C for

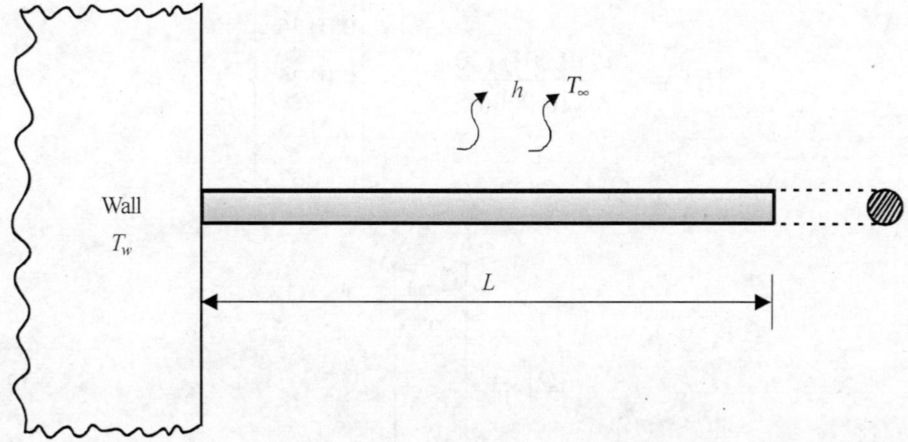

Fig. 6.20   **Temperature distrbution in a pin-fin (Example 6.7).**

aluminium; h = 20 W/m^2 °C; and T_∞ = 30°C. Assume that the tip of the fin is insulated. There is no internal heat source. We will now study how the temperature in the fin varies with time before settling to the steady state solution obtained in Example 4.10. We will use four equal length elements to model the fin.

The assembled matrices can be obtained as follows:

$$[K] = \frac{200}{0.0125}\begin{bmatrix} 1 & -1 & 0 & 0 & 0 \\ -1 & 2 & -1 & 0 & 0 \\ 0 & -1 & 2 & -1 & 0 \\ 0 & 0 & -1 & 2 & -1 \\ 0 & 0 & 0 & -1 & 1 \end{bmatrix} + \frac{(\pi)\,(0.001)\,(20)\,(0.0125)}{(6)\,(\pi)\,(0.0005)^2}\begin{bmatrix} 2 & 1 & 0 & 0 & 0 \\ 1 & 4 & 1 & 0 & 0 \\ 0 & 1 & 4 & 1 & 0 \\ 0 & 0 & 1 & 4 & 1 \\ 0 & 0 & 0 & 1 & 2 \end{bmatrix}$$

$$[C] = \frac{(2700)\,(900)\,(0.0125)}{(6)} \begin{bmatrix} 2 & 1 & 0 & 0 & 0 \\ 1 & 4 & 1 & 0 & 0 \\ 0 & 1 & 4 & 1 & 0 \\ 0 & 0 & 1 & 4 & 1 \\ 0 & 0 & 0 & 1 & 2 \end{bmatrix}$$

$$\{F\} = \frac{(\pi)\,(0.001)\,(20)}{(\pi)\,(0.0005)^2}\,(30) \begin{Bmatrix} 0.00625 \\ 0.0125 \\ 0.0125 \\ 0.0125 \\ 0.00625 \end{Bmatrix}$$

$$\{Q\} = \begin{Bmatrix} Q_{\text{wall}} \\ 0 \\ 0 \\ 0 \\ 0 \end{Bmatrix} \tag{6.306}$$

With $\Delta t = 1$ s, the nodal temperatures for the first few time steps are given in Table 6.3, and the full solution is given in Figure 6.21.

Table 6.3   Nodal Temperatures in a Pin-Fin

Time (s)	T_1	T_2	T_3	T_4	T_5
1	300	147.58	39.23	30.73	30.11
2	300	166.95	84.25	37.76	31.74
3	300	186.34	101.46	57.04	39.91
4	300	195.88	118.83	70.77	56.61
5	300	204	130.8	86.35	70.57
6	300	209.98	141.96	99.23	85.13
7	300	215.3	151.21	111.43	97.73
8	300	219.81	159.6	122.06	109.4
9	300	223.84	166.93	131.69	119.69
10	300	227.39	173.51	140.2	128.95

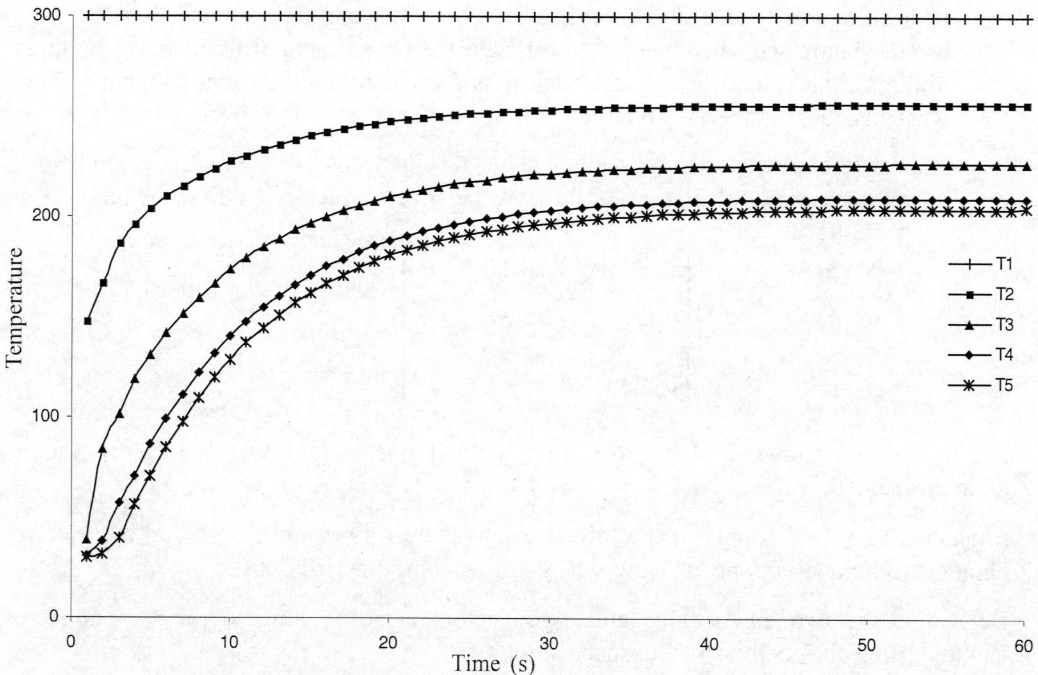

Fig. 6.21   Transient temperature distribution in a pin-fin (Four-element model of Example 6.7).

SUMMARY

In this chapter, we have discussed the formulation and solution of transient dynamic problems using FEM. We observed that the solution of transient dynamics problems is more time consuming than that of steady state (static) problems. An effective scheme (HRZ scheme) for generating lumped mass matrices has been discussed. Also, modelling of damping has been addressed. Several algorithms for solution of dynamic equilibrium equations have been covered in detail. Through the examples of structural dynamics and thermal transients, we have demonstrated that finite element formulation can be carried out essentially on the same lines as those of steady-state problems.

PROBLEMS

6.1   Consider the system shown in Fig. P6.1.

(a)   Write down the stiffness and mass matrices of the structure.

Fig. P6.1   A 3-d.o.f. spring-mass system.

(b)   It is known that the natural frequencies of the system are 1, 2 and 3 rad/s. Now use the Sturm sequence method to estimate the second natural frequency. Use three different trial values, less than, equal to, and greater than the exact value and show what happens to the $[D]$ matrix.

6.2   It is aimed at determining the largest eigenvalue and eigenvector of the matrix given below. Perform the necessary calculations following the inverse vector iteration method.

$$\begin{bmatrix} 20 & -10 & 0 \\ & 20 & -10 \\ \text{Symmetric} & & 20 \end{bmatrix}$$

6.3   For the system shown in Fig. P6.1, obtain the highest natural frequency by vector iteration method.

6.4   Explain what would happen if the lowest eigenvalue of a system is zero and the inverse iteration technique is applied. How will you overcome the difficulty?

6.5   Determine the two eigenvalues and eigenvectors corresponding to the two nonzero masses, using the method of subspace iteration.

$$[K] = \begin{bmatrix} 2 & -1 & 0 & 0 \\ -1 & 2 & -1 & 0 \\ 0 & -1 & 2 & -0 \\ 0 & 0 & -1 & 1 \end{bmatrix}, \quad [M] = \begin{bmatrix} 2 & 0 & 0 & 0 \\ 0 & 1 & 0 & 0 \\ 0 & 0 & 0 & 0 \\ 0 & 0 & 0 & 0 \end{bmatrix}$$

6.6   For the system shown in Fig. P6.6, determine the transient response over the interval 0 to t_0 when a force $P_c(t)$ acts at point C as shown.

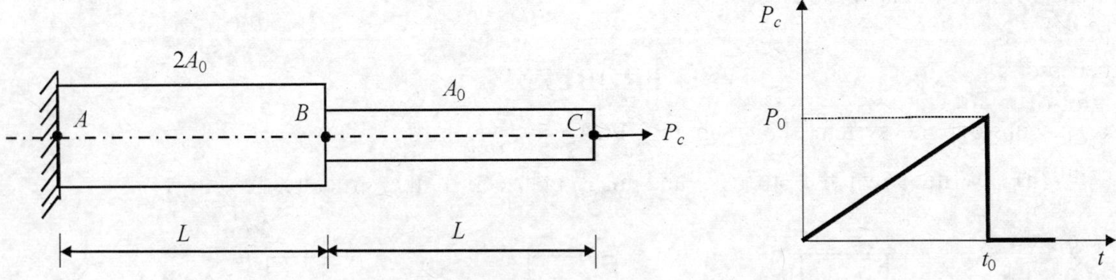

Fig. P6.6   Transient loading of a stepped bar.

CHAPTER 7

Application Examples

In this chapter, we present some examples of practical application of the various finite element concepts discussed in the earlier chapters. We will first discuss the use of beam elements in modelling the vibrations of an automotive crankshaft assembly. As an example of a two-dimensional finite element analysis, we will then discuss axisymmetric finite element analysis of pressure vessels.

7.1 Finite Element Analysis of Crankshaft Torsional Vibrations

Automotive crankshafts are subjected to fluctuating torques due to periodic strokes in the cylinders. Torsional vibrations result from the twisting reaction created in rotating shafts due to a fluctuating torque. Torsional vibration analysis is usually carried out by modelling the crankshaft and driveline components as a set of lumped masses and springs. Each throw of crankshaft is typically modelled as a set of masses and springs. Sometimes, these models account for clutch hysteresis and universal joint disturbances. Journal bearings are modelled as a set of linear springs and dashpots. The pulley and flywheel are modelled as lumped masses. These models are usually restricted to analyse torsional vibrations only. Generally, torsion modes are coupled with the bending modes in practical multicylinder engine crankshafts due to their complex geometry. Three-dimensional vibration analysis of crankshafts thus becomes necessary. Although 3d models could be used for analysis and design of real-life crankshafts, a lot of time would be consumed in the preprocessing as well as repeated analyses in a design cycle. Thus it is useful to develop a reasonably accurate one-dimensional model based on simple beam elements.

We will now describe one such model of an automotive crankshaft. Generalised beam elements (with six degrees of freedom per node) have been used to model the crankshaft. Crankshaft main bearings have been modelled as linear springs and viscous dampers for dynamic analysis. The details of the models are described in the following section.

7.1.1   Beam Element Model of Crankshaft Assembly

A typical crankshaft used in automobiles is shown in Figure 7.1. The web can be modelled as rectangular blocks as shown in Figure 7.2. The dimensions of the rectangular blocks are

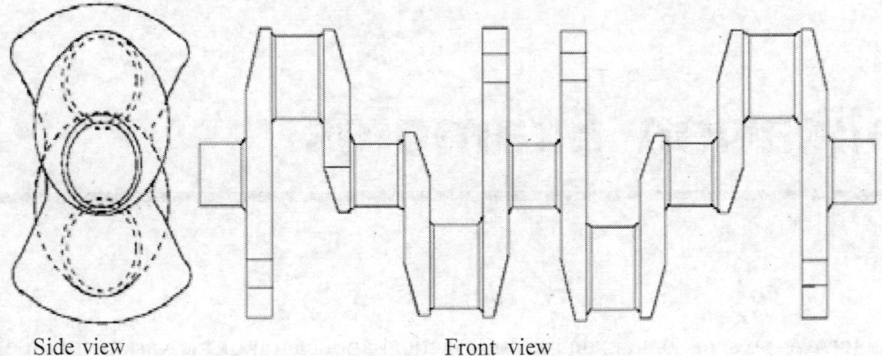

Fig. 7.1   Typical automotive crankshaft.

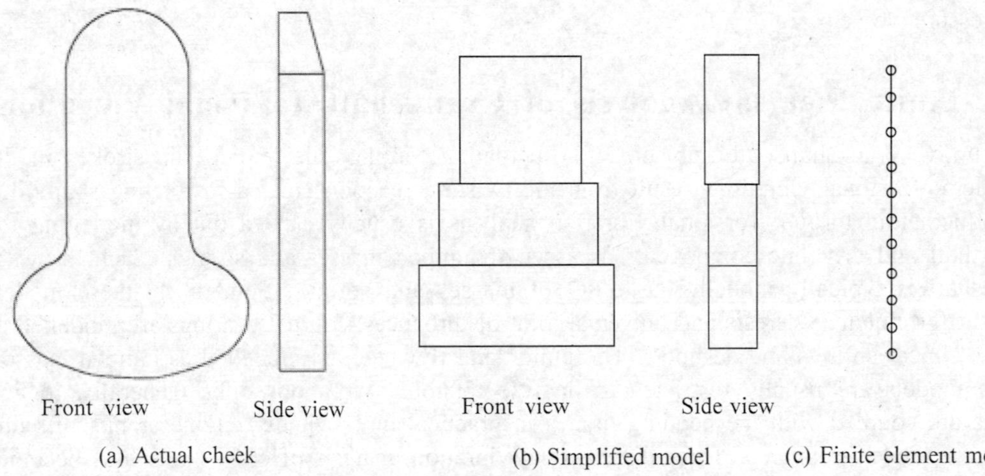

Fig. 7.2   Modelling of the cheek as a beam.

chosen such that the total area of these blocks equals the area of the web profile. This simplification enables us to use beam elements with uniform cross-section. More generally, we can estimate the various stiffnesses of the actual cross-section (viz., bending, torsional, etc.), and use these values for the beam elements. The crankpins (journals) are modelled using beam elements whose cross-sectional area and the moment of inertia are taken to be equal to the original crankpins (journals). The flywheel and pulley are assumed to be cylindrical in shape and are modelled as lumped masses.

The crankshaft main bearings are modelled as three sets of linear springs and dashpots in the directions normal to the crankshaft axis. They are attached at the middle of the respective crank-journals. It is assumed that the central set of spring-dashpot system supports 50% of the bearing load, the remaining two support 25% each because of the parabolic distribution of the oil film pressure along the crankshaft axis. The gas force and inertia force for a given cylinder are assumed to be equally supported by the adjacent crank-journals. The spring stiffness and damping coefficients are calculated using standard formulae[*]. A representative beam finite element model of the system is shown in Figures 7.3 and 7.4. For assessing the accuracy of the beam element model, a three-dimensional finite element model has also been developed as shown in Figure 7.5.

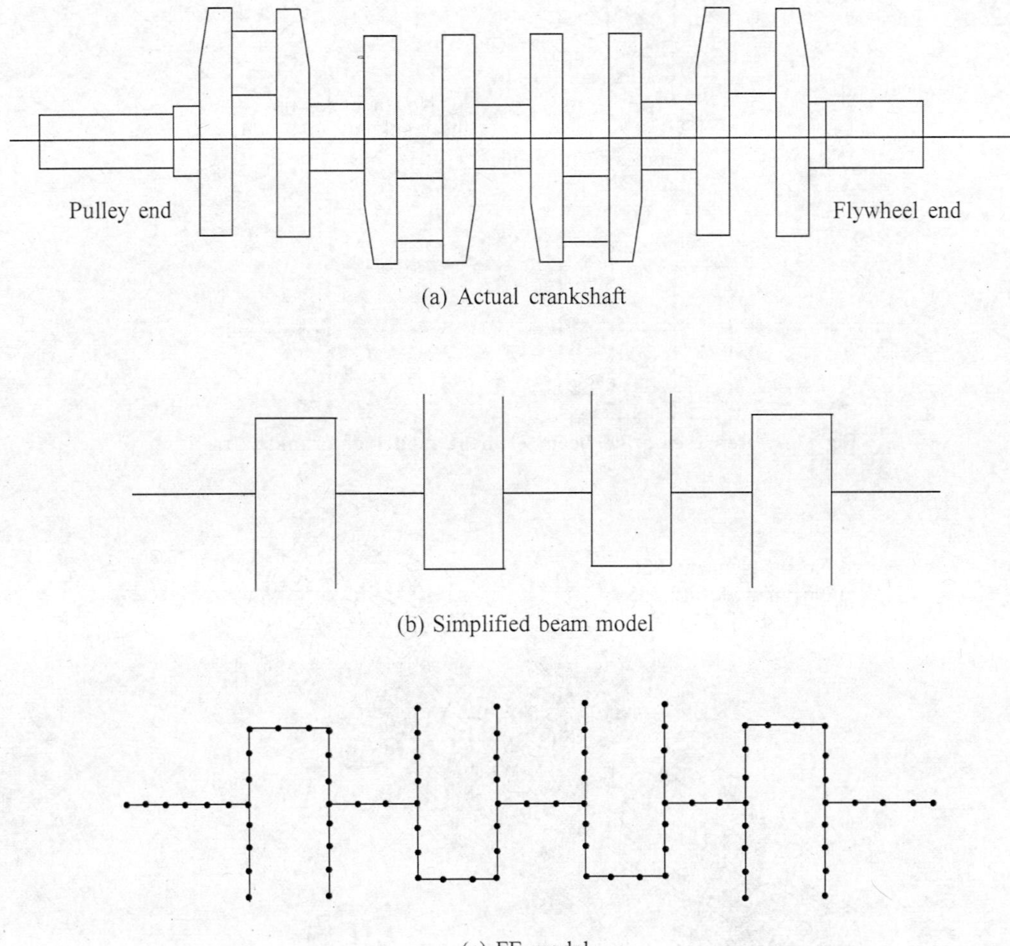

Fig. 7.3   Modelling of the crankshaft as beams.

---

*J.S. Rao, *Rotor Dynamics*, New Age International, New Delhi, 1996.

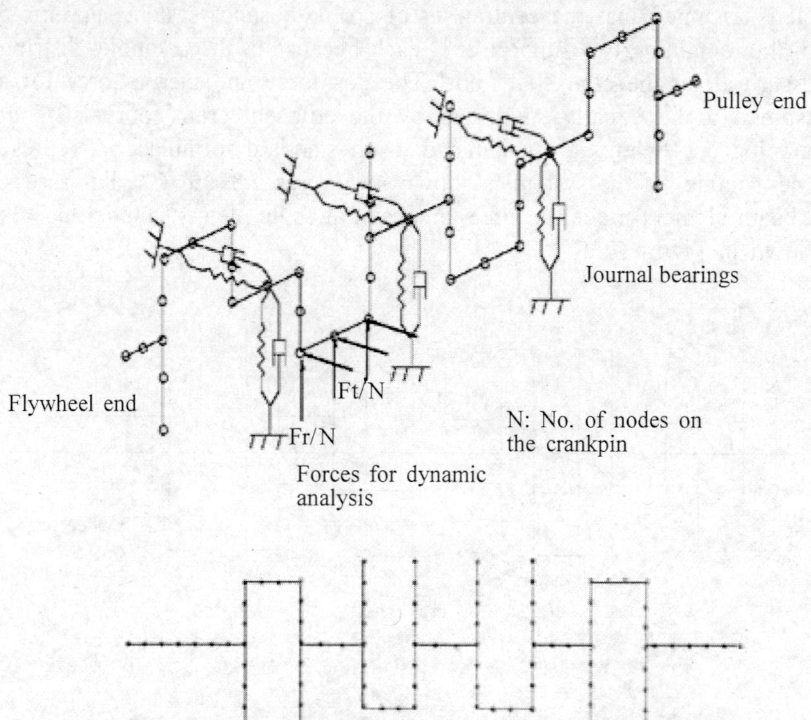

Fig. 7.4   Representative beam element model of crankshaft.

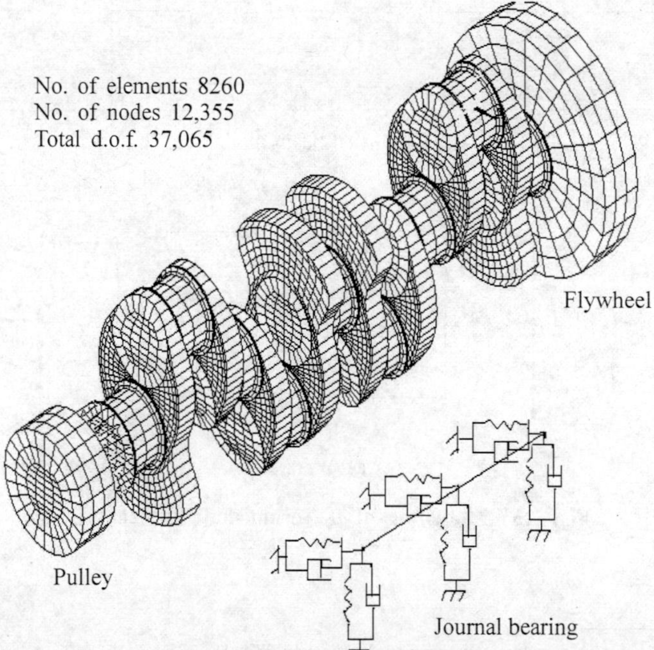

Fig. 7.5   Representative 3d-finite element model.

7.1.2   Results and Discussion

Free vibration analysis

The beam and 3-d element models are used to analyse the following five systems for their free vibration behaviour:

- System 1: A simple crankshaft (free-free) without flywheel, pulley and bearings.
- System 2: System 1 is modified by including the journal bearings.
- System 3: System 1 is modified by incorporating the flywheel and pulley.
- System 4: System 3 is modified by modelling the journal bearings as linear springs.
- System 5: System 4 is modified by modelling the reciprocating masses of the pistons. In both the 3d model and the beam model, the mass is lumped on the crankpins. This system is further used in dynamic response analysis.

The above systems are analysed in order to study the effect of journal bearings, pulley and flywheel on the natural frequencies and mode shapes of the crankshaft system. The results of free vibration analysis for these cases are compared in Tables 7.1–7.3. Conventional

Table 7.1   Comparison of Torsional Frequencies (Hz) for Beam and 3d Models

Mode No.	System 1			System 2		
	3d model	Beam model	Per cent deviation	3d model	Beam model	Per cent deviation
1	764	693	−9.3	790	710	−10.3
2	1206	1291	7.0	1280	1370	7.03
3	1941	1612	−17.0	1968	1646	−16.4
4	2129	1673	−21.4	2132	1673	−21.5
5	3023	2335	−22.8	3027	2338	−22.8

Table 7.2   Comparison of Torsional Frequencies (Hz) for Beam and 3d Models

Mode No.	System 3			System 4		
	3d model	Beam model	Per cent deviation	3d model	Beam model	Per cent deviation
1	487	562	15.4	516	606	17.5
2	1093	1172	7.2	1132	1213	7.15
3	1487	1505	1.2	1491	1513	1.5
4	1892	2207	16.7	1922	2216	15.3
5	2924	3469	18.6	2966	3472	17.1

Table 7.3   Comparison of Torsional Frequencies (Hz) for Beam and 3d Models

	System 5		
Mode No.	3d model	Beam model	Per cent deviation
1	484	346	28.5
2	1097	1286	17.2
3	1423	1487	4.5
4	1828	1951	6.7
5	2896	3228	11.5

analysis of crankshaft vibrations considers pure torsional modes. The mode shape plots (Figures 7.6–7.8), obtained using the beam element model show both the torsional and bending deflections in the model.

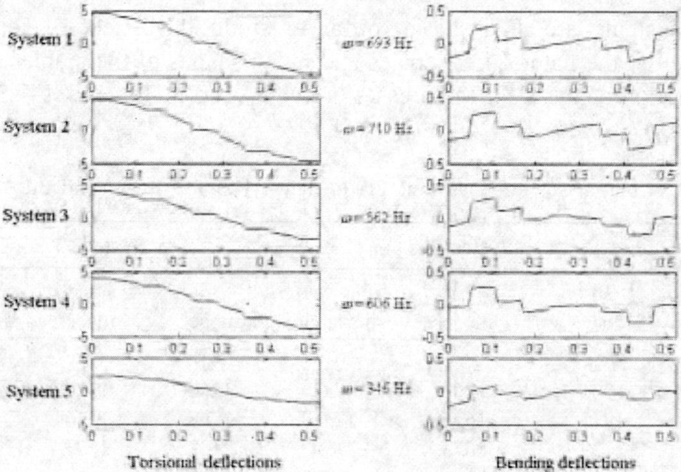

Fig. 7.6   First mode shape of crankshaft.

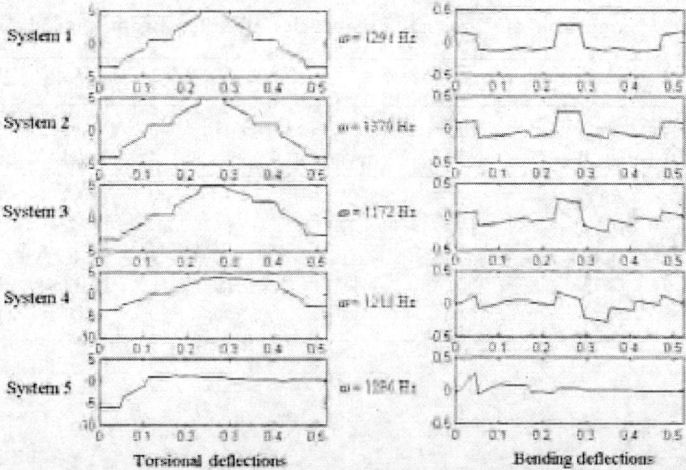

Fig. 7.7   Second mode shape of crankshaft.

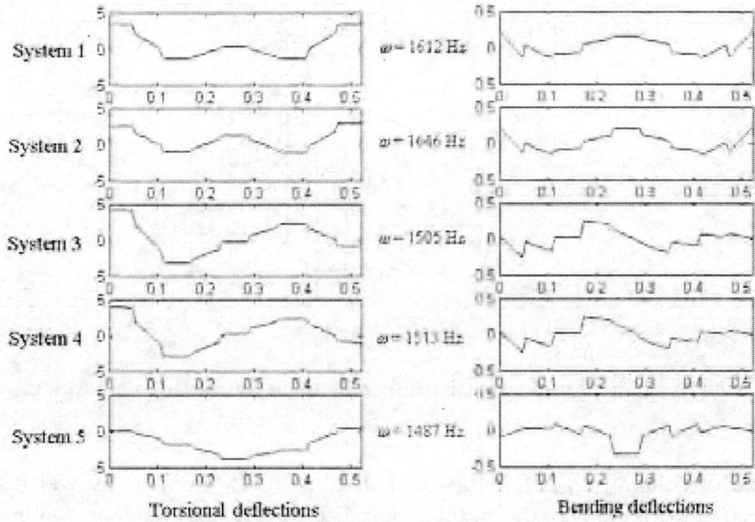

Fig. 7.8　Third mode shape of crankshaft.

Addition of the flywheel and pulley to the crankshaft significantly reduce the natural frequencies of the system, as can be observed from the results of Systems 1 and 3. The reduction in the natural frequency could be as much as 18%. It can be observed from the mode shape plots that, with the introduction of pulley and flywheel, the nodes get shifted towards the pulley, but the mode shapes do not change significantly. Incorporation of the reciprocating mass of the pistons significantly reduces the natural frequencies of System 4. The reduction in the natural frequencies is around 16% for beam model and 6% for the 3d model. In view of the coupling between the torsion and the bending modes, it is expected that the bearings will also affect the natural frequencies. From a comparison of results obtained for systems 1 and 2 (or 3 and 4), we observe that the natural frequencies may increase by as much as 6–7%. It can be observed from the mode shape plots that for a given mode shape, the bending mode shape alters due to the introduction of bearings while the torsion part remains unaffected. Hence, for accurate study of crankshaft vibrations, bearings should also be taken into consideration.

It can be observed that the beam element model overestimates (or) underestimates the resonances by as much as (16–20%). The 3-d finite element model, though requiring greater computational resources, provides accurate estimate of the resonance frequencies and also provides better insight into the coupled vibration behaviour.

7.1.3　Dynamic Response Analysis

The gas forces and the inertial forces due to the reciprocating masses will contribute to the excitation forces on the crankshaft system. Fourier analysis is performed to calculate the first 20 harmonics of the excitation forces (Figure 7.9). The radial and tangential components of the forces are applied on the crankpins as shown in Figure 7.4. These components are then used to perform harmonic analysis of crankshaft to calculate the steady state vibration behaviour. Dynamic response to individual harmonics is first obtained and then superposed to estimate the

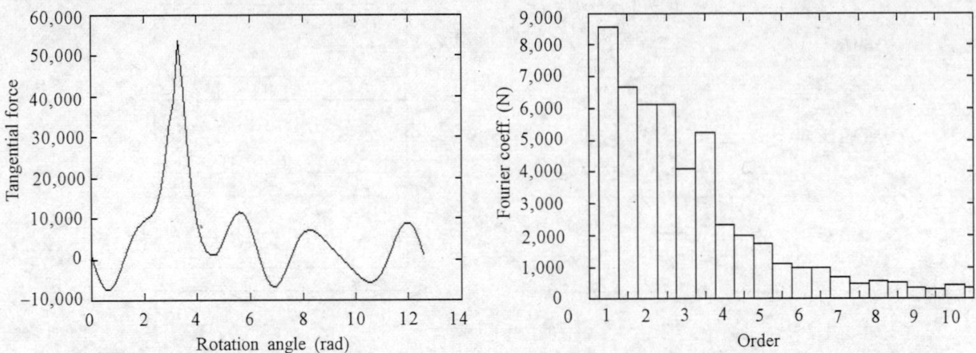

Fig. 7.9 Dynamic forces on crankshaft and their Fourier decomposition.

total response. Viscous damping in the torsional damper (2000 N s/m) as well as structural damping in the material (damping ratio = 0.01) has been taken into account while calculating the steady-state response.

The results for a given excitation order are presented in the form of stress as a function of the speed of rotation of the crankshaft. The von Mises stress at the crankpin is depicted in Figures 7.10 and 7.11.

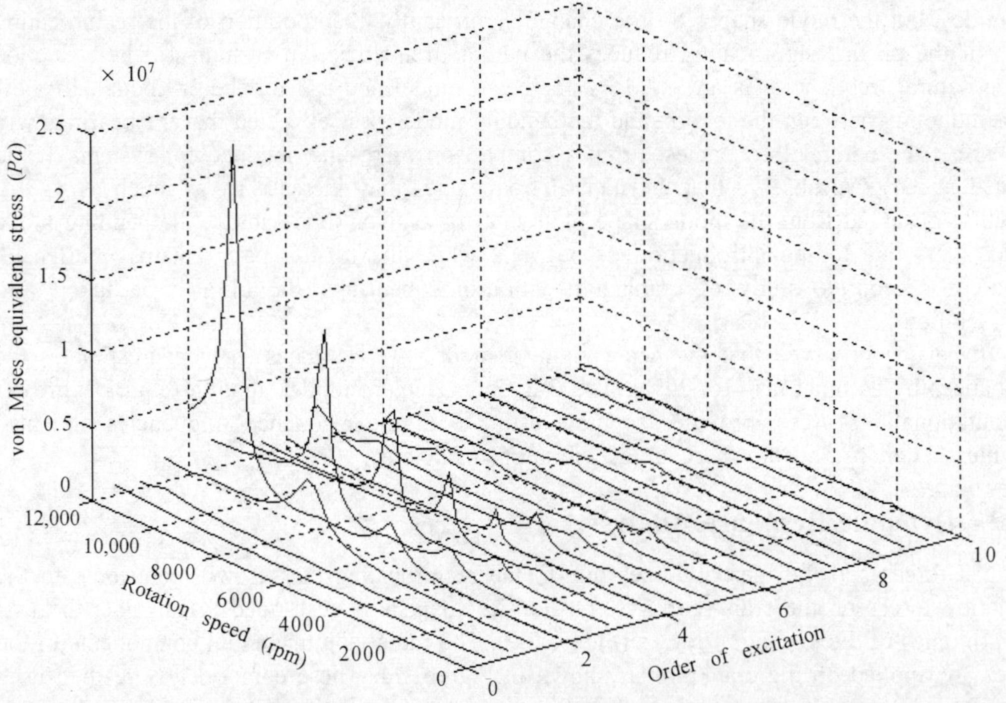

Fig. 7.10 Dynamic stresses on crankpin (beam element model).

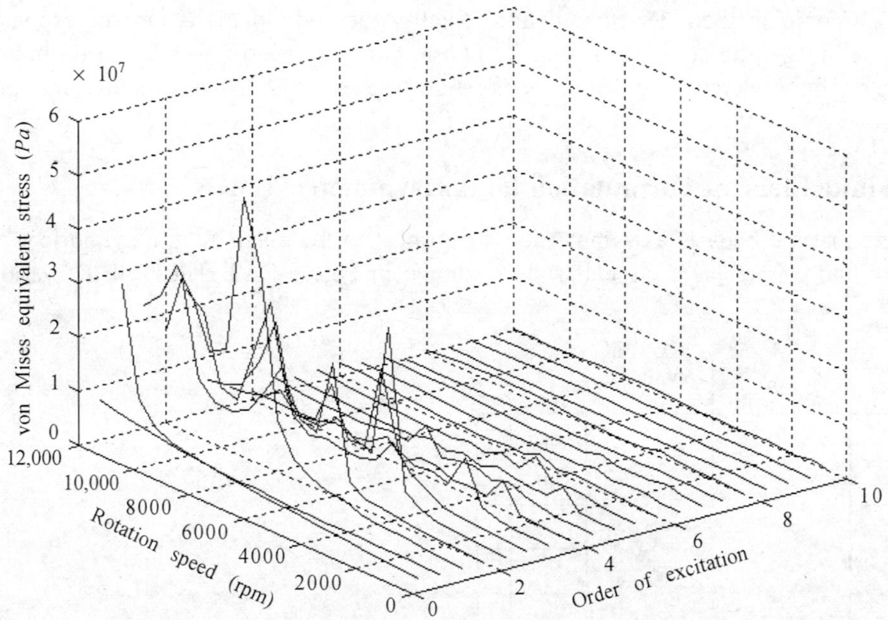

Fig. 7.11   Dynamic stresses on crankpin (3d model).

The CPU time consumed for complete dynamic analysis of 3d model on Sun Ultrasparc workstation was approximately 12 hours, while that for the beam model was approximately 12 minutes. However, it is observed that the beam element model predictions are inaccurate for the stresses even in noncritical regions such as crankpin. It is also incapable of estimating stresses at a critical location such as a fillet. Thus, even though a beam element model provides an initial guess at the natural frequencies and dynamic stresses, a full 3-d finite element model yields a much better insight into the complex mode shapes and critical stress regions of practical automotive crankshafts.

7.2   Axisymmetric Finite Element Analysis of a Pressure Vessel

Some of the commonly found structures such as pressure vessels, rocket casings, domes, submarine hulls, cooling towers, etc., possess near axisymmetric geometry. If we neglect small holes, attachments, and so on, their geometry can be viewed as perfectly axisymmetric, i.e., these are solids of revolution. They are usually made of isotropic materials and are supported all around the periphery. Thus the boundary conditions and material properties can also be approximated as axisymmetric. When subjected to an axisymmetric loading such as internal pressure/thermal gradient, the resulting deformation and stresses are also perfectly axisymmetric. In their design, however, it is crucial to consider certain non-axisymmetric loading cases such as those due to ground motion caused by earthquakes. Determination of response may then become quite complex, but can be simplified by invoking the Fourier series.

The non-axisymmetric loading is first decomposed into several Fourier harmonics; the response to each harmonic is then determined and, finally, the individual harmonic responses are superimposed to get the complete system response. We now briefly discuss the formulation of axisymmetric finite elements and their application to the stress analysis of a typical pressure vessel.

7.2.1   Finite Element Formulation for Axisymmetric Loads

We assume that the body is axisymmetric with respect to the z-axis (e.g. a cylindrical vessel), and each finite element is a circular ring as shown in Figure 7.12. Thus, all the various 2-d

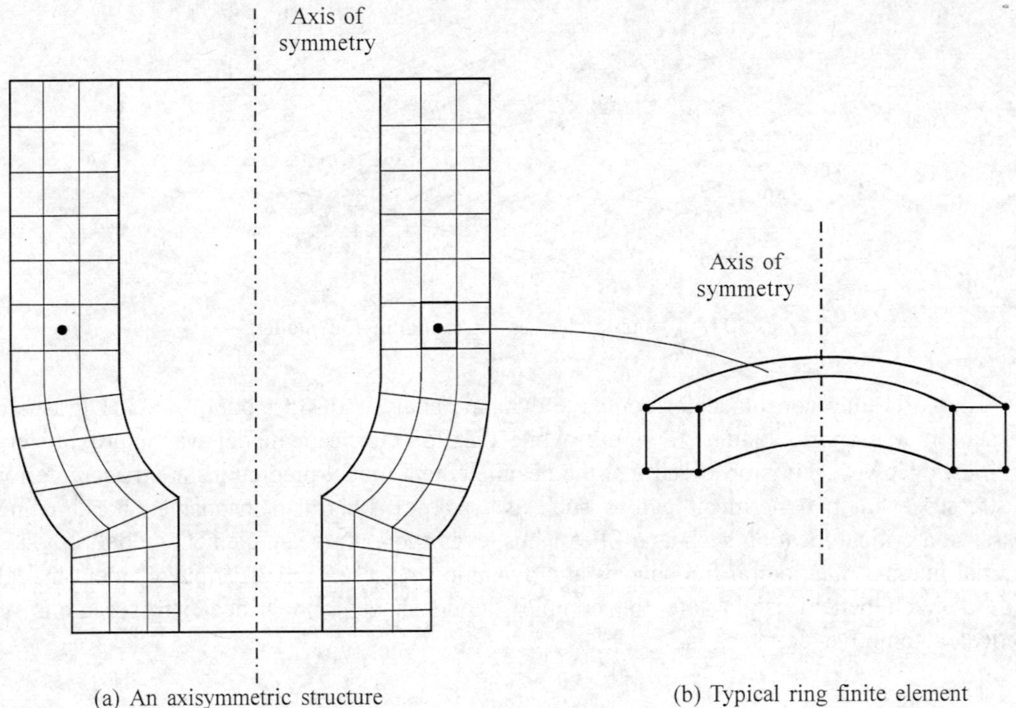

(a) An axisymmetric structure

(b) Typical ring finite element

Fig. 7.12   A typical pressure vessel.

finite elements discussed in Chapter 5, viz., three-noded triangle, six-noded triangle, four-noded quadrilateral, eight-noded quadrilateral, etc. can be formulated as axisymmetric ring elements. Each node is, in fact, a nodal circle. A typical isoparametric four node quadrilateral ring element is shown in Figure 7.12. Using standard shape functions, the displacements of any point P within the element are given, in terms of nodal displacements, as

$$u = \Sigma \, N_i u_i, \qquad w = \Sigma \, N_i w_i \tag{7.1}$$

where u and w correspond to radial and axial displacements respectively. The shape functions are the same as those given in Chapter 5 [see Eqs. (5.43)–(5.46)]. As a radial displacement cannot occur without causing a circumferential strain, we observe that there will be four nonzero independent strains unlike the case of plane stress/strain. The independent strains for axisymmetric deformation are:

$$\varepsilon_r = \frac{\partial u}{\partial r}, \qquad \varepsilon_\theta = \frac{u}{r}, \qquad \varepsilon_z = \frac{\partial v}{\partial z}$$

$$\gamma_{rz} = \frac{\partial u}{\partial z} + \frac{\partial v}{\partial r} \tag{7.2}$$

Using the shape function derivatives, we can express these strains in terms of nodal displacements as follows:

$$\{\varepsilon\} = [B]\{\delta\}^e \tag{7.3}$$

The stresses are given by

$$\begin{Bmatrix} \sigma_r \\ \sigma_\theta \\ \sigma_z \\ \tau_{rz} \end{Bmatrix} = \frac{E(1-v)}{(1+v)(1-2v)} \begin{bmatrix} 1 & \dfrac{v}{1-v} & \dfrac{v}{1-v} & 0 \\ & 1 & \dfrac{v}{1-v} & 0 \\ & & 1 & 0 \\ \text{Symmetric} & & & \dfrac{1-2v}{2(1-v)} \end{bmatrix} \begin{Bmatrix} \varepsilon_r \\ \varepsilon_\theta \\ \varepsilon_z \\ \gamma_{rz} \end{Bmatrix}$$

i.e.,

$$\{\sigma\} = [D]\{\varepsilon\} \tag{7.4}$$

The element stiffness matrix is given by

$$[k]^e = \int [B]^T [D][B] \, dv = \int [B]^T [D][B] \, (2\pi r) \, dr \, dz \tag{7.5}$$

Typically, these are evaluated using appropriate order of Gauss quadrature rule.

7.2.2 Stress Analysis of a Pressure Vessel

A typical heavy duty cylindrical pressure vessel with threaded end closures (Figure 7.13) is undertaken for linear, elastic, static stress analysis under an internal pressure of 2000 bar (i.e., approx. 200 MPa). Axisymmetric ring elements of triangular cross-section are used. Ignoring the helix angle of the thread, the problem is treated as axisymmetric.

For the typical discretisation shown in Figure 7.14 of the entire vessel, a total of about 5000 elements have been used with nearly 2750 nodes. An exploded view of the discretisation is depicted in Figure 7.15. The resulting number of equations to be solved was about 5500 (= n), with a bandwidth (= b) of about 100 for the stiffness matrix. The Cholesky scheme (LL^T decomposition) has been used to solve the equations. Typical stress (σ_z, σ_r, σ_θ) contours are shown in Figure. 7.16.

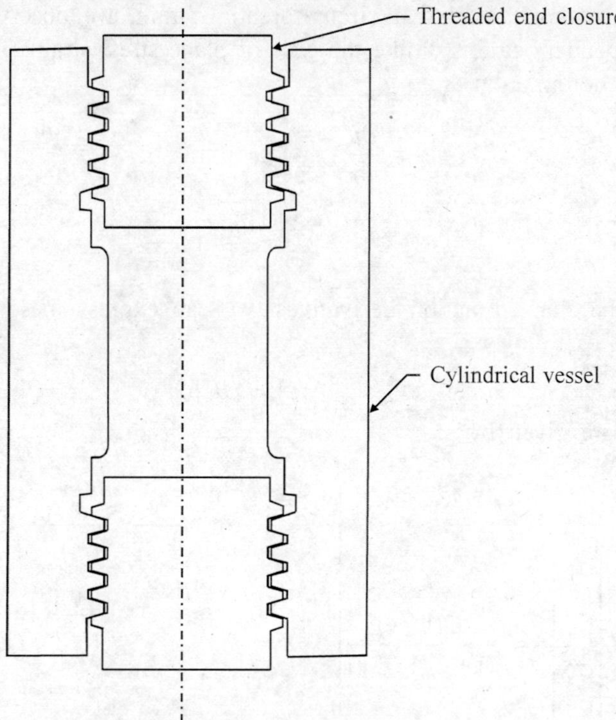

Fig. 7.13   Structural details of a typical pressure vessel.

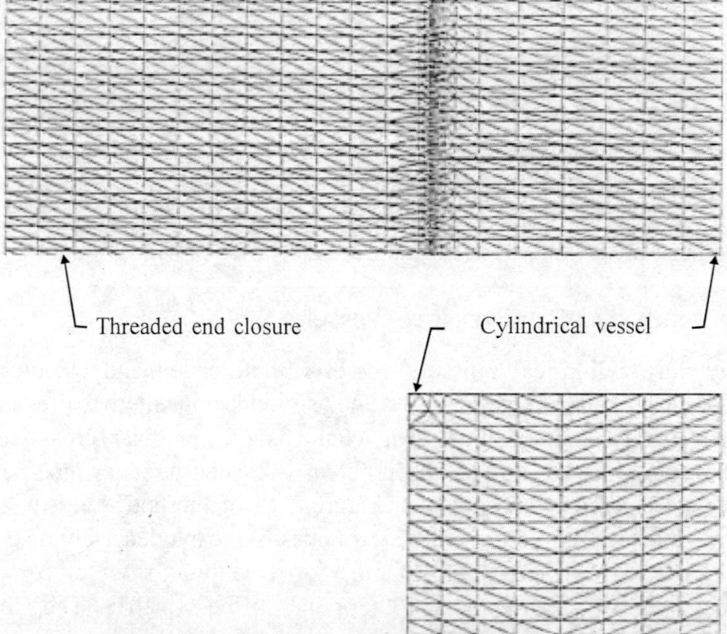

Fig. 7.14   FE discretisation of pressure vessel.

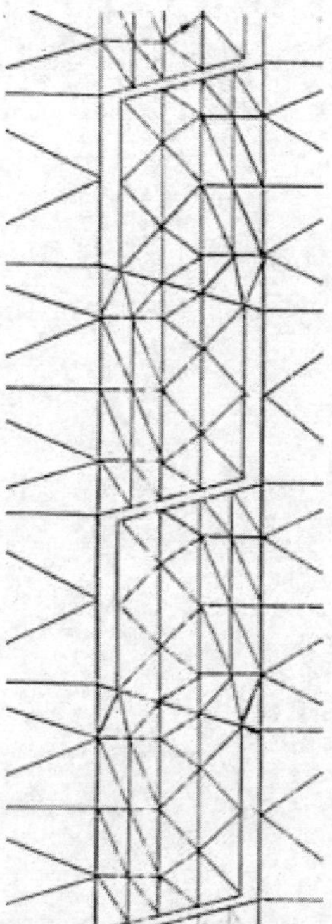

Fig. 7.15 Discretisation details at the thread.

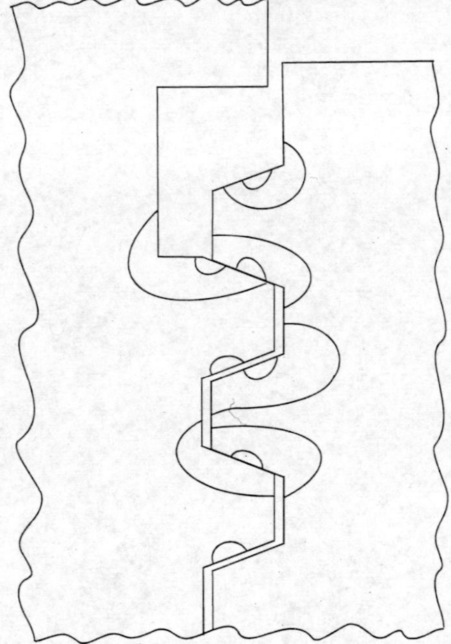

Fig. 7.16 Typical stress contours under internal pressure loading.

Suggested Mini-Project Topics

The mini-projects suggested here are aimed at giving the student a feel for the practical applications of finite element analysis. We have observed in our courses that such a term project helps significantly in reinforcing the student's understanding of the subject matter and helps build a proper perspective on the subject. These projects are expected to be carried out using any typical commercial finite element analysis software.

Project 1: Thermal Analysis of a Pressure Vessel

This project aims at conducting a thermal analysis of the pressure vessel shown in Figure MP.1. You may wish to organise your tasks in the following manner:

➢ Do the geometric modelling of the given pressure vessel. Model the pressure vessel as an axisymmetric two-dimensional structure. Also model the threads.

➢ At steady state, the wall temperatures inside and outside are T_{in} = 300°C and T_{out} = 50°C. Find the temperature distribution in the walls of the vessel and the threads.

➢ Solve for the temperature distribution when the gas inside the pressure vessel is at a temperature of 450°C. Inside the vessel assume free convection and the convective heat transfer coefficient to be 15 W/m²K. Outside, a coolant liquid is being forced and the convective heat transfer coefficient is 100 W/m²K.

➢ The coefficient of thermal expansion for the vessel is $11 \times 10^{-6}/$°C, and for the end cover material is $19 \times 10^{-6}/$°C. Find the thermal stresses at steady state in the vessel.

➢ Initially the system is at ambient temperature and suddenly the gas inside the pressure vessel is charged to 450°C. Find the temperature distribution in the vessel as a function of time.

For the material of the pressure vessel, assume Young's modulus $Y = 2.1 \times 10^{11}$ N/m², Poisson's ratio = 0.3, and density = 7800 kg/m³. Assume ambient temperature as 25°C.

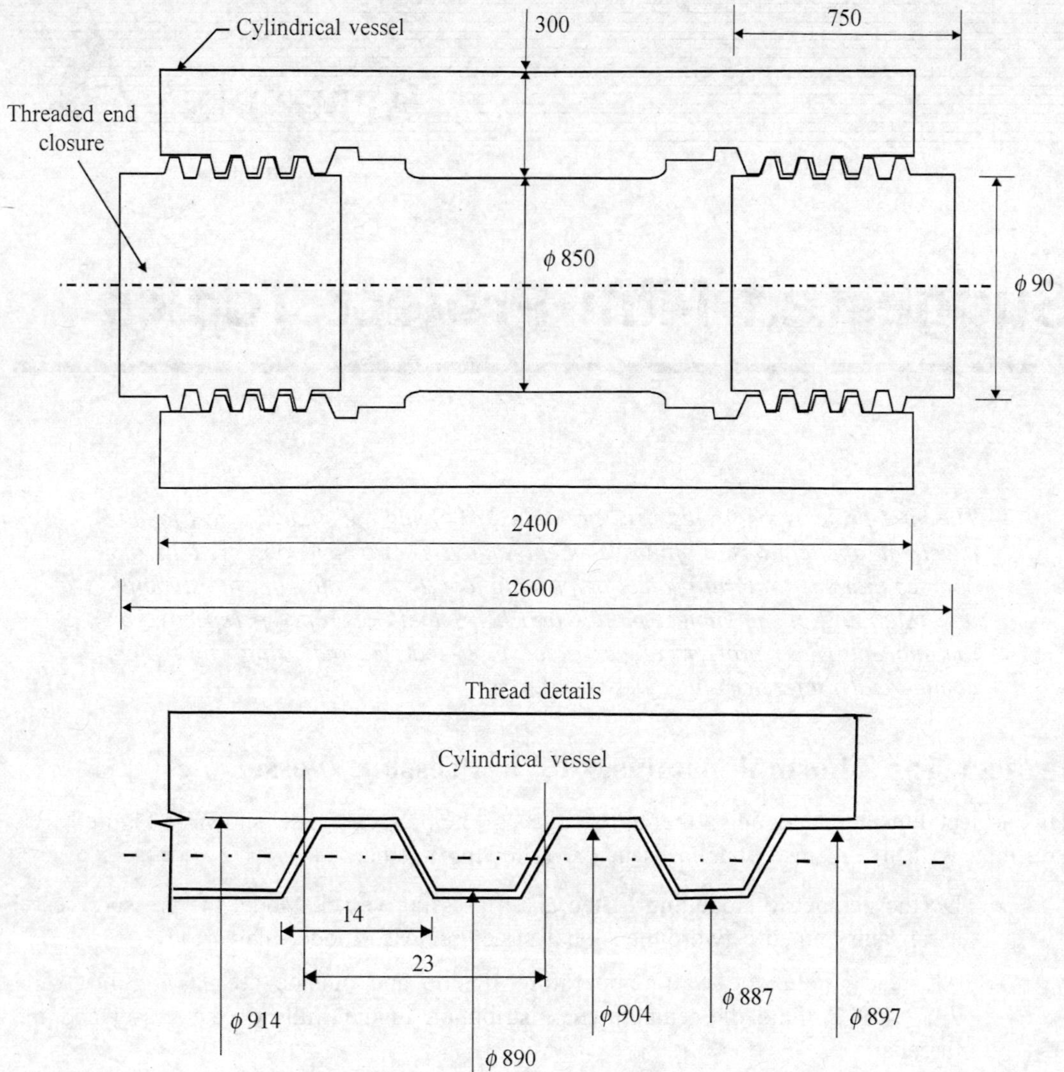

Fig. MP.1 Thermal analysis of a pressure vessel.

Project 2: Structural Dynamic Analysis of a Pressure Vessel

The objective of this project is to perform static and dynamic stress analysis of a pressure vessel. You may organise your task as follows:

> ➢ Do the geometric modelling of the given pressure vessel as axisymmetric structure.

> ➢ Let the gas pressure inside the vessel be 200 MPa. Find the stresses in the wall and the threads. How do the threads share the load?

> ➤ An explosion occurs inside the vessel. The variation of pressure with time due to this explosion is given in Figure MP.2. Perform transient analysis for the given pressure profile.

> ➤ The vessel needs to be certified for earthquake resistance. Consider a typical earthquake ground motion record and determine the structural response.

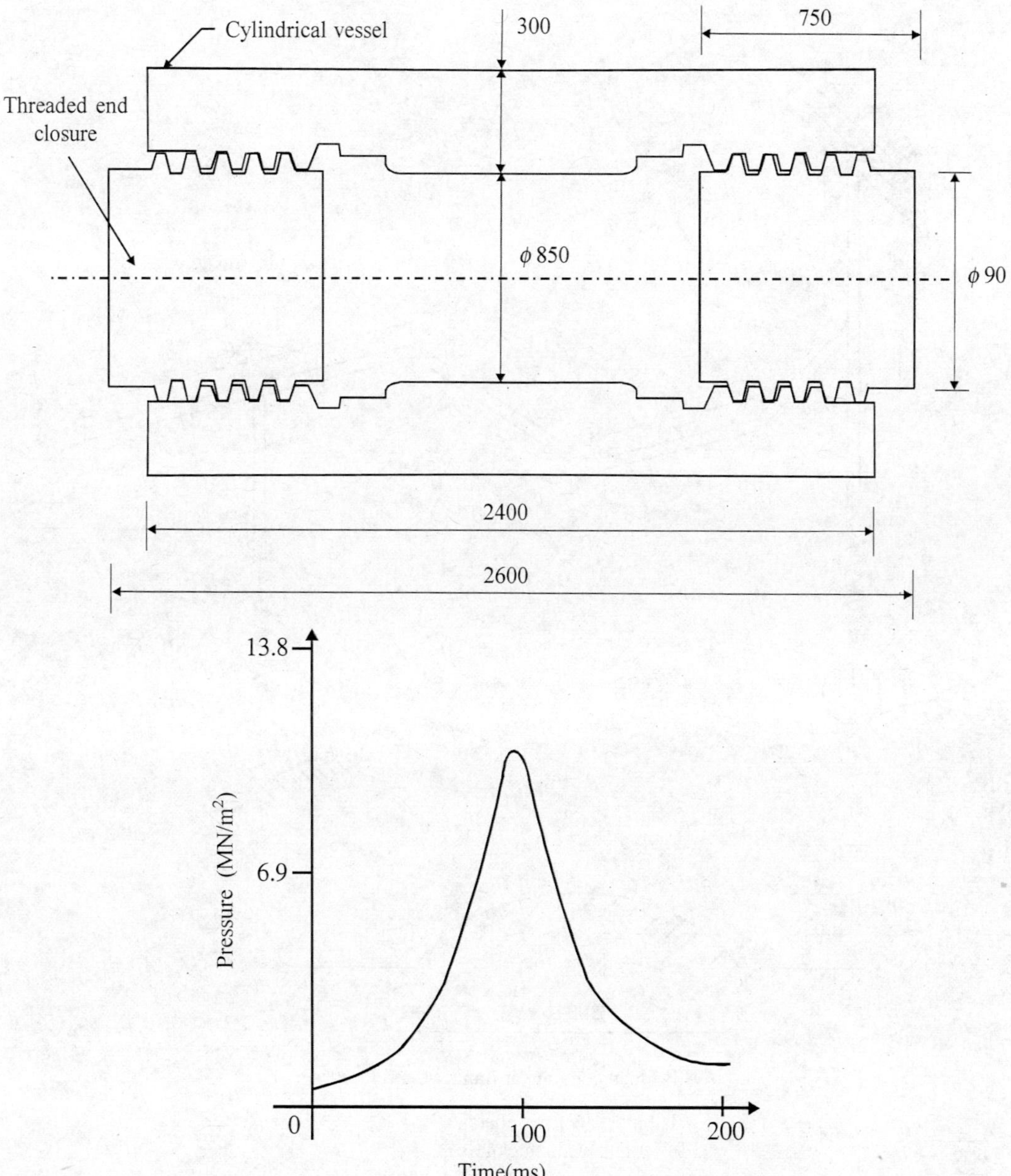

Fig. MP.2 Structural analysis of a pressure vessel.

Project 3:   Dynamics of a Scooter Frame

Figure MP.3 shows a simplified model of a typical tubular frame of a two-wheeler (Scooter). Model the system using spring, rigid mass and general beam elements, and determine the first few natural frequencies and mode shapes. Assume that the vehicle goes on a road bump as shown in the figure with a forward speed of 25 kmph. Determine the response of the system.

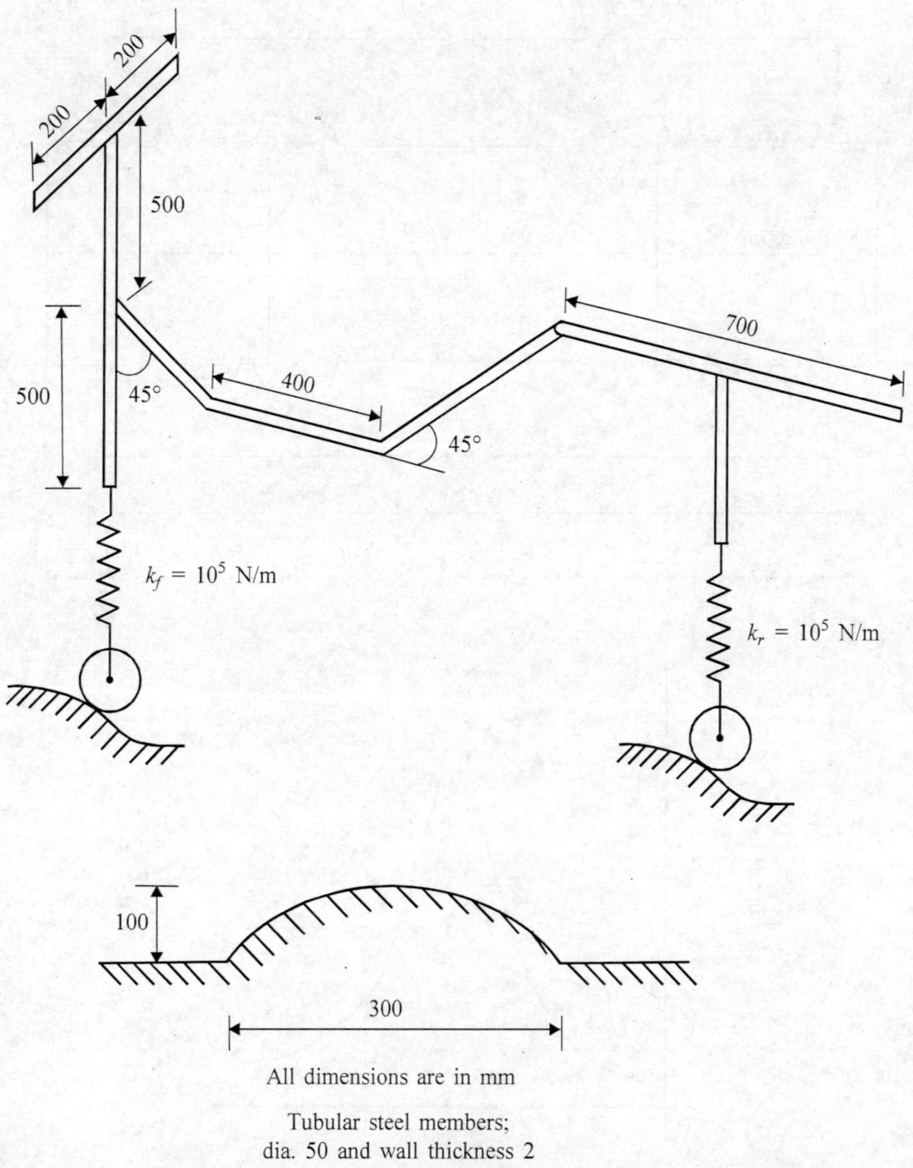

Fig. MP.3   Dynamics of scooter frame.

Project 4:   Automotive Chassis Dynamics

A typical automotive chassis is shown in Figure MP.4(a). Although an actual chassis is complicated, for the present project, we can model all the structural members of the chassis using the same cross-section.

Typical Bimensions

Side width (Rear)   = 1.5 m
Side width (Front)   = 1 m
Wheel base   = 2.5 m
End-to-end length   = 4 m
Distance between
mid-supports   = 1 m

Curb weight   = 600 kg
This mass is the sum of two discrete masses:
Engine weight   = 300 kg
    (This weight is assumed to act at point *A*)
Body frame weight   = 300 kg
    (This weight is assumed to act at point *B*,
    located 0.75 m from the rear axle.)

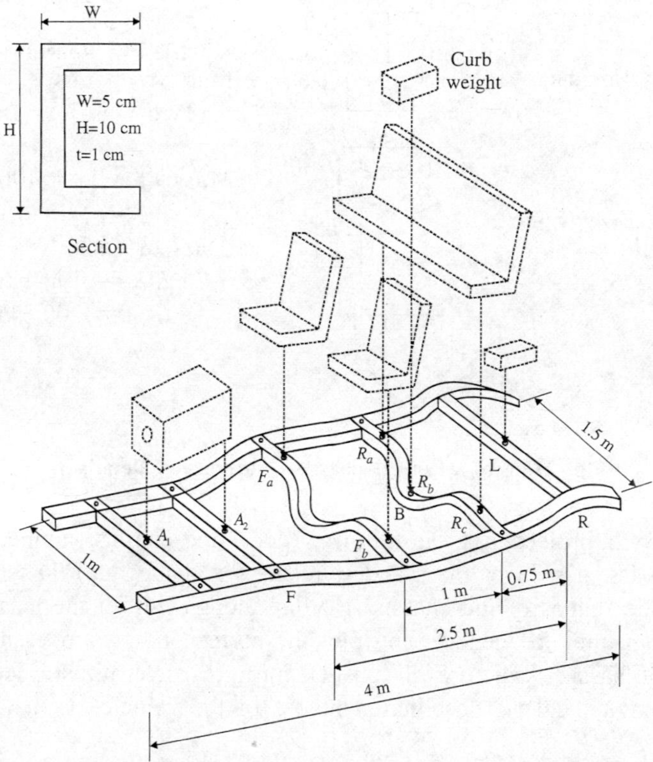

Fig. MP.4(a)   Typical Chassis—The front axle is located at *F* and the rear axle at *R*. The CG of the vehicle is at *B*.

Static Loads

Rear passenger weights (3 nos.@75 kg): Assumed to act at points Ra, Rb, and Rc.
Front passenger weights (2 nos.@75 kg): Assumed to act at points Fa and Fb.
Seat-spring stiffness K_{sf} = 90 kN/m. The luggage is modelled as a lumped mass (70 kg) acting at *L*.

Suspension

The chassis is supported over four tyres through suspension arrangements. The suspension system can be modelled as shown in Figure MP.4(b).

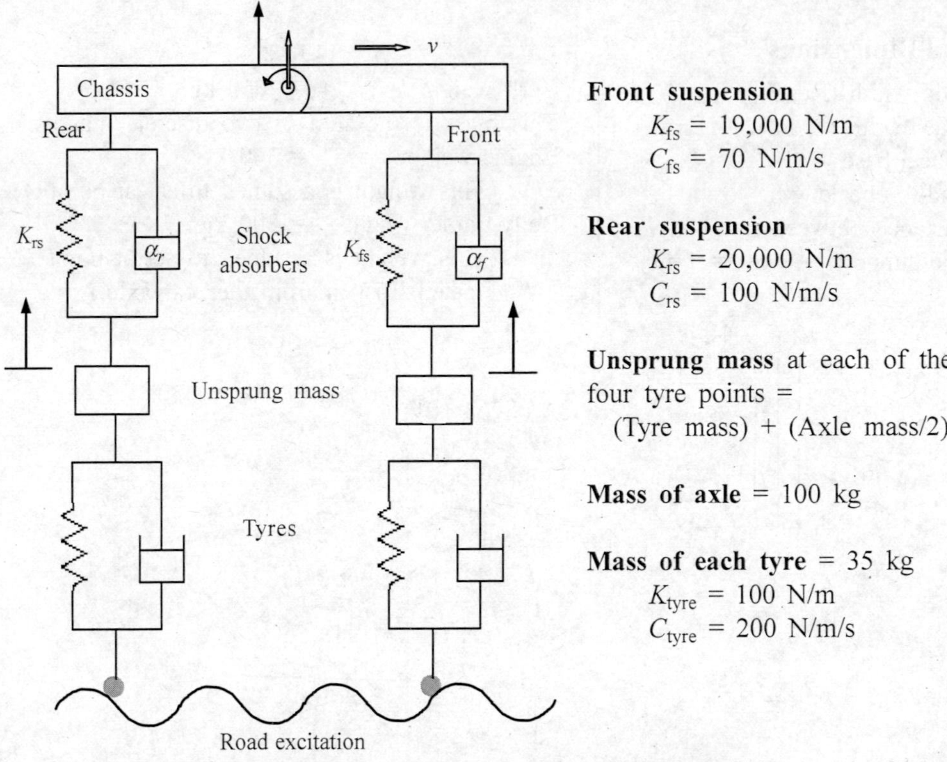

Fig. MP.4(b)   Typical chassis—Suspension details.

You may organise your project tasks as follows: Model the chassis, lumped masses (engine mass, body-frame mass, passenger mass, axle-tyre mass and luggage). Determine the natural frequencies and mode shapes of the chassis structure along with all the masses. For the first five modes, list the natural frequencies and plot the corresponding mode shapes.

Subject the (chassis structure + suspension) to deterministic road excitations. Figure MP.4(c) shows typical ride oscillation modes of the vehicle, as it navigates different road-profiles.

In order to simulate the excitation which the tyres receive from the road, we apply four displacement profiles, one each at the four tyres.

The displacements can be approximated as follows:

$$T_{FL} = A \sin(\omega t + \theta_1) \text{ [displacement at front-left tyre]}$$
$$T_{FR} = A \sin(\omega t + \theta_2) \text{ [displacement at front-right tyre]}$$
$$T_{RL} = A \sin(\omega t + \theta_3) \text{ [displacement at rear-left tyre]}$$
$$T_{RR} = A \sin(\omega t + \theta_4) \text{ [displacement at rear-right tyre]}$$

Circular frequency $\omega = (2\pi/\lambda) \times$ (vehicle speed)

Vehicle speed = 60 km/hr

Wavelength λ = 2.5 m (same as wheel base)

Amplitude of road bumps A = 0.05 m

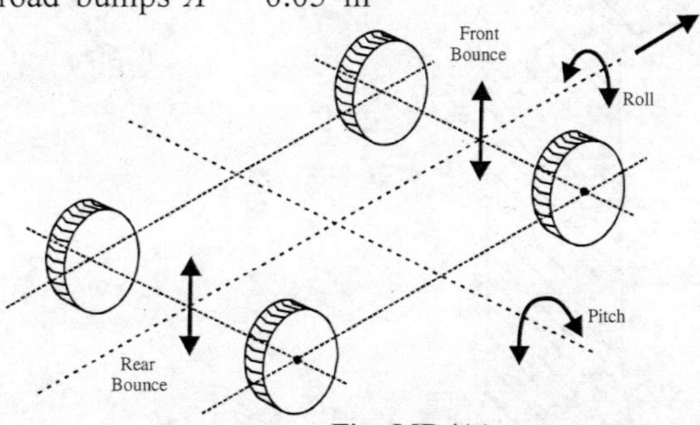

Fig. MP.4(c)

All-Four-Bounce

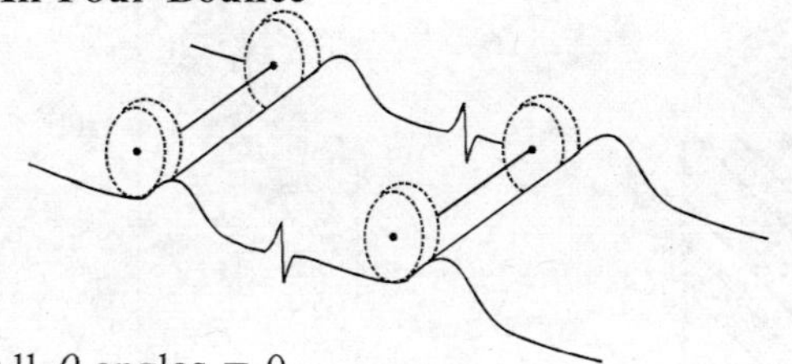

All θ angles = 0.
Both the axles heave up and down in phase.

Pitch

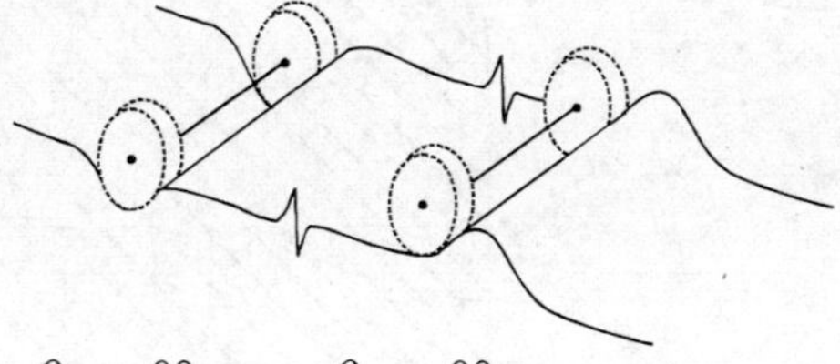

$\theta_1 = 0°,\qquad \theta_2 = 0°$
$\theta_3 = 90°,\qquad \theta_4 = 90°$

Roll

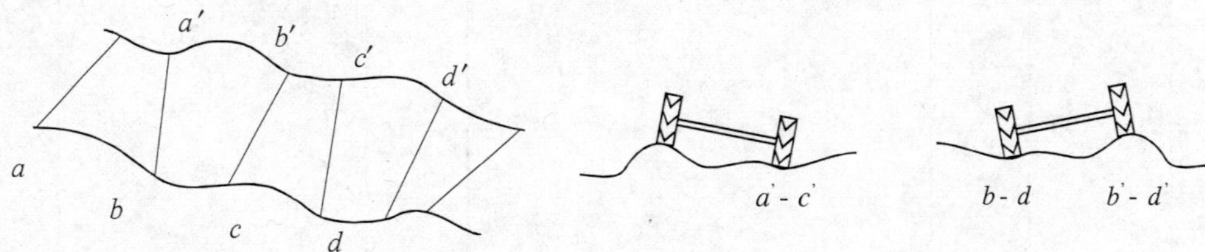

$\theta_1 = 0°,\qquad \theta_2 = 90°,\qquad \theta_3 = 0°,\qquad \theta_4 = 90°$

For the above scenarios, determine

(i) the acceleration levels and response at the passenger locations
(ii) chassis roll
(iii) chassis pitch

Project 5:   Analysis of a Turbine Disk

A typical disk used in gas turbines is shown in Figure MP.5. Perform axisymmetric finite element analysis of the disk, under centrifugal forces when the disk rotates at a steady speed N. rpm.

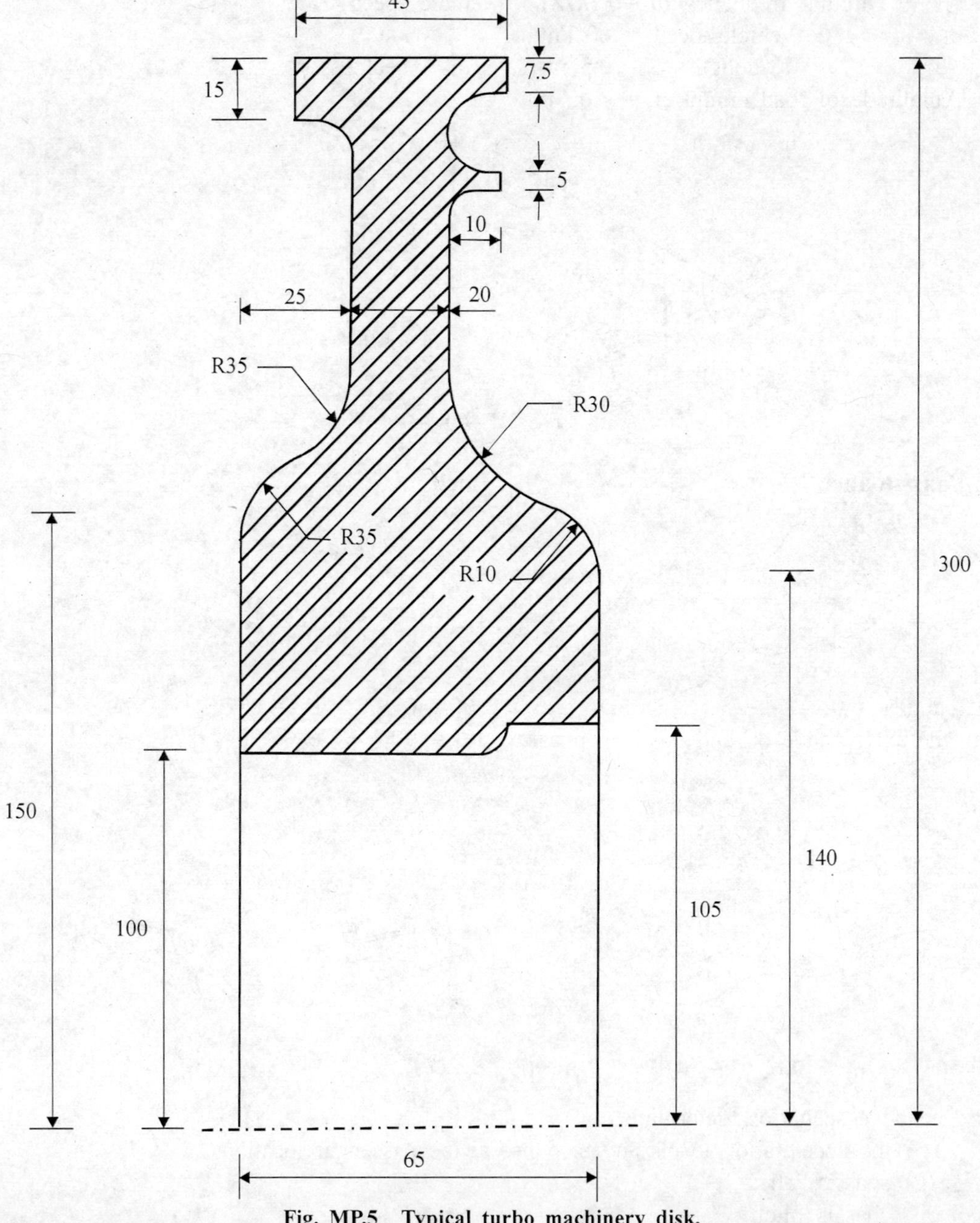

Fig. MP.5   Typical turbo machinery disk.

Project 6:   Dynamic Analysis of a Building

Consider a simplified model of a typical building as shown in Figure MP.6. Properties of concrete can be taken as follows: Young's modulus = 1.4×10^{10} N/m^2; Poisson's ratio = 0.3; density = 2240 kg/m^3; assume proportional damping (0.2 [K] + 0.4 [M]).

> ➢ Model the system using beam/plate elements and determine the first few natural frequencies and the corresponding mode shapes.

> ➢ There are two rotating machinery operating on the second floor. The specifications are as given below: weight of each machine = 5000 N; support spring stiffness = 800 kN/m; unbalanced mass on the rotor = 1 kg at an eccentricity of 0.1 m; rotational speed = 3000 rpm. Determine the steady state response of the system when these machines are in operation.

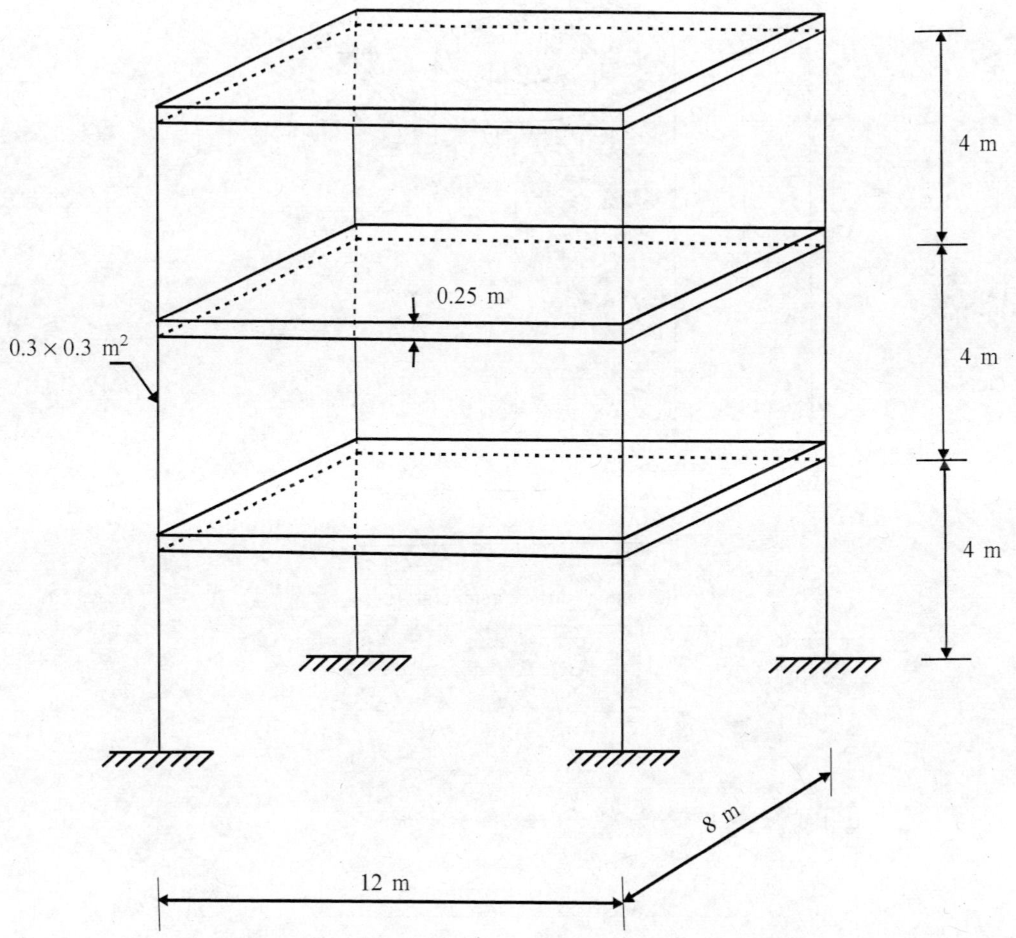

Fig. MP.6 Dynamic analysis of a typical building structure.

Project 7:   Thermal Analysis of an IC Engine Cylinder

Consider a simplified model of an IC engine cylinder shown in Figure MP.7.

> ➤ Model the system using finite elements and find the steady state temperature distribution given that $T_i = 200°C$ and $T_\infty = 30°C$. Assume $k = 45$ W/m°C and $h = 20$ W/m²°C.

> ➤ Use the typical temperature distribution over a cycle as shown in the figure and perform a transient dynamic analysis for temperature in the walls of the cylinder.

> ➤ Design appropriate fins on the cylinder block outer surface so as to enhance heat transfer. The number of fins per surface must be limited to 3, and each fin may be of thickness 3 mm and length not more than 20 mm.

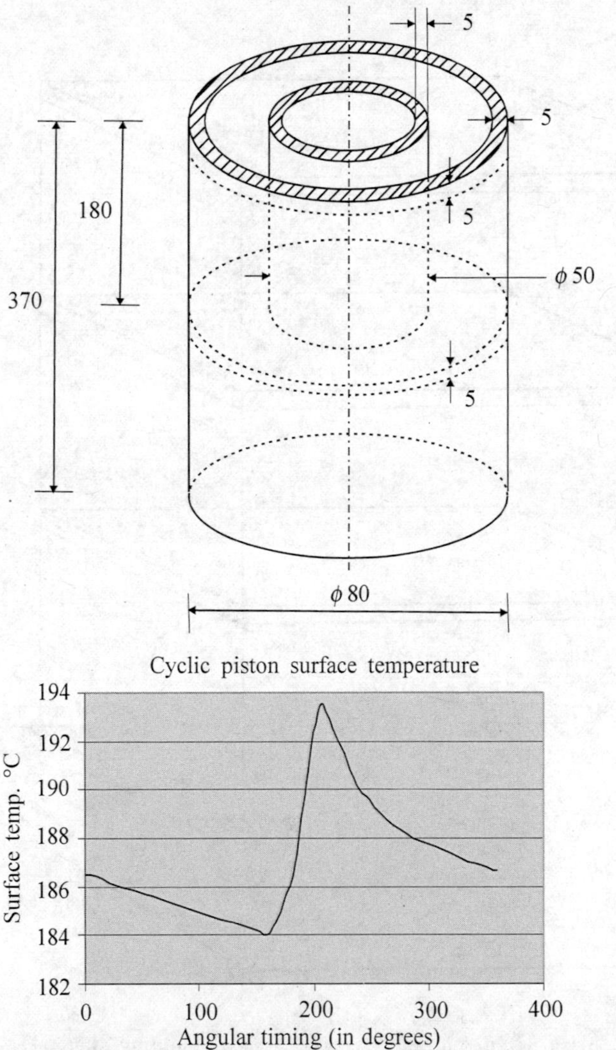

Fig. MP.7   Thermal analysis of a typical IC engine cylinder.

Project 8:   Stress Concentration

Consider the stress distribution around a small circular hole in a large plate subjected to uni-axial tension. It is well known that the theoretical stress concentration factor for this case is 3. When we consider an elliptical hole, theoretical studies show that stress concentration gets worse if we have the major axis perpendicular to the load. If the major axis is however aligned with the direction of loading, then there is significant improvement in stress distribution. However, circular holes are easier to make compared to elliptical holes. Thus, in practice, one can think of approximating the ellipse with smaller circular holes on either side of the original circular hole as shown in Figure MP.8.

> ➢ Model the plate with the central circular hole using standard two-dimensional elements (such as 8/9-noded quadrilateral; 6-noded triangle) and determine the stresses around the hole.

> ➢ Consider additional circular holes as indicated in the figure and redo the problem. Find (if necessary, by trial and error) appropriate size/location/number of the holes required to get near-uniform stress distribution.

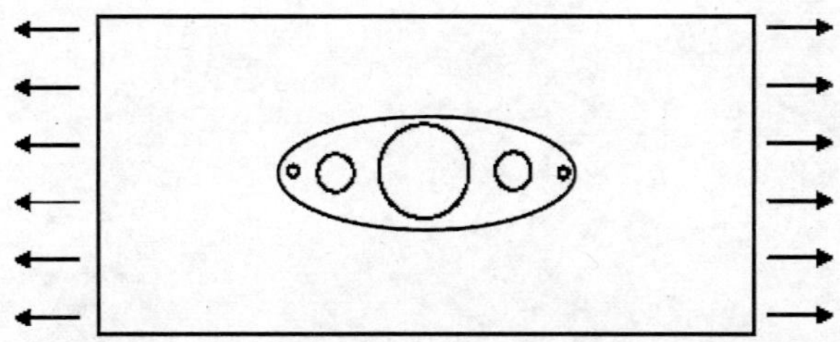

Fig. MP.8   Plate with holes.

Project 9:   Dynamics of a Hard Disk Drive Read/ Write Head Assembly

Figure MP.9 shows a simplified version of a typical Read/Write Head assembly used in computer hard disk drives. The motor is at the left end of the system while the R/W head itself is mounted at the right end. The assembly is rotated about an axis perpendicular to the plane of the paper through an appropriate angle to reach the desired location (track) on the disk. The track density, i.e. Number of tracks per mm is ever increasing and the response time desired is ever decreasing. It is therefore important that, when the assembly is moved to a desired location, the residual vibrations at the location of the R/W head be as low as possible. The system is actually very complex, with natural modes of vibration in two planes of bending and twisting, all coupled. Model the system using appropriate finite elements and determine the first few natural frequencies. Next, assume that, under a command to seek a particular track, the R/W head assembly is rotated by 60° in 10 ms through uniform acceleration/deceleration motion (i.e., triangular velocity profile). Determine the residual vibrations at the tip, carrying the R/W head. For the purpose of this analysis, assume that in the 60° position the system behaves as if its left end is clamped.

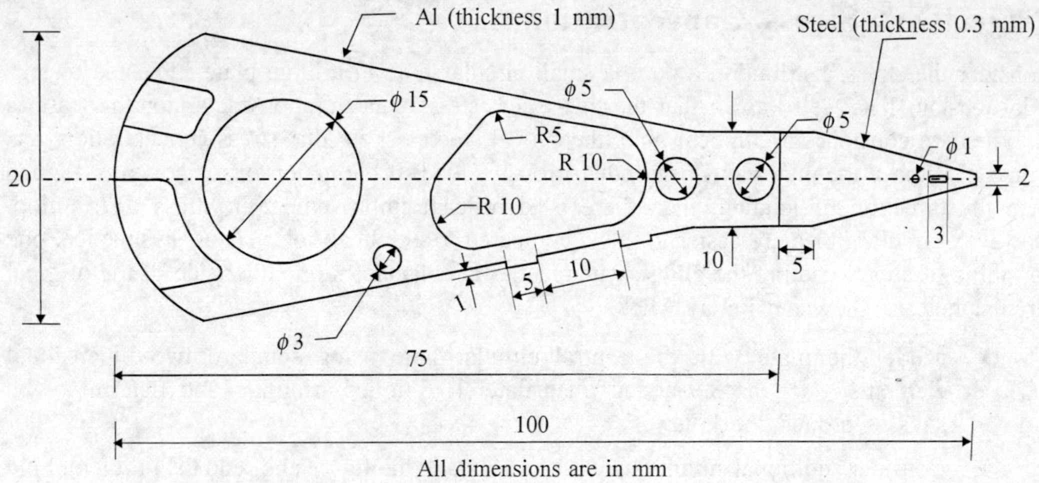

All dimensions are in mm

Fig. MP.9   Hard disc drive R/W head assembly.

Review of Preliminaries

B1.1   Matrix Algebra

A system of linear algebraic equations given by

$$a_{11}x_1 + a_{12}x_2 + \cdots + a_{1n}x_n = b_1$$
$$\vdots \qquad\qquad\qquad \vdots \qquad \vdots$$
$$a_{n1}x_1 + a_{n2}x_2 + \cdots + a_{nn}x_n = b_n$$

can be represented in compact matrix notation as

$$[A]_{n\times n}\,\{X\}_{n\times 1} = \{b\}_{n\times 1}$$

FEM helps us to convert governing differential equations into algebraic equations and hence we use matrix notation extensively. The system of equations corresponding to a finite element mesh are typically *sparse*, i.e. many coefficients a_{ij} are zero and, in particular, the equations are *banded*, i.e. nonzero coefficients are clustered around the principal diagonal as indicated in Figure B1.1.

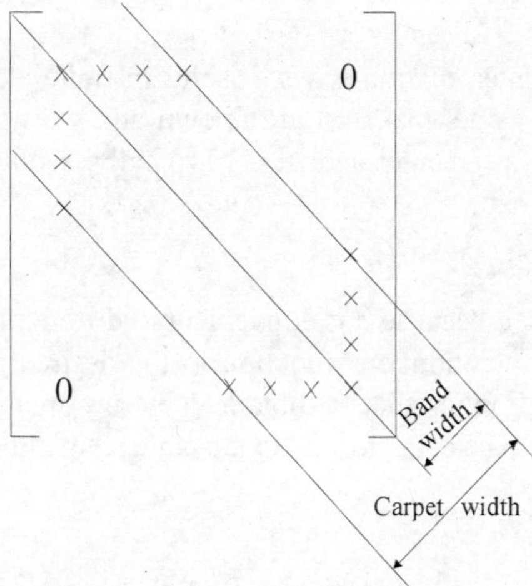

Fig. B1.1   Sparse, banded matrix.

The rank of a matrix $[A]$ is defined as the order of the largest nonzero determinant (or) the maximum number of linearly independent columns in $[A]$. Differentiation and integration of a matrix $[A]$ refers to these operators being applied on each and every element a_{ij} of the matrix. Thus,

$$\frac{\partial[A]}{\partial x} = \begin{bmatrix} \dfrac{\partial a_{11}}{\partial x} & \cdots & \dfrac{\partial a_{1n}}{\partial x} \\ \vdots & & \vdots \\ \dfrac{\partial a_{n1}}{\partial x} & \cdots & \dfrac{\partial a_{nn}}{\partial x} \end{bmatrix}$$

$$\int [A]\,dx = \begin{bmatrix} \displaystyle\int a_{11}\,dx & \cdots & \displaystyle\int a_{1n}\,dx \\ \vdots & & \vdots \\ \displaystyle\int a_{n1}\,dx & \cdots & \displaystyle\int a_{nn}\,dx \end{bmatrix}$$

A matrix eigenvalue problem is represented as

$$[A]_{n\times n}\{X\}_{n\times 1} = \lambda\{X\}_{n\times 1}$$

or

$$([A] - \lambda[I])\{X\} = \{0\}$$

The nontrivial solution set consists of n paris of λ_i, $\{X_i\}$, where λ_i, are referred to as the eigenvalues of $[A]$ and $\{X_i\}$ are the corresponding eigenvectors. If $[A]$ is real and symmetric, λ_i will be real. If, in addition, $[A]$ is positive definite, all λ_i,s are positive. The corresponding eigenvectors $\{X_i\}$ are orthogonal i.e., $\{X_i\}^T\{X_j\} = 0$ $(i \neq j)$.

B1.2   Interpolation

Understanding the process of interpolation is very useful in finite element method. We assume that at certain x_i, the function value, $f_i = f(x_i)$ are known and we wish to find an approximate estimate of the function value at an intermediate x. Typically, a polynomial $p_n(x)$ of degree n is fit to the data given such that

$$p_n(x_i) = f_i, \qquad i = 1, 2, \cdots, n+1$$

Polynomial fit is normally chosen because it is convenient to differentiate/integrate a polynomial. Also, they can approximate any continuous function to the desired accuracy. A polynomial fit p_n as above is unique, which implies that p_n obtained by any method will be the same. The various methods differ in their particular forms. One standard method is the Lagrange method. The basic idea is to write

$$p_n(x) = \sum_{i=0}^{n+1} f_i L_i(x)$$

where L_i are known as the Lagrange polynomials given by

$$L_i(x) = \frac{(x - x_1)(x - x_2) \cdots (x - x_{i-1})(x - x_{i+1}) \cdots (x - x_{n+1})}{(x_i - x_1)(x_i - x_2) \cdots (x_i - x_{i-1})(x_i - x_{i+1}) \cdots (x_i - x_{n+1})}$$

It would perhaps be intuitive that to improve the accuracy of fit, we need only increase n, the degree of the polynomial, used to fit the data. However, this is not generally true. Numerical instabilities may cause the polynomial fit $p_n(x)$ to oscillate significantly between the points x_i. As an example, let us attempt to fit the function $f(x) = \dfrac{1}{1 + x^2}$ in the interval $(-5, 5)$. The curve fit obtained using Lagrange's method, as described above, is given in Figure B1.2. The original $f(x)$ and the polynomial curve fit $p_n(x)(n = 6$ and $12)$ are compared in the figure. It is observed that significant oscillations are present in $p(x)$.

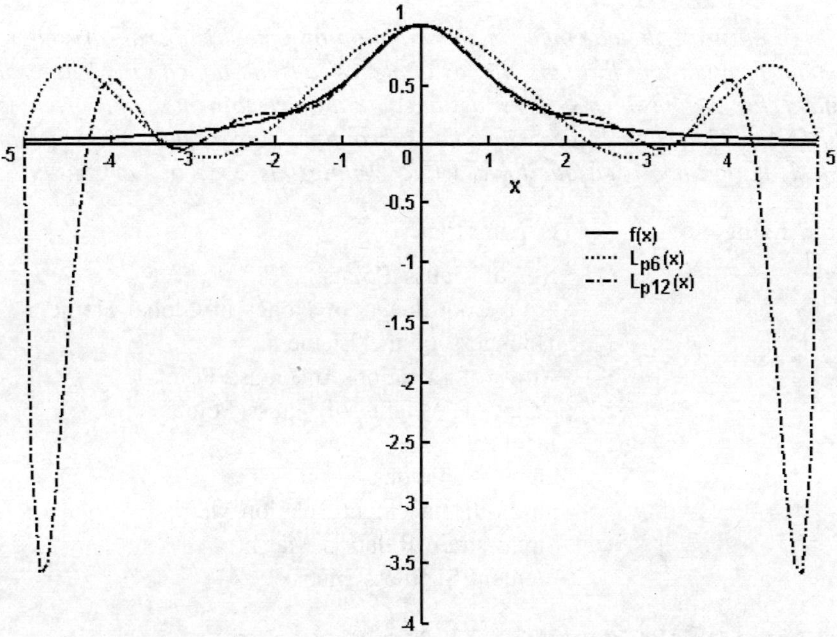

Fig. B1.2   Curve fit using a single polynomial.

Instead of a single, high degree polynomial curve fit in the entire interval $a \le x \le b$, we may subdivide the interval into many segments and fit several low degree polynomials (each valid in one segment). One such method uses cubic splines where a cubic curve is fit within each segment. Such a piecewise curve fitting is far superior to fitting a single, high degree polynomial for the entire region, as this is not subject to numerical instabilities (or) oscillations.

Typical Finite Element Program

Here we illustrate the computer implementation of typical isoparametric finite element formulation through the example of a nine-noded quadrilateral element for 2-d problems. It is assumed that a certain main program is available which interfaces with the subroutines given here. Clarity of implementation has been our focus while writing this code in Fortran-90.

Variable name	Explanation
Ngauss	No. of Gauss Points
Xl, Yl	X, Y Coordinates of Nodes in Global Frame
Thc	Thickness of the Element
Place	Array of Locations of Gauss Points
Wgt	Array of Weights of Gauss Points
Jac	Jacobean
En	Shape Functions
B	Strain-displacement Relation Matrix
D	Stress-strain Relation Matrix
Ek	Element Stiffness Matrix

```
Subroutine Quad9(Ngauss,Xl,Yl,Thc,Place,Wgt)
Implicit Double Precision (A-h, O-z)
Double Precision Jac
Common/Q9/En(9),Jac(2,2),B(3,18),D(3,3),Ek(18,18)
Dimension Xl(9),Yl(9), Btdb(18,18),Place(3,3),Wgt(3,3)

C
C     INITIALISE THE MATRICES
C
      DO 10 K=1,18
      DO 10 L=1,18
         EK(K,L)=0.0
```

```fortran
10    CONTINUE
      DO 20 I=1,3
      DO 20 J=1,3
         PLACE(I,J)=0.0
20       WGT(I,J)=0.0

         PLACE(1,2) = -0.5773502692
         PLACE(2,2) = 0.5773502692
         PLACE(1,3) = -0.7745966692
         PLACE(3,3) = 0.7745966692
         WGT(1,1) = 2.0
         WGT(1,2) = 1.0
         WGT(1,3) = 5.0/9.0
         WGT(2,2) = 1.0
         WGT(2,3) = 8.0/9.0
         WGT(3,3) = 5.0/9.0
C
C     GAUSS QUADRATURE LOOP
C
      DO 90 NA=1,NGAUSS
      PXI=PLACE(NA,NGAUSS)
      DO 80 NB=1,NGAUSS
      PET=PLACE(NB,NGAUSS)
C
C     WE HAVE NOW PICKED UP ONE GAUSS POINT WHOSE COORDINATES ARE
C     ξ (PXI), η(PET)
C     AT THIS GAUSS POINT, WE EVALUATE [J], [B] MATRICES USING SHAPE
C     ROUTINE
C
      CALL SHAPE(PXI,PET,XL,YL,DETJAC)
      DV=WGT(NA,NGAUSS)*WGT(NB,NGAUSS)*THC*DETJAC
      CALL TRAMUL(B,D,B,3,18,18,BTDB)
      DO 70 I=1,18
      DO 70 J=1,18
         EK(I,J)=EK(I,J)+BTDB(I,J)*DV
80    CONTINUE
90    CONTINUE
C
C     WE HAVE SUMMED UP OVER ALL THE GAUSS POINTS AND OBTAINED
C     ELEMENT STIFFNESS MATRIX [EK]
C
      RETURN
      END
      SUBROUTINE SHAPE(PXI,PET,XL,YL,DETJAC)
      IMPLICIT DOUBLE PRECISION (A-H, O-Z)
      DOUBLE PRECISION NXI(9),NET(9),JAC,JACINV(2,2)
      DIMENSION XL(9),YL(9),B1(3,4),B2(4,4),B3(4,18)
      COMMON/Q9/EN(9),JAC(2,2),B(3,18),D(3,3),EK(18,18)
```

```
C
C    THE SHAPE FUNCTIONS EVALUATED AT ξ (PXI) and η (PET)
C
      EN(9)=(1.0-PXI*PXI)*(1.0-PET*PET)
      EN(5)=0.5*(1.0-PXI*PXI)*(1.0-PET)-0.5*EN(9)
      EN(6)=0.5*(1.0+PXI)*(1.0-PET*PET)-0.5*EN(9)
      EN(7)=0.5*(1.0-PXI*PXI)*(1.0+PET)-0.5*EN(9)
      EN(8)=0.5*(1.0-PXI)*(1.0-PET*PET)-0.5*EN(9)
      EN(1)=0.25*(1.0-PXI)*(1.0-PET)-0.5*(EN(8)+EN(5))-0.25*EN(9)
      EN(2)=0.25*(1.0+PXI)*(1.0-PET)-0.5*(EN(6)+EN(5))-0.25*EN(9)
      EN(3)=0.25*(1.0+PXI)*(1.0+PET)-0.5*(EN(6)+EN(7))-0.25*EN(9)
      EN(4)=0.25*(1.0-PXI)*(1.0+PET)-0.5*(EN(8)+EN(7))-0.25*EN(9)


C
C    THE DERIVATIVES OF SHAPE FUNCTIONS; NXI REPRESENTS DERIVATIVE WITH ξ
C    NET REPRESENTS DERIVATIVE WITH η
C

      NXI(9)=-2.0*PXI*(1.0-PET*PET)
      NET(9)=-2.0*PET*(1.0-PXI*PXI)


      NXI(5)=-PXI*(1.0-PET)-0.5*NXI(9)
      NXI(6)=0.5*(1.0-PET*PET)-0.5*NXI(9)
      NXI(7)=-PXI*(1.0+PET)-0.5*NXI(9)
      NXI(8)=-0.5*(1.0-PET*PET)-0.5*NXI(9)


      NET(5)=-0.5*(1.0-PXI*PXI)-0.5*NET(9)
      NET(6)=-PET*(1.0+PXI)-0.5*NET(9)
      NET(7)=0.5*(1.0-PXI*PXI)-0.5*NET(9)
      NET(8)=-PET*(1.0-PXI)-0.5*NET(9)


      NXI(1)=-0.25*(1.0-PET)-0.5*(NXI(8)+NXI(5))-0.25*NXI(9)
      NXI(2)= 0.25*(1.0-PET)-0.5*(NXI(6)+NXI(5))-0.25*NXI(9)
      NXI(3)= 0.25*(1.0+PET)-0.5*(NXI(6)+NXI(7))-0.25*NXI(9)
      NXI(4)=-0.25*(1.0+PET)-0.5*(NXI(8)+NXI(7))-0.25*NXI(9)


      NET(1)=-0.25*(1.0-PXI)-0.5*(NET(8)+NET(5))-0.25*NET(9)
      NET(2)=-0.25*(1.0+PXI)-0.5*(NET(6)+NET(5))-0.25*NET(9)
      NET(3)= 0.25*(1.0+PXI)-0.5*(NET(6)+NET(7))-0.25*NET(9)
      NET(4)= 0.25*(1.0-PXI)-0.5*(NET(8)+NET(7))-0.25*NET(9)
C
C    INITIALISE [JAC] and [B]
C
      DO 10 I=1,2
      DO 10 J=1,2
10        JAC(I,J)=0.0

      DO 20 I=1,3
      DO 20 J=1,18
```

```
20          B(I,J)=0.0
C
C   FORMING THE JACOBIAN
C
      DO 30 L=1,9
      JAC(1,1)=JAC(1,1)+NXI(L)*XL(L)
      JAC(1,2)=JAC(1,2)+NXI(L)*YL(L)
      JAC(2,1)=JAC(2,1)+NET(L)*XL(L)
      JAC(2,2)=JAC(2,2)+NET(L)*YL(L)
30 CONTINUE
C
C   FINDING THE DETERMINANT OF JACOBEAN AND INVERSE OF JACOBEAN
C
      DETJAC = JAC(1,1)*JAC(2,2)-JAC(1,2)*JAC(2,1)
C
      DUM1  = JAC(1,1)/DETJAC
      JAC(1,1)= JAC(2,2)/DETJAC
      JAC(1,2)=-JAC(1,2)/DETJAC
      JAC(2,1)=-JAC(2,1)/DETJAC
      JAC(2,2)= DUM1
C
C
      DO 40 I=1,3
      DO 40 J=1,4
40       B1(I,J) = 0
      B1(1,1)=1.0
      B1(2,4)=1.0
      B1(3,2)=1.0
      B1(3,3)=1.0

      DO 50 I=1,4
      DO 50 J=1,4
50       B2(I,J)=0
      DO 60 I=1,2
      DO 60 J=1,2
60       B2(I,J)=JAC(I,J)
      DO 70 I=3,4
      DO 70 J=3,4
70       B2(I,J)=JAC(I,J)

      DO 80 I=1,4
      DO 80 J=1,18
80       B3(I,J)=0

      DO 90 J=1,9
         L=2*J
         K=L-1
         B3(1,K)=NXI(J)
```

```fortran
                B3(2,k)=NET(J)
                B3(3,L)=NXI(J)
                B3(4,L)=NET(J)
90    CONTINUE
      CALL MATMUL(B1,B2,3,4,4,B1B2)
      CALL MATMUL(B1B2,B3,3,4,18,B)
      RETURN
      END
C
C
      SUBROUTINE TRAMUL(A,B,C,L,M,N,D)
      IMPLICIT DOUBLE PRECISION(A-H,O-Z)
      DIMENSION A(L,M),B(L,L),C(L,N),D(M,N),AT(20,20),ATB(20,20)
C
C        PROGRAM TO CALCULATE  (D)=(A)T*(B)*(C)
C
         DO 10 I=1,L
         DO 10 J=1,M
10          AT(J,I)=A(I,J)
         DO 20 I=1,M
         DO 20 K=1,L
            ATB(I,K)=0.0
         DO 20 J=1,L
20       ATB(I,K)=ATB(I,K)+AT(I,J)*B(J,K)
         DO 30 I=1,M
         DO 30 K=1,N
             D(I,K)=0.0
         DO 30 J=1,L
             D(I,K)=D(I,K)+ATB(I,J)*C(J,K)
30       CONTINUE
         RETURN
         END

      SUBROUTINE MATMUL(A,B,L,M,N,C)
      IMPLICIT DOUBLE PRECISION(A-H,O-Z)
      DIMENSION A(L,M),B(M,N),C(L,N)
C
C     PROGRAM TO CALCULATE [C] = [A]*[B]
C
         DO 10 I=1,L
         DO 10 K=1,N
             C(I,K)=0.0
         DO 10 J=1,M
10           C(I,K)=C(I,K)+A(I,J)*B(J,K)
         RETURN
         END
```

Index

Assembly, 48, 117
Average acceleration method, 282, 287, 288
Axisymmetry, 146, 147, 168–169, 303, 305

Bandwidth, 207
Bar element, 90, 91
  linear, 93
  quadratic, 101
Beam element, 91, 117
Boundary condition,
  essential, 34, 38, 39
  natural, 34, 38, 39
Boundary element method, 17

Central difference method, 280
Cholesky factorisation, 208, 230, 305
Conditionally stable method, 279
Constant strain triangle (CST), 151, 173, 174, 178, 226

Damping
  factor, 235
  modelling of, 272
  Rayleigh (proportional), 275
Degrees of freedom, nodal, 48, 149
Direct integration methods, 273, 279
  explicit methods, 282
  implicit methods, 282
Discretisation 2, 3, 49, 306, 307
Displacement field, 75, 118
Duhamel integral, 275

Eigenvalue, 235, 256, 259, 263–265
  shifting, 256

Eigenvector, 256, 259, 265, 273
Element
  bar, 90, 91, 93, 101
  beam, 91, 117, 118, 238
  bilinear, 153, 180–182, 210
  conductance matrix, 59
  CST, 151, 173, 174, 178, 226
  equations, 48
  force vectors, 51
  four-noded rectangular, 152, 179
  frame, 125
  isoparametric, 156, 184, 185
  quad 4, 159, 160, 182–188, 216
  quad 8, 162, 163, 188–190, 217
  quad 9, 190–192, 217
  six-noded triangular, 153–155, 166, 192–194
  stiffness matrix, 51, 119
  sub-parametric, 156
  super-parametric, 156
  three-noded triangular, 149, 166, 171, 172
Euler–Lagrange equation, 71

Functional, 66, 68
  minimization of, 73
  stationarity of, 66, 81, 83

Gauss elimination, 107
Gram–Schmidt deflation, 267

Interpolation function (see *Shape function*)
Iteration
  forward, 265
  inverse, 265–267
  simulataneous, 268–271
  subspace, 271, 272

Jacobi method, 259–264
Jacobian, 186, 189, 191, 193, 212, 214

Lagrange
    elements, 158, 192
    equations, 240, 241, 244
    polynomial, 190
Linear acceleration method, 284–287, 287, 288
Load vector, 92
    global, 92
Local coordinate frame, 152

Magnification factor (MF), 233
Mass matrix
    consistent, 246, 250
    lumped, 246, 249
Mode superposition method, 273, 275–279

Newmark methods, 287, 288
Nodal force vector, 120–121
    consistent, 121
    lumped, 121
Nodal heat flux, 59

One-parameter solution, 18

Parasitic shear, 216
Point collocation technique, 21, 26
    collocation points, 26

Quadrature,
    Newton–Cotes, 197, 198
    Gauss, 197–204, 246
    Simpson, 196
    trapezoidal, 195
Quasi-static, 233

Rayleigh quotient, 266, 271
Rayleigh–Ritz method, 73, 75
Residual,
    boundary, 17
    domain, 1, 17
Refinement
    h-type, 220
    p-type, 220

Semi-analytical FEM, 148
Serendipity elements, 158, 192
Shape function, 2, 45, 75
Stationary total potential, principle of, 66, 73, 75
Stiffness matrix,
    element, 92, 102, 113, 129
    direct method, 85
Strain
    initial, 169, 108, 111
    plane, 146, 169, 172
Strain-displacement relation matrix, 102
Stress
    initial, 169
    plane, 146, 169, 172, 173, 175
    thermal, 108
Sturm sequence property, 256, 294
Substitution,
    backward, 209
    forward, 209

Transformation matrix, 115, 129
Trial function, 17
    piece-wise continuous, 42, 81
Trial solution (see *Trial function*)

Weighted residual,
    Galerkin, 16
    general weighted residual statement, 28–33, 236
    weak form of, 16, 33-36, 41–48, 236
Weighting function, 24
Wilson-θ method, 286